Lecture Notes in Computer Science 5723

Commenced Publication in 1973
Founding and Former Series Editors:
Gerhard Goos, Juris Hartmanis, and Jan van Leeuwen

T0223390

Helmut Horacek   Elisabeth Métais
Rafael Muñoz   Magdalena Wolska (Eds.)

Natural Language Processing and Information Systems

14th International Conference on Applications
of Natural Language to Information Systems, NLDB 2009
Saarbrücken, Germany, June 24-26, 2009
Revised Papers

 Springer

Volume Editors

Helmut Horacek
Saarland University, Dept. of Computer Science
P.O. Box 151150, 66041 Saarbrücken, Germany
E-mail: horacek@ags.uni-sb.de

Elisabeth Métais
CNAM- Laboratoire Cédric
292 Rue St. Martin, 75141 Paris Cedex 03, France
E-mail: elisabeth.metais@cnam.fr

Rafael Muñoz
Universidad de Alicante, Departamento de Lenguajes y Sistemas Informáticos
Campus de San Vicente del Raspeig, Apdo 99, 03080 Alicante, Spain
E-mail: rafael@dlsi.ua.es

Magdalena Wolska
Saarland University, Dept. of General Linguistics
P.O. Box 151150, 66041 Saarbrücken, Germany
E-mail: magda@coli.uni-saarland.de

Library of Congress Control Number: 2010924365

CR Subject Classification (1998): H.2.8, H.2, H.3, I.2, F.3-4, H.4, C.2

LNCS Sublibrary: SL 3 – Information Systems and Application, incl. Internet/Web
and HCI

ISSN   0302-9743
ISBN-10   3-642-12549-2 Springer Berlin Heidelberg New York
ISBN-13   978-3-642-12549-2 Springer Berlin Heidelberg New York

springer.com

© Springer-Verlag Berlin Heidelberg 2010
Printed in Germany

Typesetting: Camera-ready by author, data conversion by Scientific Publishing Services, Chennai, India
Printed on acid-free paper   06/3180

Preface

This volume contains the papers presented at NLDB 2009, the 14th International Conference on Applications of Natural Language to Information Systems held June 24–26, 2009, at the University of the Saarland and the German Research Center for Artificial Intelligence in Saarbrücken, Germany. In addition to reviewed submissions, the program also included contributions to the doctoral symposium held during NLDB 2009 as well as two invited talks. These talks covered some of the currently hot topics in the use of natural language for accessing information systems.

We received 51 submissions as regular papers for the main conference, 2 extra submissions as posters, and 3 short papers for the doctoral symposium. Each paper for the main conference was assigned four reviewers, taking into account preferences expressed by the Program Committee members as much as possible. Within the review deadline, we received at least three reviews for almost all submissions.

After the review deadline, the Conference Organizing Committee members and the Program Committee Chair acted as meta-reviewers. This task included studying the reviews and the papers, specifically those whose assessment made them borderline cases, and discussing conflicting opinions and their impact on the assessment of individual papers. Finally, the meta-reviewers wrote additional reviews for the few papers which received less than three reviews, as well as for papers which received reviews with considerably conflicting assessments.

In order to come up with a final decision, the meta-reviewers used a ranking list according to the weighted average scores of all papers on a scale from -3 (lowest possible) to +3 (highest possible), the reviewer's confidence being used as the weighting factor. Submissions with a score greater or equal to 1.0 (weak accept) were accepted as full papers. For this threshold, we used the unweighted average score, which makes the arrival at the precise value used conceptually motivated. In general, the differences between weighted and unweighted scores were mostly marginal; almost all submissions accepted as full papers were at least 1.0 also on weighted average score (only two scored 0.9 on this score). As short papers, we accepted submissions with scores lower than full papers, which got either three positive scores or two positive scores but no negative ones. As posters, finally, we accepted most submissions which were assessed at least slightly positive, individually deciding some borderline cases.

The final acceptance rate counting the number of full papers according to NLDB tradition was 25.5% percent (13 out of 51), quite close to the rates in at least the previous four years. In addition, eight submissions were accepted as short papers, and seven as posters, including one of the extra submissions, a system demo. Originally, two more posters were accepted, but the authors preferred to withdraw their submissions, once they were only accepted as posters. Finally, the three submissions to the doctoral symposium were all accepted.

Since the short papers were not assessed much lower than some of the full papers, we also chose a relatively small difference between the space devoted to these categories: full papers were allowed a maximum of 15, and short papers a maximum of 12 pages, while posters got only 2 pages. There was, however, a more pronounced difference between full and short papers in terms of the presentation time.

In this volume, most contributions were grouped according to their category. First, there are the two invited papers; then the majority of the contributions, long and short papers together. These papers were grouped according to topic areas. This is followed a section with the posters. Finally, there are the three doctoral symposium papers.

The invited papers both address learning of specific natural language concepts from large corpora. The paper by Hovy investigates contributions to the long-range issue of building a database out of the information found on the Web. This includes automated instances mining, metastructure harvesting, and inter-concept relation discovery. The paper by Uszkoreit et al. reports on a systematic analysis of a minimally supervised machine learning method for relation extraction grammars. These investigations led to insights on the dependencies of properties of the data and the selection of the seeds of the performance of the algorithm. Consequently, the learning method can be further improved by taking these dependencies into account.

The accepted contributions (long and short papers) covered a wide range of topics, which we classified into eight topic areas, each covering a section in this volume:

- Information Retrieval
- Term Extraction
- Information Extraction
- Classification of Text
- Classification of Documents
- Interfaces to Knowledge Bases
- Using Semantic Models of Natural Language
- Quality Assessment of Knowledge Sources

Information Retrieval. Two full papers were categorized in this section. They emphasize the role of semantic information to improve some specific forms of information retrieval. The first paper by Navarro et al. introduces evidence from natural language in multimodal fusion techniques. Through incorporating natural language annotations related to images, the performance of two widely used fusion strategies in visual information retrieval can be significantly improved. The second paper by Sorg and Cimiano applies the model of explicit semantic analysis to cross-language retrieval. The authors perform a systematic investigation through examining variations between different basic design choices, yielding considerable improvements over the original model.

Term Extraction. One full and two short papers were categorized in this section. They address a variety of applications of term extraction techniques, ranging from topic maps over schema matching to requirements analysis, mostly by combining evidence from several knowledge sources. The first one, the full paper by Ellouze et al., proposes an incremental construction of multilingual topic maps. In this construction process, it takes into account multiple knowledge sources including, as a particular factor, evidence about requests of potential users. The first of the two short papers by Coen and Xue addresses the problem of schema-matching, that is, semantic equivalences across name spaces. The algorithm presented makes use of textual information and dependency information. The second of the two short papers by Kof analyzes concept extraction techniques to be used in the construction of executable models from textual specifications. The paper obtains the first results in this new research direction through systematic comparisons in a case study.

Information Extraction. One full and one short paper were categorized in this section. They address issues of information extraction in non-standard applications. The full paper by Segura-Bedmar et al. investigates anaphora resolution for the yet untested domain of drug interaction. They achieve results similar to those obtained for other domains by using domain-specific syntactic and semantic parsers. The short paper by Had et al. investigates standard relation extraction techniques accommodated to the German language. Through an enhanced composite kernel method, they improve significantly over the poor results of standard methods, which are not well suited to free word order languages, such as German.

Classification of Text. Two full and one short paper were categorized in this section. They deal with a nice selection of special, rarely addressed topics. The first full paper by Asonov aims at effective spelling correction. By simplifying the task into only finding typographical errors rather than also correcting them, a task which falls under the responsibility of the human writer anyway, Asonov achieves improved results over other approaches. The other full paper by Reyes et al. aims at recognizing humor, a task in which computers are notoriously bad performers. The authors demonstrate that an examination of features of semantic and morpho-syntactic ambiguities enables a rather good discrimination between humor and non-humor. The short paper by Balahur and Montoyo addresses opinion mining. They apply a general annotation scheme that is suitable for a variety of domains, with quite promising results.

Classification of Documents. Two full and one short paper were categorized in this section. They cover quite a large range of techniques. The first full paper by Keim et al. aims at the extraction of discriminating and overlap terms out of a set of document classes. Their widely language-independent method is shown to outperform several competing approaches. The other full paper by Ponomareva et al. describes a fully implemented system for archiving institutional repositories. The semi-automatic approach combines automatic discovery and extraction methods with user interaction techniques to ensure the quality of the bibliographic data obtained. The short paper by He and Lin proposes a semi-supervised learning algorithm via local learning with class priors to address text classification for the domain of protein–proptein interactions. The authors demonstrate that their compound algorithm outperforms more traditional methods.

Interfaces to Knowledge Bases. Two full and one short paper were categorized in this section. The full papers address traditional interfaces to databases, the short one aims at an interface applicable to languages that are based on images. The first full paper by Cimiano and Minock investigates a quantitative grounding in empirical data. The evidence obtained from a geodatabase makes it clear that the demand on interfaces is quite high, since these systems must have some means to deal with a variety of natural language phenomena each of which appears with some frequency in the corpus. The other full paper by Giordani and Moschitti attacks the problem of mapping natural language questions onto SQL queries via learning techniques, on the basis of a corpus of pairs of natural language questions and database queries. Like the previous paper, they also use a geodatabase as their domain of application. The short paper by Sidorov et al. reports on pioneer work for a Mayan script database, the first computer-based representation for this antique sign language. The incorporated methods are widely applicable to languages based on sets of images, and the implemented system can accommodate a demand of users with varying background knowledge.

Using Semantic Models of Natural Language. Two full papers were categorized in this section. They both aim at the attainment of increased quality through the incorporation of semantics, for different application areas. The first one by Llorens et al. investigates the incorporation of semantic roles in the task of temporal expression identification. The authors demonstrate that the approach is quite valuable, especially since it is less demanding in terms of training data and development time, when compared to machine learning and knowledge-based approaches, respectively. The other full paper by Adly and Al Ansary presents an interlingua-based machine translation system, evaluated for English-Arabic translation. The elaborate evaluations in several automated metrics demonstrate that this system performs better than some well-known competitors in English-Arabic translation.

Quality Assessment of Knowledge Sources. One full and two short papers were categorized in this section. They address properties of ontologies and corpora as their target of quality evaluation. The full paper by Solskinnsbakk et al. investigates issues in verifying the quality of subsumption hierarchies. The authors formulate hypotheses about relations between classes in super/sub and sister relations, verifying these hypotheses for some widely used ontologies. The first short paper by Sabou et al. investigates the problem of evaluating the correctness of semantic relations. They use online ontologies and the Semantic Web in order to test the plausibility of a semantic relation. The other short paper by Pinto et al. investigates the quality assessment of text corpora, on the basis of corpus features, such as domain broadness and class imbalance. The formally obtained quality assessments are shown to be comparable to human judgements. The methods presented could be used, for example, to assess the quality of gold standards.

The posters included in this volume cover a wide range of topics, including navigational semantics, word level alignment, data mart schema design, spreadsheet information retrieval, weblog corpora, full text search, and knowledge management. Finally, the papers from the doctoral symposium address textual entailment, spreadsheet information retrieval, and speech interpretation.

The conference organizers are indebted to the reviewers for their engagement in a vigorous submission evaluation process. We would also thank members of the DFKI GmbH, and our various student helpers, for their help with the organization.

June 2009

Helmut Horacek
Elisabeth Métais
Rafael Muñoz
Magdalena Wolska

Organization

Conference Organization

Helmut Horacek   Saarland University, Saarbrücken, Germany
Elisabeth Métais   CNAM, Paris, France
Reind van de Riet (+)   Vrije Universiteit Amsterdam,
           The Netherlands

Program Chair

Rafael Muñoz   Universidad de Alicante, Spain

Doctoral Symposium Chair

Magdalena Wolska   Saarland University, Saarbrücken, Germany

Publicity Chair

Mokrane Bouzeghoub   Université de Versailles, France

Program Committee

Jacky Akoka   CNAM, France
Sophia Ananiadou   Manchester Interdisciplinary Biocentre, UK
Frederic Andres   University of Advanced Studies, Japan
Jing Bai   Yahoo Inc., Canada
Akhilesh Bajaj   University of Tulsa, USA
Mokrane Bouzeghoub   Université de Versailles, France
Hiram Calvo   National Polytechnic Institute, Mexico
Roger Chiang   University of Cincinnati, USA
Philip Cimiano   University of Delft, The Netherlands
Isabelle Comyn-Wattiau   CNAM, France
Antje Düsterhöft   Hochschule Wismar, Germany
Günther Fliedl   University of Klagenfurt, Austria
Alexander Gelbukh   Mexican Academy of Sciences, Mexico
Jon Atle Gulla   Norwegian University of Science and
           Technology, Norway
Udo Hahn   Friedrich-Schiller-Universität Jena, Germany
Karin Harbusch   Universität Koblenz-Landau, Germany
Harmain Harmain   United Arab Emirates University, UAE
Alexander Hinneburg   University of Halle, Germany

Helmut Horacek	Saarland University, Germany
Andreas Hotho	University of Kassel, Germany
Paul Johannesson	Stockholm University, Sweden
Epaminondas Kapetanios	University of Westminster, UK
Zoubida Kedad	Université de Versailles, France
Christian Kop	University of Klagenfurt, Austria
Manolis Koubarakis	University of Athens, Greece
Georgia Koutrika	Stanford University, USA
Nadira Lammari	CNAM, France
Jana Lewerenz	sd&m Düsseldorf, Germany
Deryle Lonsdale	Brigham Young University, USA
Stéphane Lopes	Université de Versailles, France
Heinrich C. Mayr	University of Klagenfurt, Austria
Paul McFetridge	Simon Fraser University, Canada
Elisabeth Métais	CNAM , France
Farid Meziane	Salford University, UK
Luisa Mich	University of Trento, Italy
Andrès Montoyo	Universidad de Alicante, Spain
Rafael Muñoz	Universidad de Alicante, Spain
Günter Neumann	DFKI, Germany
Jian-Yun Nie	Université de Montréal, Canada
Pit Pichappan	Annamalai University, India
Odile Piton	Université Paris I Panthé on-Sorbonne, France
Yacine Rezgui	University of Salford, UK
Reind van de Riet (+)	Vrije Universiteit Amsterdam, The Netherlands
Jürgen Rilling	Concordia University, Canada
Hae-Chang Rim	Korea University, Korea
Fabio Rinaldi	University of Zurich, Switzerland
Markus Schaal	Bilkent University, Turkey
Grigori Sidorov	National Researcher of Mexico, Mexico
Max Silberztein	Université de Franche-Comté, France
Irena Spasic	Manchester Ctr. for Integrative Systems Biology, UK
Veda Storey	Georgia State University, USA
Vijay Sugumaran	Oakland University Rochester, USA
Bernhard Thalheim	Kiel University, Germany
Krishnaprasad Thirunarayan	Wright State University, USA
Juan Carlos Trujillo	Universidad de Alicante, Spain
Luis Alfonso Ureña	Universidad de Ja'en, Spain
Panos Vassiliadis	University of Ioannina, Greece
Roland Wagner	University of Linz, Austria
Christian Winkler	University of Klagenfurt, Austria
René Witte	Concordia University, Canada
Magdalena Wolska	Saarland University, Germany

Additional Reviewers

Kristin Mittag
M. Teresa Martìn-Valdivia
Martin Atzmueller
Jose Luis Vicedo
Yulia Ledeneva
Oscar Ferrandez
Sonia Vazquez
Arturo Montejo-Raez

Table of Contents

Classification of Text

Classification of Documents

Interfaces to Knowledge Bases

Using Semantic Models of Natural Language

Quality Assessment of Knowledge Sources

Poster Papers

Doctoral Symposium Papers

Turning the Web into a Database: Extracting Data and Structure

Eduard H. Hovy

Information Sciences Institute,
University of Southern California
hovy@isi.edu

Abstract. People build databases to collect, systematize, and make available to users knowledge in a consistent and hopefully trustworthy form. But the largest data collection today, the web, is not systematic, consistent, or trustworthy, and the access techniques we use are provably inadequate. Focusing just on text, what would it take to extract information from the web, organize it, and form a database (both instances and metadata) from it? This paper discusses some of the core problems and provides examples of recent research in NLP: automated instance mining, metadata structure harvesting, and inter-concept relation discovery.

1 Introduction

The compilation and organization of information has been a human enterprise since the dawn of civilization. As technology has advanced, so too have both the scale of knowledge assembled and the methods for storing, indexing, retrieving, and displaying it. A hundred years ago, encyclopedias were the traditional way for structuring presenting the world's (formal) knowledge, and libraries were repositories of large amounts of information. However, they lacked much (most?) information useful in daily life: train timetables, films, municipal notices, restaurant reviews, and a host of other kinds of information.

Today, after the explosive growth of the web, not only is much more information accessible to a single person, but also many more forms of it, in a larger variety of media. This is not news. What is surprising, however, is that our best indexing and access technology is Information Retrieval (IR), which has been demonstrated to be only partially sufficient. The TREC competitions (TREC) of the past decade continue to show that in various settings, even the best IR technology still returns less than 50% of the extant appropriate information (recall), and less than 50% of what it does return is appropriate (precision). This is not meant as a criticism of IR technology or research as much as a comment on the sorry state of the ways we organize (or don't organize) our information in the world. In fact, given the free-form, unstructured, or perhaps even chaotic way information is posted, interconnected, and maintained on the web, IR systems are in fact doing a phenomenal job. It is amazing that we find anything at all.

H. Horacek et al. (Eds.): NLDB 2009, LNCS 5723, pp. 1–7, 2010.
© Springer-Verlag Berlin Heidelberg 2010

But can we do better? Information Retrieval is very good at supporting browsing — its delivery of a ranked list of (hopefully) relevant material can be perfect for exploring and learning. But can we devise methods that support other forms of information usage: ones that organize our information, facilitate overviews and synthesis, and reflect the myriad of ways in which our world's information is structured into knowledge?

To this end one can identify a hierarchy of levels of knowledge organization and associated technology:

- No overt organization. Current state of the web; technology exemplified by IR.
- Partial taxonomies and networks. Many online and offline metadata schemas and taxonomies, usually built by hand. Examples: Yahoo topic hierarchy (http://yp.yahoo.com/); WordNet (http://wordnet.princeton.edu/); the industrial NAICS coding system (http://www.naics.com); the Gene Ontology (http://www.geneontology.org/).
- Collaborative creation and organization of knowledge. Examples: Wikipedia (http://www.wikipedia.org/); FaceBook (http://www.facebook.com/); online newspapers; databases built by careful wrapping of websites and subsequent information extraction and automated QA, such as MIT's START system (http://start.csail.mit.edu/).
- Internal representation of semantics. Systems automatically infer and insert semantic information 'into' web content, which is the idea behind the semantic web. Technology: mainly in prototypes of isolated functionalities plus collections of ontologies and formal representation languages (http://semanticweb.org/).

None of these models is really satisfactory; all have benefits and shortcomings. For example, Yahoo's popular Yellow Pages taxonomy is informal, built in a completely data-driven way, and quite problematic, at least for many types of computational processing. Its terms are undefined and are often duplicated to name different classes at different levels of the hierarchy (e.g., classes under *Automotive* and *Motorcycles*). In contrast, Wikipedia is not hierarchicalized, but instead encourages cross-linking of articles as appropriate. It supports a small amount of standardized structuring of certain of its contents, such as *Infoboxes* with biographical information of people. But there is no preferred set of terms, no topic taxonomy, and it makes no attempt to include what can be called the 'informal' information of the world, such as personal videos, train timetables, or people's opinions about local restaurants. Regarding the semantic web, the idea is that the internal semantics of web contents will somehow – through some miraculous agency – become manifest and be included. At present, useful representations and algorithms for analyzing links and link structure are being developed. But what exactly the semantics is, and who or what the mysterious agency is, remains unclear.

Rather than argue for one model or another, one can adopt one and work on it, in hopes of making a little progress and exploring whether it could in fact become (part of) a better organization of our knowledge.

In this paper we outline the kind of work that would support the second model — the model of (semi-)automatically creating (small) taxonomic structures that organize terms and their interrelationships, as used for web content. The central question of this work is: Using text on the web, can you automatically build a domain-specific database (i.e., instance collection plus its metadata) on demand? Said differently, since term taxonomies are valuable for current (semantic and NLP) web applications, and since existing knowledge repositories have limited coverage, can one, for a given domain, learn from text its terminology (concept) structure – the instance/basic level terms, the non-instance terms, their organization, and inter-term relationships?

2   Learning Instances and Basic-Level Concepts

[1] There has been a great deal of text mining research on learning instances and basic-level concepts from web text [2]. Most of this work uses some form of surface pattern, of the kind "X such as *" introduced in [12]. Provided a term such as "animals" for X, extended web searches return animals of all kinds in the starred position. Variations of this method have been in use in a variety of projects, including IBM's WebFountain and the KnowItAll system [8] that attempts to harvest from a web a lot of instantial knowledge. Weakly supervised learning methods for term generation use co-occurrence statistics [26]; [28], syntactic information [31]; [19]; [23], lexico-syntactic contextual patterns (e.g., "resides in <location>" or "moved to <location>") [32], and local and global contexts [9].

One problem is that the pattern harvests many unwanted concepts. A recent variant of it is the so-called Doubly-Anchored Pattern DAP recently proposed by the author and colleagues in [14]: "X such as Y and *". Here the researcher provides two seed terms, such as "animals" and "lions", and harvests other animal terms in the starred position. The harvested concepts are then ranked according to various scoring functions and replaced into the Y position to extend the harvesting. As shown by Kozareva et al., this method is both more accurate than the traditional single-seed patterns and produces more results. Another advantage of this method is that it can effectively differentiate between instances and basic-level concepts, depending on the nature of the seed Y: starting with "Animal such as lions and *" delivers a different set of terms than "Animals such as Lassie and *".

---

[1] The work described in Sections 2 and 3 was performed with Dr. Zornitsa Kozareva, Prof. Ellen Riloff, and the author, while the former two were visiting USC/ISI for a year in 2007-08.

[2] Here, *basic-level concepts are terms at the Basic Level in Prototype Theory [29]; they are the most general terms of a class that can be visualized; the terms usually learned first by children; often, the shortest words in the language; for example, "dog" (not "mammal" or "collie"); "car" (not "vehicle" or "BMW 520i"). In contrast, an* instance is a term more precise than a basic-level concept; it is a single individual entity, often named or identified by unique number, suhc as "Lassie", "Asian", and "BMW 520i with registration EX740N".

3   Learning Terms for Non-instance Concepts

By providing the conceptual supertype, the researcher limits term collection to a (small) localized taxonomy below it. With many patterns of the type "moved to <location>", the learned terms might be at any taxonomic level, but as mentioned the DAP pattern picks out a specific level that forms the lower fringe of the small taxonomy. As Hovy et al. [13] show, one can invert the DAP pattern to "* such as lions and tigers" (which they call DAP^{-1}) and then harvest new terms that are conceptually superordinate to the learned instance / basic-level terms, such as "mammals" and "hunters". Where [20] uses Hearst's patterns to learn semantic class instances and class groups by acquiring contexts around the pattern, he also developed a second technique [21] that creates context vectors for a group of seed instances by searching web query logs, and uses them to learn similar instances.

4   Learning Taxonomic Structure

There has been less work on automatically learning taxonomic structure. This is a considerably more complex problem, since there are in general many different parallel taxonomic organizations that can be placed 'over' a given set of basic-level / instance terms. As Kozareva et al. [15] show, for animals one can harvest a surprising variety of potential superclass terms, including such terms as "laboratory animals", "forest dweller", "endangered species" (which are probably useful), "bait", "allergens", "seafood", "vectors", "protein", "pests" (whose utility is unclear), and "native animals" and "large mammals" (with which some further decision on strategy is required). Similar work, using the basic Hearst [12] pattern, was reported also in [27], though this included manual tagging of common and proper nouns. Snow et al. [30] learned both instances and concepts simultaneously, using an ingenious model. Other work on reorganizing, augmenting, or extending semantic concepts already contained in manually built resources includes [34] on WordNet and [24] on Wikipedia. Lexical hierarchies — what one could call 'terminology ontologies' — are built by [4]; [5]; [18].

5   Learning Other Inter-term Relationships

Once one has created the basic skeletal structure of (some portion of) the ontology, one can start providing content. Unfortunately, there is no ready source of large amounts of semantic information, formally encoded, about concept interrelationships. WordNet is probably the largest freely available classification taxonomy, but its structure is based on cognitive principles (making it locally comprehensible without solving some thorny taxonomization problems, including multiple inheritance and concept 'facets'), and it is rather incomplete.

One promising approach is to collect material related to each concept in a fairly unstructured way, and then to provide structure (in the form of inter-term or even inter-concept relations) later. It requires three major steps: 1. Collecting words related to a target word(sense); 2. Discovering the relationship between the target and each of the words collected; 3. Placing the results into an ontology of concepts (or at least word senses). The first step involves creating a 'concept family' or *concept/topic signature* for each concept in the ontology. A concept (topic) signature is defined as a head concept (term) plus a set of related concepts (terms), each with strength of association to the head; an example is { *airplane* (*engine*, 0.178) (*wing*, 0.138) (*fly*, 0.103) ... } [16]. Such concept families blur the various associations q that exist among concepts (generalization, meronymy, co-occurrence in text, etc.). They indicate topic-relatedness without specifying the precise role of each topic relative to the others. And even without explicit relations, topic signatures have proven useful, for applications as diverse as:

Multi-document summarization [17]: Create a signature for each set (topic) of input texts; create a query using the signature terms; use IR to select sentences from the given topic's texts; filter and reorder the selected sentences into a single summary. This simple method performed very well in the international DUC-01 and DUC-02 competitions organized by NIST [6]; [7].

Wordsense disambiguation [1]: Having created (from the web) a term signature for every noun synset in Wordnet, Agirre et al. performed wordsense disambiguation by comparing the context of each ambiguous noun in the SemCor corpus to the signature words of every candidate sense of the noun. They found that disambiguation using the signatures performed better than either random choice (as expected) or the WordNet baseline.

The second step is to discover the relations between the signature head and each of its terms. Much past work has focused on harvesting various relations that exist between terms, using web or domain text; this includes especially meronymy (part-whole) relation learning by [3]; [11]. Like the taxonomic relations described above, this work generally uses (surface) patterns — sequences of words — that indicate when the desired relation holds between two terms [10]. Recently, research has focused on automated methods to learn such patterns. A simple method that requires the developer to provide a few seed terms as examples of the domain and range of the desired pattern was described in [25]. A great deal of similar work has shown a variety of methods [2] [8]; [23]; etc., some comparing fully-automated methods to manual selection and tuning of patterns, such as [22].

The third step is to place the term-relation-term triples into an ontology or taxonomy. There is less work on this challenge to date. The double-anchored pattern of Kozareva et al. might be used to accomplish this task.

6   Conclusion

Extracting terms and structure from the web, and (semi-)automatically building a term taxonomy, metadata organization, or even an ontology, is a very complex undertaking. But recent work in a variety of projects focusing on information extraction can be assembled and configured into a single overall project that might make this feasible, at least in certain kinds of well-circumscribed domains.

Should we succeed in developing this kind of capability, it should be possible to provide, for certain kinds of domains, a relatively comprehensive and empirically based terminology structure/taxonomy that — importantly — is directly linked into the source texts of the domain. This can be a crucial step in realizing the dream of the semantic web. It would certainly help to provide some kind of organization over the contents of the web, making access, standardization, and cross-language operation a lot simpler.

References

1. Agirre, E., Ansa, O., Martinez, D., Hovy, E.H.: Enriching WordNet Concepts with Topic Signatures. In: Proceedings of the NAACL Workshop on WordNet, Pittsburgh, PA (2001)
2. Banko, M., Cafarella, M., Soderland, S., Broadhead, M., Etzioni, O.: Open Information Extraction from the Web. In: Proceedings of International Joint Conference on Artificial Intelligence (IJCAI), pp. 2670–2676 (2007)
3. Berland, M., Charniak, E.: Finding Parts in Very Large Corpora. In: Proceedings of the 37th conference of the Association for Computational Linguistics (ACL) (1999)
4. Caraballo, S.: Automatic Acquisition of a Hypernym-Labeled Noun Hierarchy from Text. In: Proceedings of the 37th conference of the Association for Computational Linguistics (ACL), pp. 120–126 (1999)
5. Cimiano, P., Volker, J.: Towards Large-Scale, Open-Domain and Ontology-Based Named Entity Classification. In: Proceedings of the RANLP 2005 conference, pp. 166–172 (2005)
6. DUC conference (2001), http://duc.nist.gov/
7. DUC conference (2002), http://duc.nist.gov/
8. Etzioni, O., Cafarella, M., Downey, D., Popescu, A., Shaked, T., Soderland, S., Weld, D., Yates, A.: Unsupervised Named-Entity Extraction from the Web: An Experimental Study. Artificial Intelligence 165(1), 91–134 (2005)
9. Fleischman, M., Hovy, E.H.: Fine Grained Classification of Named Entities. In: Proceedings of the international conference on Computational Linguistics (COLING), Taipei, Taiwan (2002)
10. Freitag, D.: Toward General-Purpose Learning for Information Extraction. In: Proceedings of the 36th conference of the Association for Computational Linguistics and 17th international conference on Computational Linguistics (COLING-ACL) Montreal, Quebec, pp. 404–408 (1998)
11. Girju, R., Badulescu, A., Moldovan, D.: Learning Semantic Constraints for the Automatic Discovery of Part-whole Relations. In: Proceedings of the HLT-NAACL conference (2003)
12. Hearst, M.: Automatic Acquisition of Hyponyms from Large Text Corpora. In: Proceedings of the 14th international conference on Computational Linguistics (COLING), pp. 539–545 (1992)
13. Hovy, E.H., Kozareva, Z., Riloff, E.: Toward Completeness in Concept Extraction and Classification. In: Proceedings of the conference of Empirical Methods in Natural Language Processing (EMNLP), Singapore (2009)
14. Kozareva, Z., Riloff, E., Hovy, E.H.: Semantic Class Learning from the Web with Hyponym Pattern Linkage Graphs. In: Proceedings of the 46th conference of the Association of Computational Linguistics (ACL), Columbus, OH (2008)
15. Kozareva, Z., Hovy, E.H., Riloff, E.: Learning and Evaluating the Content and the Structure of a Term Taxonomy. In: Proceedings of the AAAI Spring Symposium on Learning by Reading and Learning to Read Stanford University, CA (2009)

16. Lin, C.-Y., Hovy, E.H.: The Automated Acquisition of Topic Signatures for Text Summarization. In: Proceedings of the 18th international conference on Computational Linguistics (COLING), Strasbourg, France (2000)
17. Lin, C.-Y., Hovy, E.H.: Automated Multi-Document Summarization in NaATS. In: Proceedings of the Human Language Technology Conference (HLT), San Diego, California (2002)
18. Mann, G.: Fine-grained Proper Noun Ontologies for Question Answering. In: Proceedings of the 19th international conference on Computational Linguistics (COLING), pp. 1–7 (2002)
19. Pantel, P., Ravichandran, D.: Automatically Labeling Semantic Classes. In: Proceedings of the HLT-NAACL conference, pp. 321–328 (2004)
20. Pasca, M.: Acquisition of Categorized Named Entities for Web Search. In: Proceedings of the CIKM conference, pp. 137–145 (2004)
21. Pasca, M.: Weakly-supervised Discovery of Named Entities using Web Search Queries. In: Proceedings of the CIKM conference, pp. 683–690 (2007)
22. Patwardhan, S., Riloff, E.: Effective Information Extraction with Semantic Affinity Patterns and Relevant Regions. In: Proceedings of the joint conference on Empirical Methods in Natural Language Processing and Computational Natural Language Learning (EMNLP-CoNLL) Prague, Czech Republic, pp. 717–727 (2007)
23. Phillips, W., Riloff, E.: Exploiting Strong Syntactic Heuristics and Co-Training to Learn Semantic Lexicons. In: Proceedings of the conference on Empirical Methods in Natural Language Processing (EMNLP) (2002)
24. Ponzetto, S., Strube, M.: Deriving a Large scale Taxonomy from Wikipedia. In: Proceedings of the 22nd national conference on Artificial Intelligence (AAAI), pp. 1440–1447 (2007)
25. Ravichandran, D., Hovy, E.H.: Learning Surface Text Patterns for a Question Answering System. In: Proceedings of the 40th conference of the Association for Computational Linguistics (ACL), Philadelphia, PA (2002)
26. Riloff, E., Shepherd, J.: A Corpus-Based Approach for Building Semantic Lexicons. In: Proceedings of the 2nd conference on Empirical Methods in Natural Language Processing (EMNLP), pp. 117–124 (1997)
27. Ritter, A., Soderland, S., Etzioni, O.: What is This, Anyway: Automatic Hypernym Discovery. In: Proceedings of the AAAI 2009 Spring Symposium on Learning by Reading and Learning to Read, pp. 88–93. Stanford University, Stanford (2009)
28. Roark, B., Charniak, E.: Noun-phrase Cooccurrence Statistics for Semi-automatic Semantic Lexicon Construction. In: Proceedings of the 36th conference of the Association for Computational Linguistics (ACL), pp. 1110–1116 (1998)
29. Rosch, E.: Principles of Categorization, pp. 27–48 (1978)
30. Snow, R., Jurafsky, D., Ng, A.Y.: Learning Syntactic Patterns for Automatic Hypernym Discovery. In: Proceedings of the NIPS conference (2005)
31. Tanev, H., Magnini, B.: Weakly Supervised Approaches for Ontology Population. In: Proceedings of the 11th conference of the European Chapter of the Association for Computational Linguistics (EACL) (2006)
32. Thelen, M., Riloff, E.: A Bootstrapping Method for Learning Semantic Lexicons Using Extraction Pattern Contexts. In: Proceedings of the conference on Empirical Methods in Natural Language Processing (EMNLP), pp. 214–221 (2002)
33. TREC conferences, http://trec.nist.gov/
34. Widdows, D., Dorow, B.: A Graph Model for Unsupervised Lexical Acquisition. In: Proceedings of the 19th international conference on Computational Linguistics (COLING), pp. 1–7 (2002)

Analysis and Improvement of Minimally Supervised Machine Learning for Relation Extraction

Hans Uszkoreit, Feiyu Xu, and Hong Li

DFKI GmbH, LT Lab
Stuhlsatzenhausweg 3, D-66123 Saarbrücken
{uszkoreit,feiyu,lihong}@dfki.de

Abstract. The main contribution of this paper is a systematic analysis of a minimally supervised machine learning method for relation extraction grammars. The method is based on a bootstrapping approach in which the bootstrapping is triggered by semantic seeds. The starting point of our analysis is the pattern-learning graph which is a subgraph of the bipartite graph representing all connections between linguistic patterns and relation instances exhibited by the data. It is shown that the performance of such general learning framework for actual tasks is dependent on certain properties of the data and on the selection of seeds. Several experiments have been conducted to gain explanatory insights into the interaction of these two factors. From the investigation of more effective seeds and benevolent data we understand how to improve the learning in less fortunate configurations. A relation extraction method only based on positive examples cannot avoid all false positives, especially when the data properties yield a high recall. Therefore, negative seeds are employed to learn negative patterns, which boost precision.

1 Introduction

The charm and the power of the seed-based minimally supervised machine learning method within a bootstrapping framework has been widely recognized and frequently employed for various information extraction tasks (e.g., [8,10,6,3,12,7,4]). The approach has evolved into an empirically promising and theoretically attractive research strand, dedicated to the automatic acquisition of extraction patterns or rules from unannotated textual data (e.g., [18,6,5,3,1,2,9,15,14,16,13]). The only task-specific knowledge provided to the automatic learning process is a small set of examples of either patterns or semantic instances. Several methods have been developed that accomplish rather decent results with a minimum of effort [6,3,12,9,11]. It is mentioned in all these approaches that their seed selection is random, without any explicit criteria. In fact, some experiments conducted by [9] for named entity extraction and those by [13] for relation extraction demonstrate that certain data properties yield a high recall even with a very small number of seed examples.

[13] investigates the role of the seed selection in connection with the data properties in a careful way. Through some dedicated experiments we could obtain

H. Horacek et al. (Eds.): NLDB 2009, LNCS 5723, pp. 8–23, 2010.

and confirm some new insights on the relevant properties of seed and data. From these insights we derive proposed solutions, some of which could already be empirically validated. All measures for improving recall at the same time trigger false positives. We have observed various sources of degrading precision and propose to employ negative patterns as filters for some classes of false positives. In order to learn such negative patterns we need negative seeds. We will explain the construction of negative seeds and discuss some problems for generalizing these to effective negative rules.

The remainder of the paper is organized as follows: Section 2 describes the state of the art of approaches to seed construction and provides a systematic analysis of the properties of the seed construction and provides the first solutions to lucky seed. Section 3 shows the interaction between the complexity of the target relation, the seed construction and the performance. Section 4 presents the experimental results of the integration of negative rules. Section 5 concludes and opens discussions for the future research.

2 Magic and Challenges Around Seed

2.1 Seed Representation

With respect to the employed seeds, the bootstrapping approaches to relation extraction fall in two classes:

- pattern based seeds
- semantics (relation instance) based seeds

The former class uses simple linguistic patterns that indicate the target relation (e.g., [17,12,7]). A typical example of this tradition is the ExDisco system [17]. (1) is an example pattern of the management succession domain, e.g.,

(1) *subj(company) verb("appoint") obj(person)*

Patterns serve as structured queries for document retrieval (e.g., [12,7]). If new patterns from the retrieved documents are found, they will be used to retrieve more documents again. Each iteration in the boostrapping lerning process contains two steps: patterns (p_i) extraction and documents (m_n) retrieval as depicted in Figure 1.

In general, these patterns cannot be applied for relation extraction straightforwardly because there is no semantic role labelling information between the linguistic arguments in the patterns and semantic arguments specified in the target relations. In example (1) mentioned above, the object argument does not contain the information that it fills the role for a new person which takes over the position. Furthermore, the linguistic patterns are mostly flatter lists of grammatical functions such as a "subj-verb-obj" sequence and do not allow or even afford recursive or hierarchical structures. Therefore, these pattern-based approaches are especially suitable for simple relations that can be expressed by such simple linguistic patterns. However, complex relations with more arguments often cannot be expressed by such simple patterns. Let us consider an example relation

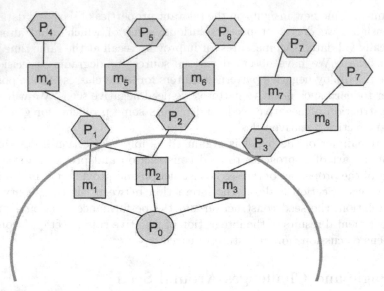

Fig. 1. Learning graph starting from pattern-based seed. p_i: patterns, m_j: textual snippets/documents.

from the prize award domain. The relation contains four arguments representing an event in which a person or an organization won a particular prize in a specific area and in a certain year:

(2) *<recipient, award, area, year>*

(3) is an example relation instance of (2), referring to an event mentioned in the sentence (4).

(3) *<Mohamed ElBaradei, Nobel, Peace, 2005>*
(4) *Mohamed ElBaradei, won the 2005 Nobel Prize for Peace on Friday for his efforts to limit the spread of atomic weapons.*

(5) is a simplified dependency tree of the parsing result of (4). The pattern-based approaches mentioned above do not provide strategies to extract hierarchical patterns like (5).

(5)

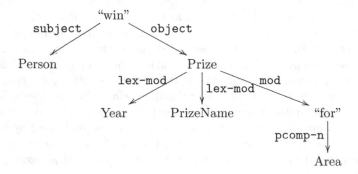

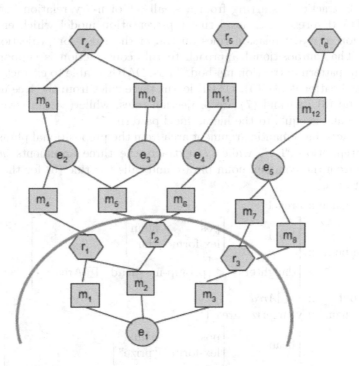

Fig. 2. Learning graph starting from semantics-based seed. e_i: relation instances; r_i: extraction rules, mj: textual snippets.

Moreover, it is difficult for domain experts but non-linguists to formulate a pattern like (5). In comparison to pattern-based approaches a seed example in the semantics-based approach is very easy to formulate, even for complex relations. It is simply an instance of the target relation, namely, a database record like (3). Users just have to provide some examples and no linguistic knowledge is needed.

In comparison to the pattern-based approach, each iteration in the learning process of the semantics based approach contains one more step, namely, the instance (e_i) extraction (e.g., [6,3]), see Figure 2. In Figure 2, the learning process starts with the instances (e.g., e_1) as seed and find textual snippets (e.g., m_1, m_2, m_3) which match the seed and then extract pattern rules (e.g, r_1, r_2, r_3). The learned pattern rules can be applied to the documents to discover new instances. The new instances can be used as seed for the next iteration again. Hence, a kind of co-training takes place here: two classifiers are applied to the textual snippets (m_i) to decide whether they mention the target relation (see [5]):

- one utilizes the semantic-instance as feature and
- another relies on the pattern rules.

Among the approaches in this class, [13] and [16] present a general framework for extracting relations of various complexity, called DARE (Domain Adaptive

Relation Extraction), starting from a small set of n-ary relation instances as "seed". DARE presents a novel rule representation model which enables the composition of n-ary relation rules on top of the rules for projections of the relation. The compositional approach to rule construction is supported by a bottom-up pattern extraction method. Thus, DARE is able to extract and build hierarchical patterns like (5). DARE learns three rules from the tree in (5), i.e., (6), (7) and (8). (6) and (7) are projection rules, while (8) can cover all four arguments and is equal to the hierarchical pattern (5).

(6) extracts the semantic argument area from the prepositional phrase headed by the preposition "for", while (7) extracts the three arguments *year*, *prize* and *area* from the complex noun phrase and calls the rule (6) for the semantic argument area.

(6) Rule name :: area_1

$$\text{Rule body} :: \begin{bmatrix} \text{head} & \begin{bmatrix} \text{pos} & \text{noun} \\ \text{lex-form} & \text{``for''} \end{bmatrix} \\ \text{daughters} & < \begin{bmatrix} \text{pcomp-n} & \begin{bmatrix} \text{head} & \boxed{1}\ \text{Area} \end{bmatrix} \end{bmatrix} > \end{bmatrix}$$

Output   :: $< \boxed{1}Area >$

(7) Rule name :: year_prize_area_1

$$\text{Rule body} :: \begin{bmatrix} \text{head} & \begin{bmatrix} \text{pos} & \text{noun} \\ \text{lex-form} & \text{``prize''} \end{bmatrix} \\ \text{daughters} & < \begin{bmatrix} \text{lex-mod} & \begin{bmatrix} \text{head} & \boxed{1}\ \text{Year} \end{bmatrix} \end{bmatrix}, \\ & \begin{bmatrix} \text{lex-mod} & \begin{bmatrix} \text{head} & \boxed{2}\ \text{Prize} \end{bmatrix} \end{bmatrix}, \\ & \begin{bmatrix} \text{mod} & \begin{bmatrix} \text{rule} & \text{area\_1} :: <\boxed{3}Area> \end{bmatrix} \end{bmatrix} > \end{bmatrix}$$

Output   :: $< \boxed{1}Year, \boxed{2}Prize, \boxed{3}Area >$

(8) is the rule that extracts all four arguments from the verb phrase dominated by the verb "win" and calls (7) to handle the arguments embedded in the linguistic argument "object".

(8) Rule name :: recipient_prize_area_year_1
Rule body ::

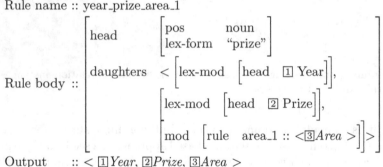

$$\begin{bmatrix} \text{head} & \begin{bmatrix} \text{pos} & \text{verb} \\ \text{mode} & \text{active} \\ \text{lex-form} & \text{``win''} \end{bmatrix} \\ \text{daughters} & < \begin{bmatrix} \text{subject} & \begin{bmatrix} \text{head} & \boxed{1}\ \text{Person} \end{bmatrix} \end{bmatrix}, \\ & \begin{bmatrix} \text{object} & \begin{bmatrix} \text{rule} & \text{year\_prize\_area\_1} :: \\ & < \boxed{4}Year, \boxed{2}Prize, \boxed{3}Area > \end{bmatrix} \end{bmatrix} > \end{bmatrix}$$

Output   :: $< \boxed{1}Recipient, \boxed{2}Prize, \boxed{3}Area, \boxed{4}Year >$

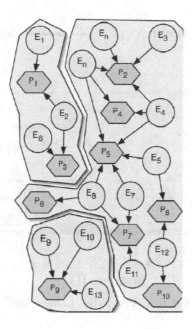

Fig. 3. Bipartie Graph of Interplay between Instances and Rules

A freely accessible online demonstration `http://dare.dfki.de` illustrates the learning process of DARE for the prize award domain. As demonstrated through the examples of DARE rules, an important advantage of this approach is that the learned rules do not only represent the linguistic patterns but also contain the relevant semantic role labels. Thus, rules learned by this strategy are real extraction rules.

2.2 Seed Magic and Lucky Seed

Although the seed plays a central role in the minimally supervised machine-learning framework, its construction is often left underspecified in the literature. Seeds are either randomly chosen or they are stipulated by users (e.g., [3,16]).

As illustrated in Figure 3, a bipartite graph can describe the connections between all mentions of instances and the patterns (or rules) that express these instances in the learning data corpus. The correct part of the learning graph is a subgraph of the bipartite graph containing all instances recognized and all the rules that were correctly constructed. If the learning starts with several instance examples, the learning graph can be composed of several subgraphs. Some of these may be small islands containing very few connections, while others appear as continents with many connections. In Figure 3, the learning graph contains a big continent and two small islands.

[13] reported on two experiments with two different domains A and B each with a distinct corpus. Corpus A contains almost 2300 documents and is more than ten size larger than corpus B. Corpus A is collected from various newspapers

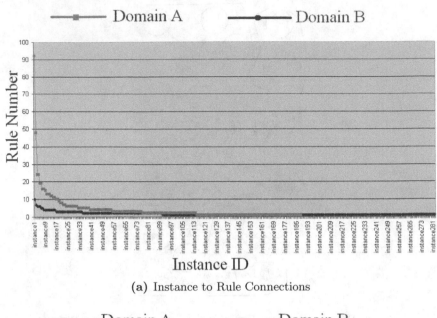

(a) Instance to Rule Connections

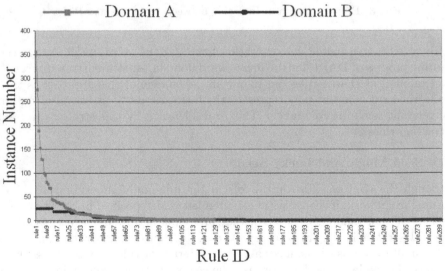

(b) Rule to Instance Relation

Fig. 4. Distribution of the connectivity degree between learned rules and instances in domains

and corpus B comes from one newspaper. The target relations for both domains are quaternary relations.

Figure 4a and Figure 4b show the distribution of the connectivity degree between learned rules and instances in the two domains.

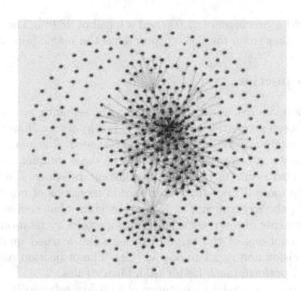

Fig. 5. Connectivity of Instances in Domain A

In Domain A, one semantic-seed is sufficient to find a large number of rules (Figure 4a) and one rule suffices to extract many instances (Figure 4b). The distribution behavior in Domain A corresponds to Zipfs law and confirms the results reported by [9], namely, the Power Law distribution. Both distributions are skewed, but in the domain A, we get some sort of scale-free graph, i.e., $P(k) \sim k^r$. In this case, the graph displays the so-called small world property. We can reach nearly all events from any seed in a few steps, even if the graph grows. The reason is simple: in order to learn the numerous less frequent patterns in the heavy tail of the distribution, we need "event hubs". But we need the less frequent patterns in order to get to many events mentioned only once.

The connected instances in domain A in the bootstrapping process can be visualized as a graph with one large component (continent) and many small ones (islands) (see Figure 5). In Figure 5, there are 138 components in the graph. The largest graph contains 298 mentioned instances, while one island contains 2 instances and 136 islands only have one instance. In order to have 100% recall, we would need 138 seed examples for Domain A corpus, of which 137 would have to miraculously match the instances on the islands. However, one seed example in the biggest component can already achieve a maximal recall of 68%.

Thus any instance on the continent is a lucky seed. Such seed triggers the apparent magic reported in the literature: for domain A, a precision of 80.59% and a recall of 62.9% could be achieved with only one example as seed.

The connectivity behavior in Domain B is completely different from Domain A. We find a boring distribution: most instances are just mentioned a single time. Patterns and instances have a very small degree of connectivity (Figure 4a and Figure 4b). Thus, more instances are needed as seed to discover enough patterns. Even with 20 seed examples, its precision is 48.4% and recall 34.2%.

55 examples yield a precision of 62.0% and a recall of 48.0%. The chances for a lucky seed are rather small for such corpus. Thus the magic fails in this case.

3   Seed Properties

In the last section, we discussed the advantages of semantic seeds and also the relationship between properties of the data and lucky seeds. However, a pattern seed has more descriptive content than a semantic seed. A semantic seed is simply a list of semantic arguments. As we know, the same combination of semantic arguments can also occur in mentions of different relations. The simpler the combination, the more likely is its occurrence in mentions of multiple relationships. In [13,15], the experiments in Domain A show that sentences matching all or most arguments of a semantic seed for a quaternary relation are the best candidates for mentions of the target relation. The learned quaternary rules exhibit best precision and recall among all rules. The projection rules with three arguments again perform much better than binary rules.

Let $r = < a_1, a_2, ..., a_n >$ be an instance of an n-ary relation R.

Then T_r, the tuple-mention set of r, is the set of all segments (i.e., sentences) in which mentions of all arguments are detected. T_r' is a subset of T_r containing exactly those segments that actually intensionally refer to the instance of r. The ratio T_r'/T_r we call distinctiveness. The more distinctive a seed, the better for precision.

If we have a large number of correct seed instances available, we can use all of them in order to secure a good start of the learning process. If we only have a small set, we need to think about ways to increase the seed fast during the first step without already collecting too many false positives.

Assume we want to build a database of married couples from the web. If we use as seeds some pairs of the form:

   <wife: "*Robin Wright*", husband: "*Sean Penn*">,

then the seed might also trigger many mentions of other events or relations involving the two persons, i.e., *meeting, co-starring, fighting* and *separating*.

An effective method for increasing the distinctiveness is the extension of the n-ary relation by increasing the arity of the relation and accordingly of the seed. If wisely selected, the addition of another relevant argument for the first step of the bootstrapping can make the relation more distinctive. In subsequent steps, the original nary relation will be among the selected projections of the initial n+1-ary relation.

Another strategy for increasing the distinctiveness is to use seed examples whose participants are less likely to appear in reports about other relation types besides the target relation. If we select as a seed pair two participants who rarely appear in the data (in our case in the news) we may not get the desired degree of redundancy. Thus we have to restrict this strategy to one participant, e.g., select a seed couple in which only one partner enjoys the desired level of popularity.

We are now going to back up these proposals by measured evidence. To this end we stay with the detection of married couples. We consider the following parameters:

a) arity of the seed example
b) size of the seed set
c) distinctiveness of the seed example

In case of a), two options are taken into account

1. <person1, person2>
2. <person1, person2, marriage year>

In the case b), we assume that including more pairs exhibiting the same relation will help to find the relevant patterns fast. For the case c), we selected couples of which one partner does not appear in the press except as spouse of the more popular partner. We also picked people who had not been married before or after to further increase the distinctiveness of the seed examples. In our experiment we vary the three factors.

In order to avoid data sparseness and unlucky seeds, we choose only prominent persons from the Wikipedia. 313 persons were selected, belonging to the English Wikipedia categories "Times person", "presidents of US", "best drama actor Golden Globe", "best actor of Academy award", "best actress of Academy award" and "best support actor Golden Globe". Our data setup contains 313 documents. We extract 11733 sentences which contain mentionings of two persons. We conducted one experiment for each of the six seed configurations:

(1) arity 2, size 1, more distinctive
(2) arity 2, size 3, more distinctive
(3) arity 3, size 3, more distinctive

To explain our attempt to increase distinctiveness: The seeds contain couples whose spouses married only once, and for which one spouse is not mentioned in other contexts in the Wikipedia, i.e., the US president "Andrew Johnson" and his wife "Eliza McCardle".

(4) arity 2, size 3, less distinctive
(5) arity 3, size 3, less distinctive

Each person in the above marriage couples have married at least twice and both partners are involved in different mentioned relations, e.g., "Audrey Hepburn" and "Mel Ferrer".

(6) arity 3, size 6, mixed distinctiveness

The seed contains three more distinctive and three less distinctive couples with their respective wedding years. Figure 6 illustrates the recall and the precision results with the above six seed configurations. The most striking result is the evidence showing that more distinctive examples achieve much better precision and better recall than less distinctive seeds. Raising the arity for increased distinctiveness results in slight improvements of precision and recall. Increasing the number of seed examples does not help with the Wikipedia data, while the size of the seed set plays a central role for the experiment with Domain B mentioned in Section 2 ([14]) whose corpus exhibits less redundancy and connectivity.

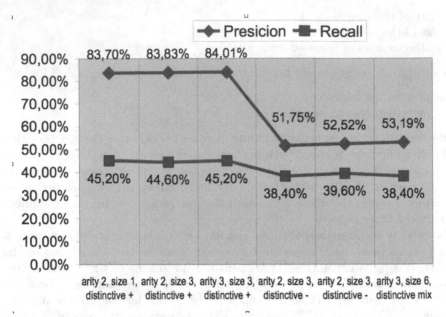

Fig. 6. Arity, size and distinctiveness

4   Negative Seed for Negative Rules

An ideal learning graph contains all instances of the target relation that are mentioned in the data and therefore probably also all inducible rules. As mentioned in Section 3, the ambiguity potential of the semantic seed is one of the error sources, in particular, those of the simple relations. In the experiment with Domain A ([14]), a systematic evaluation has been conducted for errors. It turned out that modality (17.6%) and rules (14.7%) are two major error resources in addition to factual errors in the content and parsing errors.

Facts or events can be embedded in the scope of modalities that either deny or weaken the truth of the statements. The relevant mentions of the relation may occur within the scope of negation particles, modal verbs, modal adverbials propositional attitude verbs, and other modifications affecting the truth value. In example (9) from Domain corpus A, the facts embedded within the scope of "speculation" are wrong.

(9)   *The talk has included speculation [that North Korean leader Kim Jong Il and South Korean President Kim Dae-jung might win the Nobel Peace Prize for their step toward reconciliation, the most promising sign of rapprochement since the Korean war ended with a fragile truce in 1953]*

Learned rules can be classified into four categories: good, useless, dangerous and bad. Good rules extract only correct instances. Useless rules are rules that do not contribute any extractions of new instances in the bootstrapping process. Bad rules are those that discover only wrong instances. In Domain A, the bad rules only resulted in 5% of the error instances. Dangerous rules are the rules, which extract both correct and incorrect instances. They exhibit the highest error spreading potential and they caused most of the errors in our experiments. Often they detect semantically related relations whose extensions overlap with the target relation. In the prize award domain, for instance, the patterns headed by the verb "nominate" extract many award winners, because they had also been nominated before being selected. In the marriage domain, some high-ranked dangerous rules are headed by the semantically independent verbs "meet" and "play" because these are two other reported relations in which the spouses often stand. A tempting strategy for preventing bad and dangerous rules is to apply negative rules learned from negative examples as filters of the learned instances. There are two semantically different negative seed sorts

1. dangerous instances:
   seed examples that also represent other relations with the same argument combination as the target relation;
2. wrong facts or events:
   seed examples that are explicitly mentioned as wrong instances in the corpus.

The dangerous instances already help us to learn dangerous rules, while the wrong facts or events will hopefully lead us to learn modalities affecting the truth value. In the following, we present two bootstrapping methods that integrate learning of negative rules. By the first algorithm (Method 1), we learn at first the negative rules from the negative seed and then integrate the negative rules in each positive rule learning iteration, see the following detailed description. It is a conservative method because it learns the negative rules from the initial negative seed set. The number of learned rules grows with the cardinality of the seed set.

Algorithm 1. Method 1 - Single Step Negative Rule Learning

Given: PositiveSeeds, NegativeSeeds

```
//get negative patterns
NegativePatterns = getPaterns(NegativeSeeds);

//bootstrapping for positive seeds
while (PositiveSeeds not empty) {

  patterns = getPatterns(PositiveSeeds);

  //remove the negative patterns
  patterns = patterns - NegativePatterns ;

  PositiveSeeds = getInstance(patterns);

}
```

Algorithm 2. Method 2 - Double Negative Rule Learning
___
Given: PositiveSeeds, NegativeSeeds

//get negative patterns
NegativePatterns = *getPaterns*(NegativeSeeds);

//double bootstrapping
while (PositiveSeeds not empty) {

   PositivePatterns = *getPatterns*(PositiveSeeds);
   PositivePatterns = PositivePatterns NegativePatterns;
   PositiveSeeds = *getInstance*(PositivePatterns);

   //learning for negative patterns
   NegativeSeeds = *getInstance*(NegativePatterns);
   NegativeSeeds = NegativeSeeds PositiveSeeds;
   NegativePatterns += getPatterns(NegativeSeeds);

}
___

The second method (Method 2) is an extension of method 1. It allows the bootstrapping of negative seeds. The negative seeds will be the set of instances learned by the negative patterns minus the positive seeds. This method is an eager approach, since the new negative seed can potentially contain positive instances that are not member of the known positive seed. This extended notion of negative seed will cause incorrect negative rules.

We first conduct the experiments in the marriage domain. We consider the following relations as dangerous: "meet", "work-with", "affair", "siblings" and "friendship" and construct a negative seed set of these relations for our learning system. Table 1 shows the experiment results of the six options with the two negative methods in comparison to the previous results shown in Figure 6. The results in Table 1 show that the integration of negative seeds has improved the precision values, in particular for the less distinctive seed examples. Method 2 works a little better than method 1 for precision.

Table 2 presents the number of correct instances extracted before and after integration of the negative seeds. Method 1 slightly reduces the number of correct instances whereas Method 2 really damages recall. This means that the eager approach has also deleted positive seeds during the bootstrapping process.

The second experiment applying Method 1 uses the same corpus for the extraction of prize and award events. The negative seed is an instance of the event "nomination". The results are listed in Table 3. The precision has considerably improved without hurting recall.

For constructing the second kind of negative seed, namely, explicitly negated facts and events mentioned in the corpus, a corpus analysis is needed. We take the corpus for Domain A and try to find the negative examples of Nobel Prize winners. "Rudy Giuliani" is among the frequently mentioned persons who also co-occur with the concept "Nobel Prize".

Table 1. Precision with Negative Seeds

methods	without negative	with negative method 1	with negative method 2
arity 2, size 1, more distinctive	83,70%	96,23%	97,50%
arity 2, size 3, more distinctive	83,83%	96,17%	97,44%
arity 3, size 3, more distinctive	84,01%	96,71%	97,44%
arity 2, size 3, less distinctive	51,75%	96,19%	96,43%
arity 3, size 3, less distinctive	52,52%	96,02%	96,56%
arity 3, size 6, mixture	53,19%	95,91%	96,43%

Table 2. Correct Results with Negative Seeds

methods	without negative	with negative method 1	with negative method 2
arity 2, size 1, distinctive	226	204	117
arity 2, size 3, distinctive	223	201	152
arity 3, size 3, distinctive	226	204	152
arity 2, size 3, indistinctive	192	202	135
arity 3, size 3, indistinctive	198	169	138
arity 3, size 6, mixture	192	164	135

Table 3. Negative Seed for prize and award domain

methods	without negative	with negative
precision	69,48%	74,89%
recall	42,82%	42,82%

The mentioning of "Rudy Giuliani" and the "Nobel Prize" is embedded in the scope of the modal adjective "possible":

(10) *It's also possible [that O.J. Simpson will find the real killer, that Bill Clinton will enter a monastery and that Rudy Giuliani will win the Nobel Peace Prize.]*

In fact, modalities can be expressed in a variety of ways, e.g., by a noun such as "speculation" in (9), or by modal adjective or adverb such as "possible" like (10) or "never", or by some ctional contexts provided by films or novels. The linguistic structures embedded in the modality scopes are highlighted by us with brackets. (10) poses an additional challenge because of irony. Sentence (11) introduces a fictional Nobel Prize winner, "Josiah Bartlett", announced by a TV program. Thus, world knowledge is needed here to resolve the modality.

(11) *In NBC "West Wing," [we get President Josiah Bartlett, a Nobel Prize Winner]*

Our current rule-inducing algorithm cannot yet identify and learn the relevant constructions outside the actual mention of the relation instance because this would require a more sophisticated generalization step. Work on extending our rule generalization is in progress.

5   Conclusion and Future Work

Curiosity was our original motivation for the systematic analysis of the effects of data and seed properties for the performance of the learning method. After realizing that performance varied drastically among different domains and data sets, we wanted to understand the reasons for this variation. The deeper understanding of the interaction of target relation, seeds and data then lead us to some hypotheses concerning strategies for improving the method. Most of these strategies could already be tested, some of them with positive results.

The implementation and validation of the other proposed strategies require considerable additional efforts, which are now on our plans for future research. Among them are the learning of negative patterns and rules from a small subset of the relation together with its complement. In the case of Prize Awards, Nobel Prize Winners constitute such a closed subset, since all laureates are known. For many other domains such a set can easily be constructed. An example for marriages of movie celebrities is a set of known non-couples among frequently co-occurring pairs of movie stars.

Acknowledgement

The presented research was partially supported through the project TAKE funded by a grant from the German Federal Ministry of Education and Research (FKZ: 01IW08003) and through the project KomParse funded by the ProFIT programme of the Federal State of Berlin and the EFRE programme of the European Union.

References

1. Abney, S.: Bootstrapping. In: Proceedings of the 40th Annual Meeting of the Association for Computational Linguistics, pp. 360–367 (2002)
2. Abney, S.: Understanding the Yarowsky algorithm. Computational Linguistics 30(3), 365–395 (2004)
3. Agichtein, E., Gravano, L.: Snowball: Extracting relations from large plain-text collections. In: Proceedings of the 5th ACM International Conference on Digital Libraries (DL 2000), San Antonio, TX (June 2000)
4. Blohm, S., Cimiano, P.: Using the Web to Reduce Data Sparseness in Pattern-based Information Extraction. In: Kok, J.N., Koronacki, J., Lopez de Mantaras, R., Matwin, S., Mladenič, D., Skowron, A. (eds.) PKDD 2007. LNCS (LNAI), vol. 4702, pp. 18–29. Springer, Heidelberg (2007)
5. Blum, A., Mitchell, T.M.: Combining labeled and unlabeled sata with co-training. In: COLT, pp. 92–100 (1998)

6. Brin, S.: Extracting patterns and relations from the world wide web. In: WebDB Workshop at 6th International Conference on Extending Database Technology, EDBT 1998 (1998)
7. Greenwood, M.A., Stevenson, M.: Improving semi-supervised acquisition of relation extraction patterns. In: Proceedings of the Workshop on Information Extraction Beyond The Document, Sydney, Australia, July 2006, pp. 29–35. Association for Computational Linguistics (2006)
8. Hearst, M.A.: Automatic Acquisition of Hyponyms om Large Text Corpora. In: Proceedings of the Fourteenth International Conference on Computational Linguistics (1992)
9. Jones, R.: Learning to Extract Entities from Labeled and Unlabeled Text. PhD thesis, University of Utah (2005)
10. Riloff, E.: Automatically generating extraction patterns from untagged text. In: Proceedings of Thirteenth National Conference on Artificial Intelligence (AAAI 1996), pp. 1044–1049. AAAI Press/MIT Press (1996)
11. Stevenson, M., Greenwood, M.A.: A Semantic Approach to IE Pattern Induction. Ann. Arbor. 100 (2005)
12. Sudo, K., Sekine, S., Grishman, R.: An improved extraction pattern representation model for automatic IE pattern acquisition. In: Proceedings of ACL 2003, pp. 224–231 (2003)
13. Xu, F.: Bootstrapping Relation Extraction from Semantic Seeds. Phd-thesis, Saarland University (2007)
14. Xu, F., Uszkoreit, H.: Minimally supervised learning of relation extraction rules using semantic seeds. In: A seminar talk at the National Center for Text Mining (NaCTeM) (May 2007)
15. Xu, F., Uszkoreit, H., Li, H.: Automatic event and relation detection with seeds of varying complexity. In: Proceedings of AAAI 2006 Workshop Event Extraction and Synthesis, Boston (July 2006)
16. Xu, F., Uszkoreit, H., Li, H.: A seed-driven bottom-up machine learning framework for extracting relations of various complexity. In: Proceedings of the 45th Annual Meeting of the Association of Computational Linguistics (ACL 2007), pp. 584–591 (2007)
17. Yangarber, R.: Scenarion Customization for Information Extraction. Dissertation, Department of Computer Science, Graduate School of Arts and Science, New York University, New York, USA (2001)
18. Yarowsky, D.: Unsupervised word sense disambiguation rivaling supervised methods. In: Proceedings of the 33rd annual meeting on Association for Computational Linguistics (ACL), Morristown, pp. 189–196 (1995)

Using Evidences Based on Natural Language to Drive the Process of Fusing Multimodal Sources

Sergio Navarro, Fernando Llopis, and Rafael Muñoz

Natural Language Processing and Information Systems Group,
University of Alicante, Spain
{snavarro,rafael,llopis}@dlsi.ua.es
http://gplsi.dlsi.ua.es

Abstract. This paper focuses on the proposal and evaluation of two multimodal fusion techniques in the field of Visual Information Retrieval (VIR). These proposals are based on two widely used fusion strategies in the VIR area, the multimodal blind relevance feedback and the multimodal re-ranking strategy. Unlike the existent techniques, our alternative proposals are guided by the evidence found in the natural language annotations related to the images. The results achieved by our runs in two different ImageCLEF tasks, 3rd place in the Wikipedia task [1] and 4th place within all the automatic runs in the photo task [2], jointly with the results obtained in later experiments presented in this paper show us that the use of conceptual information associated with an image can improve significantly the performance of the original multimodal fusion techniques used.

1 Introduction

The wide variety of digital formats on the Internet and the boom of multimedia content create the need to develop and/or adapt tools for finding information with these new characteristics such as video and image among others. Currently, the well known commercial multimedia search engines such as Youtube[1] or Flickr[2] only base their searches on the text related to the image or video indexed. The development of such tools is beyond the scope of the VIR research field and specifically beyond the scope of this paper. We can say that the VIR is a specific area within the Information Retrieval (IR), which in fact initially used traditional IR systems without any specific adaptation to the VIR, performing searches only using the annotations related to the images. Thus, the collections used by VIR systems are composed of images and their related annotations describing their content.

Historically in VIR area there were two approaches used to carry out the IR of images: In the beginning of the VIR in the late 70s, VIR systems were based on the image annotations, therefore, these were Text-Based VIR (TBIR)

---

[1] *http://www.youtube.com*
[2] *http://www.flickr.com*

H. Horacek et al. (Eds.): NLDB 2009, LNCS 5723, pp. 24–35, 2010.

systems. Later in the early 90's, in an attempt to overcome the dependence of TBIR systems from the existence of textual annotations to perform the indexing of an image, the image Content Based VIR (CBIR) systems appear [3].

Finally, in recent years as the technologies used by CBIR systems matured, a third approach to tackle the problem of the VIR emerged, these systems combine textual and image based technologies. In this context are organized competitions like ImageCLEF[3] which is a specific VIR task which takes place within the framework of the annual competitions of the CLEF[4] campaigns. These competitions aims the development of multimodal systems using image collections with their related short annotations.

In this paper we initially work with two multimodal combination techniques widely used in VIR area. These techniques have shown in different competitions better results than other techniques only based on one source of information, the image content or the image annotation [4][5]. Moreover, we suggest two versions of these techniques which, unlike the original ones, use heuristics based on the evidences found in the natural language annotations of the images in order to drive the multimodal fusion process.

The first of the techniques used is the multimodal blind relevance feedback. Specifically in this paper we focused on two textual blind relevance feedback strategies which can be used on the one hand, as blind relevance feedback techniques in TBIR systems, and on the other hand, as a multimodal fusion strategy on VIR systems based on textual and image content information. The compared strategies are Probabilistic Relevance Feedback (PRF) [6] and Local Context Analysis (LCA) [7]. PRF is widely used as a blind relevance feedback technique in text based systems [4] and as a multimodal fusion strategy [5] (MMPRF). Its multimodal version selects for the query expansion those best ranked terms from the annotations related with the top-ranked images returned by a CBIR. Besides, LCA is a conservative blind relevance feedback strategy less popular than PRF in VIR area. Indeed, the only precedent that exists of its use as a multimodal fusion strategy (MMLCA) in VIR area was presented in our participation in the 2008 edition of ImageCLEFphoto task [2]. From the good results achieved by MMLCA in our participation in the above task, in this paper we focused on the comparison of PRF and LCA methods as blind relevance feedback techniques in the TBIR area. Our goal was to experiment with the largest number of available query sets and image collections in order to find evidences which explain the good results achieved by MMLCA.

The second multimodal fusion technique evaluated is the standard Multimodal Re-ranking (MMRR) which is based on the fusion of the rankings returned by a TBIR and a CBIR system. While the MMRR strategy uses the relevance value or the rank position returned by each system for each image in order to work out the final relevance value of an image, our proposed TF-IDF Multimodal Re-ranking (TF-IDF MMRR) technique, also uses TF-IDF as a measure of the quality of the text related to an image. TF-IDF MMRR uses it to determine

---

[3] http://www.imageclef.org

[4] http://www.clef-campaign.org

when it should not use the relevance value returned by the CBIR system for the final relevance value of an image.

The TBIR system used is IR-n [8], it is a passages based system which showed better results than most of the TBIR systems based on documents which participated in different ImageCLEF tasks [9,1].

This paper is structured as follows: Firstly, it presents the main characteristics of the IR-n system focusing on the multimodal fusion techniques used, then it moves on to describe the collections used, the experiments and the results obtained. Finally, it presents conclusions and further work.

2   The IR-n System

To perform the experiments we used IR-n, an information retrieval system based on passages. Such systems treat each document as a set of passages; each passage defines a portion of text from the document. Unlike systems based on documents, passages based systems give greater relevance to those documents where the query terms appear in closer positions to each other [8].

Stemmers and stopword lists are used to determine which information in a document is to be used for retrieval. On the one hand, the stopword list for each language contains words whose presence in a document is not considered relevant to determine when a document is relevant for a query even if these words appear in a query. Stemmers, on the other hand, are used to obtain the root of a word, thus eliminating any suffixes or prefixes, for indexing and search purposes. For a list of the stemmers and stopwords used by IR-n, see www.unine.ch/info/clef.

IR-n allows choosing several weighting models; those allow the quantification of the similarity between a text (a complete document or a passage in a document) and a query. The values returned by the weighting model are based on the terms that are shared by the annotations and the query and on the discriminatory importance of each term.

2.1   Blind Relevance Feedback

In the TBIR area PRF is the most widespread blind relevance feedback technique, some examples of its use by ImageCLEF participants are the following ones [5] [4]. This strategy considers the top-documents returned in a ranking as relevant documents, extracting from these documents those terms which have a higher frequency in the top-documents than in the whole collection.

Although, this technique can deter retrieval, in case most of the top-ranked documents are not relevant, results in TREC and CLEF conferences show that is an effective technique[7]. In fact, almost all the systems which used textual blind relevance feedback which participated at ImageCLEF 2007 used PRF, some examples are [5] [4].

A more conservative approach that does not assumes that all the top-ranked documents returned by the system are relevant is LCA. This strategy as well as PRF is based on the frequency of the terms occurrence in the top-ranked documents, but unlike PRF, LCA attempts to avoid including terms from the top-ranked non-relevant documents selecting those terms within those documents which have a greater number of co-occurrences with the query terms.

The main limitation of the co-occurrence method is that the authors of documents can express the same concepts as users do in a query, but in the different terms. It would disables any possibility of a term matching between the query and the document [10]. The problem could be even greater in short texts collections like image annotations collections, due to the fact that these collections have a lower number of relations reflected between terms than the ones reflected in collections with longer texts.

Experiments conducted by [7] showed that performance is more independent to differences in the number of documents used for the expansion for LCA than for PRF. It demonstrated the LCA effectiveness for discarding non relevant documents between the top-ranked documents.

Under a multimodal viewpoint, the previous works which used textual blind relevance feedback as multimodal fusion strategy used PRF. We did not find previous works using LCA in a multimodal way except our contribution to the state of the art described in this paper and its evaluation in our participation in the 2008 edition of the ImageCLEFphoto task.

Our system performs the two multimodal blind relevance feedback strategies (MMPRF and MMLCA) using the top-ranked documents returned by the CBIR system as input for the term expansion strategy (PRF or LCA).

Point out that the IR-n architecture allows us to use query expansion based on either the most relevant passages or the most relevant documents.

2.2 Multimodal Re-ranking Strategy

This strategy involves the merging of the list returned by a TBIR system and the list returned by a CBIR system. That is done giving a different weight to the normalized relevance value or ranking position for a document in each list. We have included the standard re-ranking strategy and also a variation of it in order to try to improve it and to compare their behavior.

The re-ranking strategy fuses the IR-n list and the CBIR list in order to obtain one final list with documents ranked by relevance, the final relevance (FR). The merging process is done by giving different importance to the visual relevance (VR) given for a document in the visual list and the textual relevance (TR) given by the textual IR system:

$$FR(d) = TR(d) * wText + VR(d) * wImg \qquad (1)$$

- where d is a document.
- where VR is a normalized value of the relevance value returned by the CBIR for a document.
- where TR is a normalized value of the relevance value returned by the textual IR system for a document.

Despite this strategy usually improves the results, it usually adds a great number of non relevant images in the ranking. That is due to the low coverage that a

CBIR system usually obtains for its results. Thus, when the system uses images from the CBIR list there is a high probability to be selecting a non relevant image.

In an effort to overcome this problem we propose a modification of the re-ranking strategy having in mind the following assumptions: on the one hand, the list based on image annotations is more confident than the list based on images and on the other hand we assume that TF-IDF is a suitable way to measure the quantity and the quality of a text.

In order to reduce to the minimum the number or non relevant images used from the image based list, we establish a TF-IDF threshold (2). The images which annotations have a TF-IDF value over this threshold are skipped.

$$threshold = (MaxTFIDF * tImg) \tag{2}$$

- $MaxTFIDF$ is the maximum TF-IDF value found in the image list.
- $tImg$ is a user defined parameter which indicates the percentage value respect the $MaxTTIDF$ in order to work out the threshold.

Thus, in the TF-IDF MMRR formula (3) the system uses the *threshold* in order to avoid the risk of use the CBIR relevance values for those images which annotations have enough quantity and quality of terms to perform a suitable TBIR.

$$FR(d) = \begin{cases} TR(d) + VR(d), & \text{if threshold} > TFIDF(d) \\ TR(d), & \text{else} \end{cases} \tag{3}$$

- where $TFIDF(d)$ is the TF-IDF value of the text document d related to an image.

3   Experiments

For our experiments we chose the image collections and the query sets used in different editions of the ImageCLEFphoto and ImageCLEFWikipediaMM tasks. Table 1 shows the collection names, the tasks and editions in which the collection were used, the language, the number of documents (N. Docs), the word average per document (Word Avg.) and finally, the phrase average per document (Phra. Avg).

Below the collections are discussed in detail:

- **St Andrews (Photo 2004)** [11]. It is a black and white historical photograph collection, the image annotations have a high quality due to its authors are experts in the field. These are composed by 8 fields in clear text, which are: a short and a long title, location, description, date, photographer, notes and categories. The last two fields are particularly rich in information about contextual information of the photograph (information which cannot be extracted by a human only looking at the photo). This collection has the annotations with highest quality and greatest length of the collections used in our experiments.

Table 1. Image Collections

Collection	Language	N Docs	Word Avg	Phra Avg
St Andrews (photo 2004,2005)	English	28.133	60.7	4.18
IAPR TC-12 (photo 2006)	English	20.000	27.46	2.32
IAPR TC-12 (photo 2007,2008)	English	20.000	12.93	2.6
WikipediaMM (2006,2007,2008)	English	151.519	20.03	1,7

– **IAPR TC-12 (photo 2006 y photo 2007 y 2008)** [12,13]. This is a collection of images from around the world made almost all by a travel agency. A difference with the St. Andrews collection is that the IAPR TC-12 collection uses color images and XML annotations with the following fields: title, description, notes, location and date.

For the 2006 photo task edition is provided a version of the IAPR TC-12 collection which was previously preprocessed, in order to reduce the quality of the annotations, as follows: a 70% of the images annotations contain all the fields, a 10% do not contain the description field, a 10% do not contain neither description nor title fields, and finally a 10% do not have related annotations.

In the 2007 task edition was provided a version of the collection to which the images annotations contain all the fields except the description, which on the other hand is the richest in semantic information. This reduces the amount of textual information available in this collection in comparison to the 2006 collection.

– **WikipediaMM (INEX 2006 y 2007, WikipediaMM 2008)** [14]. This image collection has images and English annotations from the Wikipedia. It was used in the 2006 and 2007 editions of the INEX MM[5] task and in the 2008 edition of the WikipediaMM task.

The annotations use the XML Wikipedia format which includes fields with the image filename and the descriptions added by the authors or other users. Looking some samples, we observed that the annotation have noisy information as strings without meaning added by users. In addition, we observed that most of the filenames contain highly relevant information about the content of the images.

Regarding the photo task query sets, it should be pointed out that while the 2004 query set (25 queries) and the 2005 query set (28 queries) is different, the query set used for 2006 and 2007 editions is the same for the two editions (60 queries), only changing the characteristics of the collection. In general, looking

---

[5] http://inex.is.informatik.uni-duisburg.de

at the number of terms and photos that compound these queries we can consider
them as an ideal testbed for evaluating multimodal techniques. Since that they
allow us to focus on the evaluation of multimodal techniques without take into
account real but external problems such as the lack of pictures provided by the
user.

The query sets used with WikipediaMM collection were provided in the INEX
MM and WikipediaMM tasks. They are characterized by using a large number
of queries (70 queries) which are obtained through a collaborative model in
which are involved the task participants. Queries are composed by a title and
optionally by an image and a set of visual concepts from a predefined list of
visual concepts. This variability in the availability of visual information in the
queries and the fact that the title of the WikipediaMM 2008 queries has an
average of 2.64 terms per query makes this task a very realistic testbed within
the VIR area.

Below is a description of the input parameters of the system used in the
experiments:

- **RR Strategy**: It indicates if the system has to use a multimodal re-ranking
  strategy or not, and which one (MMRR or TF-IDF MMRR).
- **Blind Relevance Feedback (FB)**: It indicates which blind relevance feedback is used (PRF, LCA, MMPRF or MMLCA).
- **Blind Relevance Feedback Parameters**: If E has value 1, this denotes
  blind relevance feedback based on passages is used. But, if E has value 2,
  the blind relevance feedback is based on documents. Moreover, N denotes
  the number of passages or documents that the blind relevance feedback will
  use from the textual ranking, $Ncbir$ denotes the number of documents that
  the multimodal blind relevance feedback will use from an image based list
  and finally, T indicates the number of terms to add to the original query.

For the experiments we used Divergence From Randomness (DFR) [15] as the
weighting schema. We have taken this decision based on the training results
obtained in our participation in the 2007 edition of the ImageCLEFphoto task.
Moreover, the passage size and DFR parameters used are those which returned
best results for each collection without using blind relevance feedback. Finally,
the CBIR used for the experiments was FIRE[6] [16].

The following sections describe the experiments performed and the results
obtained for each proposed strategy. Firstly, for the MMLCA strategy we focused on carrying out textual experiments using LCA in order to explain the
good results achieved by MMLCA in our participation in the ImageCLEF-
photo 2008 task. Finally, for the multimodal re-ranking strategy, we focused
on extending the multimodal experiments conducted in our participation in
this task, using configurations which did not involve other VIR techniques,
in order to clearly discern the effects of using our proposed TF-IDF MMRR
technique.

---

[6] http://thomas.deselaers.de/FIRE

Table 2. St. Andrews, 2004

E	N	T	PRF	LCA
			0.7262	0.7262
2	45	5	**0.7361**	0.7321
2	60	5	0.7129	**0.7368**
2	65	5	0.7180	**0.7375**
1	45	10	**0.7385**	0.7104
1	40	10	**0.7387**	0.7047
2	40	5	0.7395	**0.7404**
2	40	10	**0.7436**	0.7146

Table 3. St. Andrews, 2005

E	N	T	PRF	LCA
			0.3493	0.3493
2	5	15	**0.3545**	0.3436
1	25	5	**0.3559**	0.3499
2	70	5	0.3246	**0.3566**
2	80	5	0.3206	**0.3580**
2	5	5	0.3455	**0.3596**
2	10	5	0.3533	**0.3607**
2	50	5	0.3307	**0.3607**

3.1 Blind Relevance Feedback Experiments

The strategy followed in the experiments was to use a wide range of values for the blind relevance feedback parameters in order to figure out the most suitable values for each technique, collection and query set used. The ranges of values used was from 5 to 100 documents for the number of documents used in the query, and from 5 to 100 terms for the number of selected terms for expansion. Moreover, all the experiments were also done with both blind relevance feedback modalities supported by IR-n (based on passages and based on documents).

The following tables show the results for the best seven runs with query expansion and the best run without query expansion. The data are presented in increasing order of best MAP (with PRF or LCA technique).

Table 2 and Table 3 show the results of the experiments with the St. Andrews collection; we can see that the same collection with different query sets determined contradictory precision results for the baselines, obtaining for the 2004 experiment precision values dramatically higher than the results obtained for baseline runs with the other query sets. It explains that for the experiments with a high precision baseline (2004 query set) the results achieved are slightly higher using PRF, while for the experiments with low precision (2005 query set) the results are significantly better using LCA.

Table 4 and Table 5 show the results achieved for the 2006 and 2007 collection; we can see how the annotations size reduction affects to the precision results of the experiments. Furthermore, we see that for these experiments where the baseline precision is low, LCA achieves better results than PRF. The explanation for this behavior is that in a low precision ranking there are a great number of non relevant documents between the top-ranked ones, what specially hurts the results achieved by PRF.

Moreover, the results show that in general the number of documents which LCA is able to manage in order to find terms for the query expansion is higher than the number of documents managed successfully by PFR without decreasing the precision.

Finally, we compare the results of these experiments with the performance achieved by the runs which used blind relevance feedback in our participations

Table 4. IAPR TC-12, 2006   **Table 5.** IAPR TC-12 No-Descrip., 2007

E	N	T	PRF	LCA
			0.1988	0.1988
2	20	10	0.2146	**0.2191**
1	10	10	0.2174	**0.2218**
2	10	10	0.2186	**0.2222**
2	40	10	0.2025	**0.2226**
1	5	5	0.2119	**0.2233**
2	25	5	0.2004	**0.2241**
2	**5**	**5**	0.2126	**0.2242**

E	N	T	PRF	LCA
			0.1544	0.1544
2	25	10	0.1811	**0.1834**
1	5	5	**0.1849**	0.1781
1	50	10	0.1675	**0.1884**
2	90	5	0.1452	**0.1898**
2	50	5	0.1551	**0.1911**
2	85	5	0.1444	**0.1918**
2	**45**	**5**	0.1598	**0.1923**

in the 2008 edition of the WikipediaMM task [1] (Table 6) and in the Image-CLEFphoto 2008 task [2] (Table 7). The collection used on that edition of ImageCLEFphoto was the IAPR TC-12 Non-Descriptions collection and a subset of the query set used in the 2007 edition.

Our results in the WikipediaMM task show that the blind relevance feedback techniques used worsened the baseline. However is interesting to see that LCA showed more robust behavior than PRF on a experiment where the use of blind relevance feedback is not suitable.

Regarding the 2008 edition of the ImageCLEFphoto task the objective of that edition was to evaluate the diversity of the results in the top 20 images of the ranking returned by the participant systems. With this objective the relevnace assessments were divided in clusters per query, and a measure of the coverage of the different cluster between the top 20 images of the ranking (CR20) was defined to evaluate the diversity degree of the results. In addition, F-measure was used as a balanced measure of the P20 measure (precision at the 20 documents retrieved) and the CR20 measure.

Comparing the results showed in this work with the 2008 ImageCLEFphoto results we note that despite the 2008 ImageCLEFphoto task used the same collection and a subset of the queries used in the 2007 edition, the accuracy achieved by the baseline on the 2008 participation reached higher precision values than the 2007 baseline. Thus, for the 2008 subset, the presence of a higher number of relevant documents between the top-ranked documents, used for the query expansion, explains that PRF slightly improved the precision obtained by LCA. Furthermore, it is observable that the best result was achieved by MMLCA based on the ranking returned by a CBIR system; such systems are characterized by low coverage results. Emphasize that in our 2008 participation our MMLCA run was the only one of our blind relevance feedback runs which did not hurt the diversity results and, in turn the one which reached our best precision result, obtaining the 4th best MAP and the 5th best P20 of the 1039 automatic executions sent by the participants. All this considering that we was the only group that did not use the narrative of the query in any of the submitted runs. It is important to point out this fact, since that the participants who submitted official runs using the narrative field of the topic and not using it showed that simply using the

Table 6. WikipediaMM 2008 Results

FB	E	N	T	MAP
-				0.2700
LCA	1	75	5	0.2614
PRF	2	70	5	0.2321

Table 7. ImageCLEFPhoto 2008 Blind Relevance Feedback Results

FB	E	N	Ncbir	T	MAP	P20	CR20	F-Mea.
-					0.2362	0.2936	0.3113	0.3022
LCA	1	5	0	5	0.2656	0.3154	0.2737	0.2931
PRF	2	5	0	5	0.2699	0.3244	0.2816	0.3015
MMPRF	2	5	5	5	0.2856	0.3744	0.2576	0.3052
MMLCA	2	0	20	5	**0.3436**	**0.4564**	0.3119	0.3706

title and the first sentence of the narrative their runs doubled their precision results and increase significantly the diversity values in comparison with their runs which only used the title [17].

Finally, it is also observed that while the best MMPRF run only used the 5 top-ranked image annotations returned by the CBIR system and the 5 top-ranked annotations returned by the TBIR system, the best MMLCA run only used the annotations returned by the CBIR system and, in turn it was able to manage the greatest amount of image annotations (20 images).

3.2 Multimodal Re-rankig Experiments

The experiments carried out in order to evaluate the re-ranking techniques were an extension of the experiments performed in the 2008 edition of the Image-CLEFphoto task. These experiments, unlike the experiments performed in our participation in the ImageCLEFphoto task, used the re-ranking techniques with the baseline system instead of combining them with other techniques that may affect the results. The values used for the parameters of the re-ranking strategies evaluated are the same ones which we used in our participation in this task, Table 8 shows the results obtained.

We can see that the TF-IDF MMRR technique obtained our highest diversity result; it indicates that this strategy was able to include within the 20 top-ranked images a higher proportion of relevant images than the standard re-ranking does.

Table 8. ImageCLEFPhoto 2008 Multimodal Re-ranking Results

RR Strategy	MAP	P20	CR20	F-Mea.
-	0.2362	0.2936	0.3113	0.3022
MMRR wText=0.6 wImg=0.4	0.2780	0.3833	0.3923	0.3877
MMRR TF-IDF tImg=0.6	0.2731	0.4026	**0.4273**	**0.4146**

4   Conclusion and Future Work

On the one hand, the experiments performed show a clear relationship between the precision of the ranking used for the blind relevance feedback technique and the performance of the relevance feedback technique used. Be aware of this relationship allows us to say that LCA is a robust strategy which fits perfectly with the low precision rankings which are usually returned by VIR systems, especially those returned by CBIR systems. This observation is supported by the fact that LCA is able to take advantage of the query term co-occurrences within the annotations related to the visual ranking, which is a meaningful relevance hint, due to those annotations has been obtained by a CBIR system only using visual characteristics of the image.

On the other hand, the MMRR TF-IDF technique has improved considerably the precision and the diversity of the results returned by the system allowing it to harness the characteristics of the image annotations in order to drive the multimodal combination process through the reduction of the inherent risk of use a CBIR system.

Finally, due to the ease of extracting data related with conceptual information from natural language sources in comparison with the complexity of obtaining it from other media, and analyzing the results achieved with both proposed techniques, we conclude, that in general the quality of the information extracted from natural language sources can add significantly improvement when it is used to drive the multimodal combination process.

In future works we want to study different ways to take advantage of a number of natural language processing tools in order to improve our proposed multimodal combination techniques driven by natural language evidences.

Acknowledgment

This research has been partially funded by the Spanish Government within the framework of the TEXT-MESS (TIN-2006-15265-C06-01) project and by European Union (EU) within the framework of the QALL-ME project (FP6-IST-033860).

References

1. Navarro, S., Muñoz, R., Llopis, F.: A Textual Approach based on Passages Using IR-n in WikipediaMM Task 2008. In: Online Working Notes, CLEF 2008 (2008)
2. Navarro, S., Llopis, F., Muñoz, R.: Different Multimodal Approaches using IR-n in ImageCLEFphoto 2008. In: Online Working Notes, CLEF 2008 (2008)
3. Grubinger, M.: Analysis and Evaluation of Visual Information Systems Performance. PhD thesis, Engineering and Science Victoria University (2007)
4. Díaz-Galiano, M., García-Cumbreras, M., Martín-Valdivia, M., Montejo-Raez, A., Urea-López, L.: Sinai at imageclef 2007. In: Working Notes of the 2007 CLEF Workshop, Budapest, Hungary (September 2007)
5. Gao, S., Chevallet, J.P., Le, T.H.D., Pham, T.T., Lim, J.H.: Ipal at imageclef 2007 mixing features, models and knowledge. In: Working Notes of the 2007 CLEF Workshop, Budapest, Hungary (September 2007)

6. Robertson, S.E., Sparck Jones, K.: Relevance weighting of search terms. Journal of the American Society for Information Science 27(3), 129–146 (1977)
7. Xu, J., Croft, W.B.: Improving the effectiveness of information retrieval with local context analysis. ACM Trans. Inf. Syst. 18(1), 79–112 (2000)
8. Llopis, F., Vicedo, J.L., Ferrández, A.: IR-n System at CLEF-2002. In: Peters, C., Braschler, M., Gonzalo, J. (eds.) CLEF 2002. LNCS, vol. 2785, pp. 291–300. Springer, Heidelberg (2003)
9. Navarro, S., Muñoz, R., Llopis, F.: A Multimodal Approach to the Medical Retrieval Task using IR-n. In: Online Working Notes, CLEF 2008 (2008)
10. Liu, H., Junzhong Gu, Z.L.: Improving the Effectiveness of Local Context Analysis Based on Semantic Similarity. In: 2007 International Conference on Convergence Information Technology, ICCIT 2007 (2007)
11. Reid, N.: The photographic collections in st andrews university library. Scottish Archives 5, 83–90 (1999)
12. Clough, P., Grubinger, M., Deselaers, T., Hanbury, A., Müller, H.: Overview of the imageclef 2006 photographic retrieval and object annotation tasks. In: Peters, C., Clough, P., Gey, F.C., Karlgren, J., Magnini, B., Oard, D.W., de Rijke, M., Stempfhuber, M. (eds.) CLEF 2006. LNCS, vol. 4730, pp. 579–594. Springer, Heidelberg (2007)
13. Grubinger, M., Clough, P., Hanbury, A., Müller, H.: Overview of the ImageCLEF-photo 2007 photographic retrieval task. In: Peters, C., Jijkoun, V., Mandl, T., Müller, H., Oard, D.W., Peñas, A., Petras, V., Santos, D. (eds.) CLEF 2007. LNCS, vol. 5152, pp. 433–444. Springer, Heidelberg (2008)
14. Tsikrika, T., Kludas, J.: Overview of the wikipediaMM task at ImageCLEF 2008. In: Peters, C., Giampiccol, D., Ferro, N., Petras, V., Gonzalo, J., Peñas, A., Deselaers, T., Mandl, T., Jones, G., Kurimo, M. (eds.) Evaluating Systems for Multilingual and Multimodal Information Access. LNCS, vol. 5706, pp. 539–550. Springer, Heidelberg (2009)
15. Amati, G., Van Rijsbergen, C.J.: Probabilistic Models of information retrieval based on measuring the divergence from randomness. ACM TOIS 20(4), 357–389 (2002)
16. Deselaers, T., Keysers, D., Ney, H.: Features for image retrieval: An experimental comparison. Information Retrieval 11(2), 77–107 (2008)
17. Demerdash, O.E., Kosseim, L., Bergler, S.: CLaC at ImageCLEFPhoto 2008. In: Online Working Notes, CLEF 2008 (2008)

An Experimental Comparison of Explicit Semantic Analysis Implementations for Cross-Language Retrieval

Philipp Sorg[1] and Philipp Cimiano[2]

[1] Institute AIFB, University of Karlsruhe
sorg@kit.edu
[2] Web Information Systems Group, Delft University of Technology
p.cimiano@tudelft.nl

Abstract. Explicit Semantic Analysis (ESA) has been recently proposed as an approach to computing semantic relatedness between words (and indirectly also between texts) and has thus a natural application in information retrieval, showing the potential to alleviate the vocabulary mismatch problem inherent in standard Bag-of-Word models. The ESA model has been also recently extended to cross-lingual retrieval settings, which can be considered as an extreme case of the vocabulary mismatch problem. The ESA approach actually represents a class of approaches and allows for various instantiations. As our first contribution, we generalize ESA in order to clearly show the degrees of freedom it provides. Second, we propose some variants of ESA along different dimensions, testing their impact on performance on a cross-lingual mate retrieval task on two datasets (JRC-ACQUIS and Multext). Our results are interesting as a systematic investigation has been missing so far and the variations between different basic design choices are significant. We also show that the settings adopted in the original ESA implementation are reasonably good, which to our knowledge has not been demonstrated so far, but can still be significantly improved by tuning the right parameters (yielding a relative improvement on a cross-lingual mate retrieval task of between 62% (Multext) and 237% (JRC-ACQUIS) with respect to the original ESA model).

1 Introduction

The quest for a more "semantic" retrieval of information items (documents, videos, music etc.) still represents one of the more challenging research directions in information retrieval today. There have been many approaches so far aiming at incorporating "semantics" into the retrieval process. Prominent examples are those that use thesauri for query expansion. These thesauri can be either manually created as in the case of WordNet [1] or derived from the local document collection (see e.g. [2]). Other approaches integrate semantic relatedness or semantic similarity between words into the retrieval process [3]. Finally, other approaches aim at a concept-based retrieval, where such concepts can be either computed implicitly from the document collection, as in Latent Semantic Indexing [4] or given explicitly by external resources such as WordNet [5].

One very successful approach in the latter direction which has attracted a lot of attention in recent years is the Explicit Semantic Analysis (ESA) model by Gabrilovich

H. Horacek et al. (Eds.): NLDB 2009, LNCS 5723, pp. 36–48, 2010.

and Markovitch [6]. In essence, ESA indexes documents with respect to the Wikipedia article space (as "conceptual" space), indicating how strongly a given word in the document (and by aggregation also the whole document) is associated to a specific Wikipedia article. Gabrilovich and Markovitch instantiate a geometric framework in which each word is represented as a vector of Wikipedia articles and similarity is calculated using the cosine measure, where the $tf.idf$ value of a word in a given Wikipedia article is used as weight of the corresponding dimension in the vector. As a word can be associated to many articles (with different weights), ESA alleviates the vocabulary mismatch problem [7] inherent in the BOW model, where every word corresponds exactly to one dimension, the dimensions being orthogonal. In the ESA model, two words or texts can be semantically related in spite of not having any word in common (but associated to similar Wikipedia articles).

In this paper, we put the ESA model under scrutiny and empirically analyze variants of the original ESA model, both looking at alternatives for calculating the association between Wikipedia articles and words as well as examining alternative retrieval models, in particular based on language modeling approaches as well as probabilistic models. We investigate these variants in the context of a cross-language retrieval task following a cross-lingual extension of ESA (CL-ESA) (see [8] and [9]). We evaluate the ESA variants with respect to the well-known mate retrieval task, i.e. given a parallel corpus, retrieving for each document its parallel document in another language as in [10]. We report experiments on two parallel datasets, the Multext dataset as well as the JRC-ACQUIS corpus on three languages: English, French and German.

Our results show on the one hand that the choice of some parameters (in particular the association strength but also the retrieval model) can have a significant impact and, on the other hand, that, while the settings adopted in the original ESA model are reasonable, its performance can be significantly increased by changing some of the parameters. To our knowledge, there has been no empirical analysis and comparison between different implementation choices before.

The paper is structured as follows: in the following Section 2 we present the ESA model in the standard (monolingual) version (as described in [6]) as well as the cross-lingual formulation along the lines of [8], both for the sake of completeness and to facilitate the understanding of this paper. In Section 3 we then first introduce a generalization of the ESA model which makes explicit the choices that it leaves open and discuss various alternatives for these choices. In Section 4 we then experimentally analyze and present the results of the different variants on a cross-lingual mate retrieval task.

2   Explicit Semantic Analysis (ESA)

2.1   Classical (Monolingual) ESA

Explicit Semantic Analysis (ESA) [6] attempts to index or classify a given document d with respect to a set of explicitly given external categories. It is in this sense that ESA is explicit compared to approaches which aim at representing texts with respect to latent topics or concepts, as done in Latent Semantic Analysis (LSA) (see [4,11]). Gabrilovich and Markovitch have outlined the general theory behind ESA and in particular described its instantiation to the case of using Wikipedia articles as external categories. We will basically build on this instantiation as described in [6], which we briefly summarize in the following.

In essence, Explicit Semantic Analysis takes as input a document d and maps it to a high-dimensional real-valued vector space. This vector space is spanned by a Wikipedia database $\mathcal{W}_k = \{a_1, \ldots, a_n\}$ in language L_k such that each dimension corresponds to an article a_i. This mapping is given by the following function: $\Phi_k : D \to \mathbb{R}^{|\mathcal{W}_k|}$ with

$$\Phi_k(d) := \langle as(d, a_1), \ldots, as(d, a_n) \rangle$$

The function as expresses the *association strength* between d and the Wikipedia article a_i. In the original ESA model, as is defined by sum of $tf.idf$ values of all words of $d = \langle w_1, \ldots, w_s \rangle$ in the article a_i multiplied by tf in d:

$$as(d, a_i) := \sum_{w_j \in d} tf_d(w_j) tf.idf_{a_i}(w_j)$$

Essentially the Semantic Interpreter applying ESA described in [6] computes the function Φ. As output we thus get a vector representing the strength of association of a document d with respect to the articles in Wikipedia $\mathcal{W}_k$. These vectors can then be used to assess the similarity between documents at a conceptual level (e.g. using cosine similarity between the articles indexed with respect to the Wikipedia articles, i.e. the vectors yielded by the Φ-function) and have thus a natural application in information retrieval tasks, which we are concerned with in this article.

In the following section, we present the extension to ESA called CL-ESA (Cross-language Explicit Semantic Analysis), which represents a relatively straightforward extension of ESA to a cross-lingual setting presented before [9,8].

2.2 Cross-lingual ESA (CL-ESA)

It has been shown recently that, when instantiating ESA for Wikipedia, one can rely on Wikipedia's language links to transform ESA vectors in one language to another one[1]. This is done by mapping each dimension corresponding to article a in Wikipedia $\mathcal{W}_a$ to the dimension corresponding to article b in Wikipedia $\mathcal{W}_b$ so that there exists a language link from a to b. In the following we therefore assume the existence of a mapping function $m_{a \to b} : \mathcal{W}_a \to \mathcal{W}_b$ that maps articles according to the language links to articles in another language. This function is only defined for articles having a language link to the Wikipedia in the target language. To overcome this restriction we will use only that subset of the Wikipedia articles having unique language links to all languages considered, such that the function is actually a bijection. We will describe the Wikipedia subset used in more detail in Section 4.

Given a document $d \in D$ in language L_a, CL-ESA allows to index this document with respect to any of the other languages $L_1, \ldots, L_n$ by transforming the vector $\Phi_a(d) = \langle d_{a_1}, d_{a_2}, \ldots \rangle$ into a corresponding vector in the vector space that

---

[1] Cross-language links are those that link a certain article to a corresponding article in the Wikipedia database in another language.

is spanned by the Wikipedia articles in the target language. This mapping function $\Psi_{a \to b} : \mathbb{R}^{|\mathcal{W}_a|} \to \mathbb{R}^{|\mathcal{W}_b|}$ is calculated as follows:

$$\Psi_{a \to b}(\Phi_a(d)) := \langle d_{m_{b \to a}(b_1)}, d_{m_{b \to a}(b_2)}, \ldots \rangle$$

where b_j are the articles of Wikipedia $\mathcal{W}_b$. This means that $\Psi_{a \to b}(\Phi_a(d))$ is the ESA representation of d with respect to Wikipedia $\mathcal{W}_b$ based on the ESA representation of d with respect to Wikipedia $\mathcal{W}_a$ and the language links between $\mathcal{W}_a$ and $\mathcal{W}_b$.

Given the above settings, it should be straightforward to see how the actual retrieval works. The cosine between a query q_a in language L_a and a document d_b in language L_b is calculated as:

$$cos(q_a, d_b) := cos(\Psi_{a \to b}(\Phi_a(q_a)), \Phi_b(d_b))$$

In our settings the query vector is thus mapped to the target language and compared to documents in the target language. This thus gives us an elegant retrieval model which is uniform across languages. A prerequisite for this model is certainly that we know the language of the query and of the different documents in order to know which mapping Ψ should be applied.

3   ESA Variants

We first present the generalization of the ESA model, making the choices for different parameters explicit. This will provide a uniform model to investigate the impact of different parameters on the ESA model. Then, we present the specific alternatives for the different choices that we have experimentally compared in Section 4.

3.1   Generalization

A cross-lingual retrieval model based on ESA can be generalized as follows (q_a is a query and d_b a document in the collection):

$$rel(q_a, d_b) := rel(\Pi(\Psi_{a \to b}(\Phi_a(q_a))), \Pi(\Phi_b(d_b)))$$

with

$$\Phi(d) := \boldsymbol{d} = \langle as(d, a_1), \ldots, as(d, a_{|\mathcal{W}|}) \rangle$$

The relevant parameters to be instantiated are:

- **Dimension Projection Function** Π: For most implementations of ESA, it is impossible to work with all of the dimensions for which the association strength is greater than 0 (for pragmatic reasons related to efficiency of computation). Therefore, most approaches index a text only with respect to a subset of the relevant dimensions.
- **Association Strength Function** as: The so called association strength function quantifies the degree of association between a document d and a category a_j.
- **Relevance Function / Retrieval Model** rel: Concerning the retrieval model, while the cosine (thus assuming a geometric retrieval model) has been used, other alternatives are possible here.

- **Category System**: ESA relies on the fact that there is some external category system with respect to which words and texts can be indexed. While Wikipedia has been used in most implementations, the originators of ESA have also tested on an alternative category system: the Open Directory Project (ODP)[2], achieving worse results than with Wikipedia. Though the choice of the category system is also crucial, in this work we will rely on the Wikipedia-based implementation as in the context of our cross-lingual retrieval experiments we directly exploit the language links of Wikipedia to map between languages.

This offers a generalized framework for the ESA model allowing different parameters to be explored and to analyze their impact. We will discuss particular implementations of the above functions for which we will also provide experimental evaluation in Section 4.

3.2 Dimension Projection

We will consider the following variants for the dimension projection function Π that have been considered in previous literature (but never been analyzed systematically). As notation we will refer to d_i as the i-th dimension of the ESA vector of d which is the association strength of d to the article a_i. The function α_d defines an order on the indices of the dimensions according to descending values such that $\forall i, j : i < j \rightarrow d_{\alpha(i)} \geq d_{\alpha(j)}$, e.g. $d_{\alpha(10)}$ is the 10-th highest value of d.

1. **Absolute**, with $\Pi_{abs}^m(d)$ being the projected vector by restricting d to the m dimensions with highest values:, i.e. $\alpha(1), \ldots, \alpha(m)$ (as in [9] and [8])
2. **Absolute Threshold**, with $\Pi_{thres}^t(d)$ being the projected vector by restricting d to the dimensions j with values $d_j \geq t$ (as in [12])
3. **Relative Threshold**, with $\Pi_{rel}^t(d)$ being the projected vector by restricting d to the dimensions j with values $d_j \geq t * d_{\alpha(1)}, t \in [0..1]$, thus restricting it to those values above a certain fraction of the highest-valued dimension
4. **Sliding Window**, with $\Pi_{window}^{t,l}(\Phi(d))$ being the projected vector by restricting d to the first i dimensions according to the order α_d for which the following condition holds: $d_{\alpha(i-l)} - d_{\alpha(i)} \geq t * d_{\alpha(1)}, t \in [0..1]$ (as in the original ESA model [13])

A relevant question is certainly how to set the parameters m and t. We address this in the experiments by first fixing a reasonable value for m in Π_{abs}^m. In order to be able to compare the different approaches, we choose the parameter t in such a way that the number of non-zero dimensions of the projected ESA vectors of all documents in the datasets amounts to m on average. The parameter l was set to 100 as in [6].

3.3 Association Strength

In the following we will describe the different choices of the association strength function $as(d, a_i)$ between documents and articles determining the values of the ESA vector d. These functions are based on the term vectors of d and a_i. As notation we use $|\mathcal{W}|$ as the number of articles, $|a_i|$ as number of terms in article a_i, $tf_d(w)$ ($tf_{a_i}(w)$) as the

---

[2] http://www.dmoz.org

term frequency of w in document d (article a_i), $rtf_{a_i}(w) = tf_{a_i}(w)/|a_i|$ as the relative term frequency and $af(w)$ as the number of articles containing term w in Wikipedia $\mathcal{W}$.

1. **TF.IDF**: The most widely used version of the $tf.idf$ function:

$$as_{tf.idf} := \sum_{w \in d} tf_d(w)\ rtf_{a_i}(w) \log \frac{|\mathcal{W}|}{af(w)}$$

2. **TF.IDF***: A modified $tf.idf$ version ignoring how often the terms occur in document d:

$$as_{tf.idf^*} = \sum_{w \in d} rtf_{a_i}(w) \log \frac{|\mathcal{W}|}{af(w)}$$

3. **TF** : An association function only based on term frequencies (ignoring inverse document frequencies):

$$as_{tf} = \sum_{w \in d} tf_d(w) rtf_{a_i}(w)$$

4. The **BM25** ranking function as defined by Robertson et al. [14] with parameters set to the following standard value: $k_1 = 2, b = 0.75$.
5. The **Cosine** similarity between the tf and $tf.idf$ vectors $\mathbf{d} = \langle tf_d(w_1), tf_d(w_2), \ldots \rangle$ and $\mathbf{a}_i = \langle tf.idf_{a_i}(w_1), tf.idf_{a_i}(w_2), \ldots \rangle$:

$$as_{cos} = \frac{< \mathbf{d}, \mathbf{a}_i >}{\|\mathbf{d}\| \|\mathbf{a}_i\|}$$

Note that we have also experimented with versions of the above where the tf_{a_i} instead of rtf_{a_i} values were used, yielding in all cases worse results with a performance degradation of about 75% in all cases. For this reason, we do not present the results with the tf_{a_i} versions of the above functions in detail.

3.4 Relevance Function

The relevance function $rel(q, d)$ defines the score of a document $d \in D$ for a given query q and is used to rank the documents in the retrieval process. In this multilingual setting, the function is defined on the translated and projected ESA vector $\hat{q} := \Pi(\Psi(\Phi(q)))$ of query q and the projected ESA vector $\hat{d} := \Pi(\Phi(d))$ of document d (see section 2.2).

Analogous to the Bag-of-Words model the ESA vectors can be seen as Bag-of-Articles model for a document d. The term frequency of $a_i \in \mathcal{W}$ is defined as $tf_d(a_i) := \hat{d}_{a_i}$, the document frequency $df(a_i)$ is the number of documents in D with $\hat{d}_{a_i} > 0$. Based on this model different relevance functions defined for text retrieval can by applied to the ESA vectors.

- The **Cosine** similarity of query and document vectors (used by all ESA implementations known to us):

$$rel_{Cosine} = \frac{< \hat{q}, \hat{d} >}{\|\hat{q}\| \|\hat{d}\|}$$

- **TF.IDF:** The TF.IDF function transfered to the Bag-of-Articles model:

$$rel_{tf.idf} = \sum_{a \in W} tf_q(a) rtf_{d_i}(a) idf(a)$$

$$= \sum_{a \in W} \hat{q}_a \frac{\hat{d}_a}{\sum_{a^* \in W} \hat{d}_{a^*}} \log \frac{|D|}{df(a)}$$

- **KL-Divergence:** Many recent text retrieval systems use relevance functions based on the theory of language modeling. In order to be able to apply these approaches to our setting we define the conditional probability of an article given a document as follows:

$$P(a|d) := \frac{\hat{d}_a}{\sum_{a^* \in W} \hat{d}_{a^*}}$$

This definition of the conditional probability originates from the bag-of-words model and is inspired by [15], where it is also described how these probabilities can be used to define a ranking function based on the Kullback-Leibler divergence [16], which measures the difference between the query and the document model (leading ultimatively to the negative sign in the formula below). Transferred to our model this results in the following retrieval function:

$$rel_{KL} = -D_{KL}(q\|d) \cong - \sum_{a \in W} P(a|q) \log P(a|d)$$

- **LM:** An alternative approach is to use the conditional probability $P(q|d)$ as relevance function. This distribution can be converted using the conditional distributions of documents given articles, Bayes law and the a priori probability of articles $P(a) = \frac{df(a)}{|D|}$:

$$rel_{LM} = P(q|d) = \sum_{a \in W} P(q|a) P(a|d)$$

$$\cong \sum_{a \in W} \frac{P(a|q)}{P(a)} P(a|d)$$

4 Experiments

Our experiments have been carried out in an iterative and greedy fashion in the sense that we start form the original ESA model as a baseline, then iteratively varying different parameters and always fixing the best configuration before studying the next parameter. At the end of our experiments we will thus be able to assess the combined impact of the best choices on the performance of the ESA model.

To prove the significance of the improvement of our best settings (projection function Π_{abs}^{10000}, association strength function TF.IDF*, cosine retrieval model) we carry out paired t-tests (confidence level 0.01) comparing the best settings pairwise with all other results for all language pairs on both datasets. Results where the differences are **not** significant with respect to **all** other variants at a confidence level of 0.01 are marked with "X" in Figure 1 to 4.

4.1 Datasets and Evaluation Measures

For the ESA implementation we used the English, German and French Wikipedia database[3]. As we rely on the language links to map the ESA vectors to other languages, we only chose articles that are linked across all three languages. This means that the mapping function $m_{a \to b}$ used for CL-ESA is defined for all articles and is a bijection between the Wikipedia subsets for all language pairs considered. Altogether we used 166,484 articles in every language.

To evaluate the performance of the CLIR system we performed mate retrieval on two well known parallel corpora: The Multext corpus derived from the Multext project[4] consisting of 2,783 question and answer pairs, and the JRC-ACQUIS corpus[5] consisting of 15,464 documents. For both datasets documents in one language were taken as queries to search the documents in another language. In this case automatic evaluation is possible as the relevant document, i.e. the translation of the query, is known in advance. All mentioned collections were prepared using common IR-like preprocessing steps including elimination of stopwords, special characters and extremely short terms (length < 3) and stemming.

As evaluation measures we used TOP-k accuracy, i.e. the number of queries for which the mate was found in the top k documents, and Mean Reciprocal Rank, which measures the average position of the mate documents (all standard measures in information retrieval). As the observed effects were constant across measures, we only present TOP-1 accuracy in Figures 1 to 4. For experiments on the Multext corpus we used all documents (2,783) as queries to search in all documents in the another languages. The results for language pairs were averaged for both retrieval directions (e.g. using English documents as queries to search in the German documents and vice versa). For the JRC-ACQUIS dataset we randomly chose 3000 parallel documents as queries (to yield similar settings as in the MULTEXT scenario) and the results were again averaged for language pairs. This task is harder compared to the experiments on the Multext corpus as the search space now containing 15,464 documents is bigger by a factor of approximately 5, which explains the generally lower results on the JRC-Acquis dataset.

4.2 Results

In the following we discuss the results of the different variations of the CL-ESA model:

Projection Function. We first used different values for the parameter m in the projection function Π_{abs}^m. The results in Figure 1 showed that $m = 10,000$ is a good choice for both datasets.

On the basis of this result, we investigated different projection functions. In order to be able to compare them, we set the different threshold values t such that the projected

---

[3] Snapshot of 03/12/2008 (English), 06/29/2008 (German) and 06/25/2008 (French).

[4] http://aune.lpl.univ-aix.fr/projects/MULTEXT/

[5] http://wt.jrc.it/lt/Acquis/

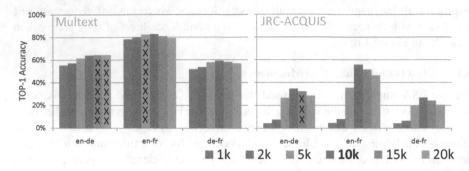

Fig. 1. Variation of m in Π_{abs}^m using the TF.IDF* association function and cosine retrieval model

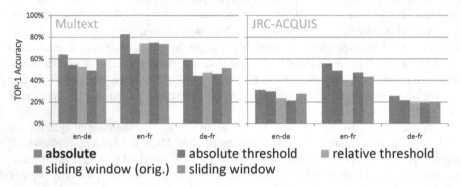

Fig. 2. Variation of the projection function Π using the TF.IDF* association function and cosine retrieval model

ESA vectors had an average number of approx. 10,000 non-zero dimensions. An exception is the function *sliding window (orig.)* where we used the parameters described in [13]: $t = 0.05$ and $l = 100$. Using an absolute number of non-zero dimensions yielded the best results (see Figure 2), the difference being indeed significant with respect to all other variants. Thus, we conclude that neither the settings of the original ESA approach (sliding window) nor in the model of Gurevych et al. (fixed threshold) are ideal in our experimental settings. For the remaining experiments we thus fix the absolute dimension projection function with 10,000 articles ($\Pi_{abs}^{10,000}$).

Association Strength. The results in Figure 3 show that the functions TF.IDF (used in the original ESA model) and TF.IDF* perform much better compared to the other functions. The better performance of TF.IDF*, which ignores the term frequencies in the queries, was indeed significant w.r.t. all other alternatives for all language pairs considered on both datasets. We thus conclude that the settings in the original ESA model are reasonable, but, surprisingly, can be improved by ignoring the term frequency of the words in the document to be indexed. The low results using the TF function show that IDF is an important factor in the association strength function. Otherwise the normalization of the TF.IDF values (= Cosine function) reduces the retrieval performance substantially.

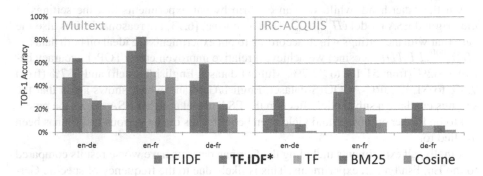

Fig. 3. Variation of the association strength function as using the projection function $\Pi_{abs}^{10,000}$ and cosine retrieval model

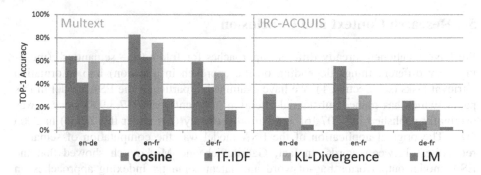

Fig. 4. Variation of the retrieval model using $\Pi_{abs}^{10,000}$ and TF.IDF*

Retrieval Model. The variations of the retrieval model lead to the result that the cosine function, which is used by all ESA implementations known to us, constitutes indeed a reasonable choice. All other models perform worse (the difference being again significant for all language pairs on both datasets), which can be seen at the charts in Figure 4, especially on the JRC-ACQUIS dataset.

4.3 Discussion

Our results show on the one hand that ESA is indeed quite sensitive to certain parameters (in particular the association strength function and the retrieval model), the choices for which can have a large impact on the performance of the approach. For example, using a tf_{a_i} values instead of rtf_{a_i} (which is length normalized) values in the association strength function decreases performance by about 75%. Unexpectedly, abstracting from the number of times that a word appear in the query document (using TF.IDF*) improves upon the standard TF.IDF measure (which takes them into account) by 17% to 117%. We have in particular shown that all the settings that are ideal in our experiments are so indeed in a statistically significant way (with the exception of the number of dimensions taken into account).

On the other hand, while we can confirm by our experiments that the settings in the original ESA model ($\Pi_{window}^{0.05,100}$, TF.IDF, cosine) [6,13] are reasonable, it is also the case that with the settings which according to our experiments are ideal on both datasets ($\Pi_{abs}^{10,000}$, $TF.IDF^*$, cosine) we achieve a relative improvement in TOP-1 accuracy between 62% (from 51.1% to 82.7%, Multext dataset, English/French) and 237% (from 9.3% to 31.3%, JRC-ACQUIS dataset, English/German), which shows again that the settings can have a substantial effect on the ESA model and that ESA shows the potential to be further optimized and yield even better results on the various tasks it has been applied to.

Finally, all experiments including the German datasets have worse results compared to the English/French experiments. This is likely due to the frequency of specific German compounds in the datasets, which lead to a vocabulary mismatch between documents and Wikipedia articles. However an examination of this remains for future work.

5   Research Context and Conclusion

We have mentioned already different approaches for folding in "semantics" (meaning very different things depending on the approach in question) into information retrieval tasks (see Section 1). We have examined in particular the ESA model in this paper, which has gained substantial attention in recent years [17,3,9,8] since it was originally published in 2007 [6] and partially already (not under this name) in 2005 [18]. The original application of the ESA model was the computation of semantic relatedness between words. In fact, Gabrilovich and Markovitch showed that the ESA model outperforms bag-of-word and latent semantic indexing approaches on this task. ESA has been also exploited in text classification approaches [18,17,19] where it has been already shown that an appropriate dimension selection function has significant influence on the performance of the ESA model. ESA has been also applied with reasonable success to information retrieval settings [3], in particular cross-language retrieval settings [9,8]. In this paper we have generalized the original ESA model and made explicit the degrees of freedom that it offers and highlighted the different choices that various implementations have adopted. The starting point for our investigation has been the observation that none of the above works has examined the various possible choices systematically due to the fact that they have focused on different aspects and this was not their main research question. In any case, if the ESA model continues to be applied successfully to various text-centered tasks, a systematic investigation of the impact of different choices seems definitely necessary. We have provided such an analysis in the context of a cross-lingual mate retrieval task (presenting results on two datasets), showing which choices have or don't have a large impact and confirmed that the settings of the original ESA model are indeed reasonable, something which to our knowledge has never been shown, but can still be improved for cross-lingual retrieval settings. Our results are clearly limited to the type of cross-lingual mate retrieval task that we have considered and an avenue for future work could be the investigation of the choices under consideration for monolingual or more general (non-mate-retrieval like) cross-lingual retrieval tasks, text classification or semantic relatedness computation. The examination of optimal settings constitutes an interesting topic for future investigation which can help to shed additional light on the ESA approach. Furthermore, our generalization of ESA can help to guide such investigations in the future, providing a common framework for comparisons.

Acknowledgments

This work was funded by the German Research Foundation (DFG) under the Multipla project (grant 38457858).

References

1. Richardson, R., Smeaton, A.: Using wordnet in a knowledge-based approach to information retrieval. In: Proceedings of the BCS-IRSG-Colloquium (1995)
2. Schütze, H., Pedersen, J.: A cooccurrence-based thesaurus and two applications to information retrieval. Information Processing and Management 33(3), 307–318 (1997)
3. Gurevych, I., Müller, C., Zesch, T.: What to be? - electronic career guidance based on semantic relatedness. In: Proceedings of ACL (2007)
4. Deerwester, S.C., Dumais, S.T., Landauer, T.K., Furnas, G.W., Harshman, R.A.: Indexing by latent semantic analysis. Journal of the American Society of Information Science 41(6), 391–407 (1990)
5. Gonzalo, J., Verdejo, F., Chugur, I., Cigarran, J.: Indexing with wordnet synsets can improve text retrieval. In: Proceedings of the COLING/ACL 1998 Workshop on Usage of WordNet for NLP, pp. 38–44 (1998)
6. Gabrilovich, E., Markovitch, S.: Computing semantic relatedness using wikipedia-based explicit semantic analysis. In: Proceedings of IJCAI, pp. 1606–1611 (2007)
7. Furnas, G., Landauer, T., Gomez, L., Dumais, S.: The vocabulary problem in human-system communication. Communications of the ACM 30(1), 964–971 (1987)
8. Sorg, P., Cimiano, P.: Cross-lingual information rerieval with explicit semantic analysis. In: Working Notes of the Annual CLEF Meeting (2008)
9. Potthast, M., Stein, B., Anderka, M.: A wikipedia-based multilingual retrieval model. In: Proceedings of ECIR, pp. 522–530 (2008)
10. Littman, M., Dumais, S., Landauer, T.: Automatic Cross-Language Information Retrieval using Latext Semantic Indexing. In: Cross-Language Information Retrieval, pp. 51–62. Kluwer, Dordrecht (1998)
11. Dumais, S., Letsche, T., Littman, M., Landauer, T.: Automatic cross-language retrieval using latent semantic indexing. In: Proceedings of the AAAI Symposium on Cross Language Text and Speech Retrieval (1997)
12. Müller, C., Gurevych, I.: Using wikipedia and wiktionary in domain-specific information retrieval. In: Peters, C., Deselaers, T., Ferro, N., Gonzalo, J., Jones, G.J.F., Kurimo, M., Mandl, T., Peñas, A., Petras, V. (eds.) Evaluating Systems for Multilingual and Multimodal Information Access. LNCS, vol. 5706, pp. 219–226. Springer, Heidelberg (2009)
13. Gabrilovich, E.: Feature Generation for Textual Information Retrieval using World Knowledge. PhD thesis, Israel Institute of Technology, Haifa (2006)
14. Robertson, S.E., Walker, S., Jones, S., Hancock-Beaulieu, M., Gatford, M.: Okapi at trec-3. In: Proceedings of TREC (1994)
15. Zhai, C.X., Lafferty, J.D.: Model-based feedback in the language modeling approach to information retrieval. In: Proceedings of CIKM, pp. 403–410 (2001)

16. Lee, L.: Measures of distributional similarity. In: Proceedings of ACL (1999)
17. Egozi, O., Gabrilovich, E., Markovitch, S.: Concept-based feature generation and selection for information retrieval. In: Proceedings of AAAI (2008)
18. Gabrilovich, E., Markovitch, S.: Feature generation for text categorization using world knowledge. In: Proceedings of IJCAI (2005)
19. Gupta, R., Ratinov, L.: Text categorization with knowledge transfer from heterogeneous data sources. In: Proceedings of AAAI, pp. 842–847 (2008)

CITOM: Incremental Construction of Topic Maps

Nebrasse Ellouze[1,2], Nadira Lammari[1], Elisabeth Métais[1], and Mohamed Ben Ahmed[2]

[1] Laboratoire Cedric, CNAM
292 rue Saint Martin, 75141 Paris cedex 3, France
{metais,lammari}@cnam.fr
[2] Ecole Nationale des Sciences de l'Informatique, Laboratoire RIADI
Université de la Manouba, 1010 La Manouba
{nebrasse.ellouze,mohamed.benahmed}@riadi.rnu.tn

Abstract. This paper proposes the CITOM approach for an incremental construction of multilingual Topic Maps. Our main goal is to facilitate the user's navigation across documents available in different languages. Our approach takes into account three types of information sources: (a) a set of multilingual documents, (b) a domain thesaurus and (c) all the possible questioning sources such as FAQ and user's or expert's requests about documents. We have been validating our approach with a real corpus from the sustainable construction domain.

Keywords: Topic Map (TM), incremental construction, enrichment, multilingual documents, thesaurus, user requests.

1 Introduction

Topic Maps [1] are used for structuring contents and knowledge provided from different information sources and different languages. They are defined as semantic structures which allow organizing all the subjects they represent. They are intended to enhance navigation and improve information search in these resources.

The amount of information sources available today is very huge and continuously increasing, in this case, it is impossible to create and maintain manually a Topic Map to represent and organize all these information. Many Topic Maps building approaches can be found in the literature [2]. However, none of these approaches takes as input multilingual documents content. In addition, although Topic Maps are basically dedicated to user's navigation and information search, no one approach takes into consideration users requests in the Topic Map building process.

In this paper, we propose CITOM, an incremental approach to build a multilingual Topic Map. The resulting Topic Map gives a user the possibility to acquire knowledge from documents written in languages different from his native language.

Having a content of multilingual documents, CITOM takes into consideration two other information sources which are: a domain thesaurus and all the possible questioning sources such as FAQ, user or expert requests related to the source documents, phone discussions and consultations with people working in the application domain.

CITOM aims at providing a global Topic Map as a semantic structure linking and organizing concepts in various languages taking into account the specificity of multilingual

H. Horacek et al. (Eds.): NLDB 2009, LNCS 5723, pp. 49–61, 2010.

contexts. In fact, a term may not have a semantic equivalent in all treated languages. This is very frequent when we consider documents provided from different cultures.

In our approach, we propose also to assign to each topic, a list of meta properties initialized when the Topic Map is created. These meta properties reflects a topic relevance according to its usage when users explore the Topic Map. They are also used in the Topic Map pruning process, especially to delete all the topics considered as non pertinent.

The paper will be structured as follows: In section 2, we present the main characteristics of our Topic Map Model. In section 3, we describe the different steps of our proposed approach for multilingual Topic Map construction. Section 4 is dedicated to the pruning process of the generated Topic Map. In section 5, we present a case study from the sustainable construction domain to illustrate our approach. At least, in section 6, we conclude and give some perspectives for this work.

2   Characteristics of the Topic Map Model

Topic Maps are an ISO standard (ISO13250) included into the ODM (Ontology Meta-Model Definition) by the OMG community in order to provide a standard TM-UML model. Figure 1 shows an extract of the ODM Topic Map Meta-Model. A Topic Map specification is also proposed by « Topic Maps.Org » consortium called XTM (XML Topic Maps).

A Topic Map is a semantic structure which allows organizing and representing knowledge from information resources (documents, databases, videos, etc). The main concept of Topic Maps is *topic* which represents the subject being referred to. A topic may have a base name and variant names. A Topic Map can contain many topics that can be categorized into topic types. A topic may be linked to one or more information sources that are deemed to be relevant to the topic. Such links are called *occurrences* of the topic. An *association* is a link element, showing relationships between topics. The role played by a topic in an association is one of the topic characteristics.

Semantic links between topics allow navigation in the Topic Map structure. As mentioned in [3], there is no limitation in the definition of links in a Topic Map. They are specified by the Topic Map designer according to its requirements, the knowledge to be represented by the Topic Map and the domain application.

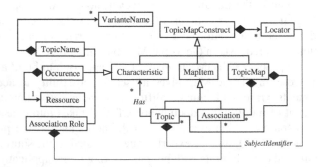

Fig. 1. An extract of the ODM Topic Map Meta-Model

Several works have been proposed to classify semantic links between terms and concepts. For example, the work proposed by ANSI (American National Standards Institute) in ANSI/NISO Z39.19-2005, they define three types of links: (1) equivalence links like synonymy, (2) hierarchical links (generalization/specialization) and (3) other links called "associative links" such as "cause/effect link". In [4] Storey and Purao propose a multi-layered ontology to classify the semantics of relationship verb phrases.

In our CITOM approach, we produce, in addition to the links between topics and resources already defined by the Topic Map standard (called "occurrences") two types of links which are: (a) ontological and structural links and (b) usage links. Ontological and structural links are defined as specialization/generalization links, composition links and associative links as described in the ANSI/NISO Z39.19-2005 standard. Associative links can be identified by analyzing the source documents using Natural Language Processing techniques. Usage links are hyper links that we have called "is an answer to" (hyper link questions/answers) between the question represented as a topic and all the associated answers. In this case, we propose also to link a question to all the keywords that compose this question using hyper links called "is composed of".

The TM standard provides also the notions of *scope* (context) and *facets*. The scope is defined as a descriptor to contextualize topics corresponding to a vision of certain users. It provides the means for indicating when each topic name, role and occurrences are appropriate. Facets are used to complete information about an occurrence linking topics to documents by assigning attributes-values pairs to each occurrence. In our case, we propose to explore these two concepts to take into account the multilingual context. We propose to define a scope for each language in the Topic Map. We attribute to each topic a list of names in each language. We propose also to use facets to filter documents according to their language, for that, we add to each occurrence linking topics to documents an attribute to indicate the document language. Language attribute can be used later in the system as selecting criteria which means: the user gives the query in his native language and the system should return first all related documents in the user language and then documents in other languages.

At least, we propose to extend the TM-UML model by adding to the topic characteristics, a list of meta properties that reflects a topic pertinence along the time. These meta properties are initialized when the Topic Map is created, they reflect a topic relevance according to its usage by Topic Map users. They are also explored in the Topic Map pruning process, especially to delete all the topics considered as non pertinent.

3 Our Approach

Many approaches have been proposed to build Topic Maps [2]. These approaches take as input different data types: structured documents, databases, unstructured documents and semi-structured data, and propose different techniques to generate Topic Maps from these sources. Some of these works take as input XML documents and propose to apply automated processes to leverage these documents [5][6]; other approaches propose to map directly RDF metadata to Topic Maps [7][8]. Some other works propose to use learning techniques and Natural Language Processing techniques to extract topics and associations from textual documents [9][10]. Learning methods can be applied with different automation levels: manual, semi-automatic or automatic. Some research works are dedicated to cooperative Topic Map building involving different actors in the construction process [11][12][13][14].

In all existing approaches, we note that the multilingual aspect is not handled except Kasler's approach [15] which takes into account English and Hungarian texts. Also, all these approaches do not propose to explore different types of information sources in the TM construction process.

Having a content composed of multilingual textual documents, our approach CITOM takes into consideration two information sources which are: a domain thesaurus and all the possible questioning sources such as FAQ, user or expert requests related to the source documents, phone discussions and consultations with people working in the domain.

The main idea of our approach is to build, in an incremental way, a Topic Map TM_i corresponding to a set of documents $D=\{d_1, d_2 \ldots d_i\}$ by enriching the Topic Map TM_{i-1} associated to the set of document $D-\{d_i\}$. This enrichment of TM_i is realized by integrating the Topic Map associated to a document d_i into TM_{i-1}. Each phase of a Topic Map building process from document d_i and its integration into the global Topic Map use not only the source document but also the domain thesaurus and a set of questions related to the document and extracted form the questioning sources.

CITOM aims at providing a global Topic Map as a semantic structure linking and organizing concepts in various languages taking into account the particularity of multilingual contexts. In fact, a term may not have a semantic equivalent in all treated languages. This is very frequent when we consider documents provided from different cultures.

CITOM is an incremental approach in the sense that it produces a Topic Map that has been evolved, during the building process, gradually, with the construction of the resources at every introduction of a new document in the content or new questions frequently asked about the documents. This contributes in the Topic Map evolution process.

The general algorithm to build the global Topic Map is the following:

Algorithm 1. The general algorithm of our approach

Inputs: *A set of multilingual documents, a domain thesaurus and all questioning sources related to the source documents (experts questions, users requests, FAQ, phone discussions and interviews with people working in the domain, etc).*

Output: *A global topic map*

    Action 1. *Build the root of the global Topic Map. We mean by root the topic which has the domain name in different languages.*

    Action 2. *Select all the questioning sources related to the source documents.*

For *each document in the multilingual document base* ***do:***

    Action 3. *Extract a list of topics and associations from document i.*

    Action 4. *Enrich the Topic Map with new ontological and structural links extracted from the domain thesaurus.*

    Action 5. *Enrich the Topic Map using the questions related to the document.*

    Action 6. *Validate the resulting Topic Map by domain experts.*

    Action 7. *Enrich the global Topic Map with Topic Map i created from document i.*

End

The Topic Map validation step consists in defining the semantics of some links, adding or deleting some topics and/or relations. This phase is realized thanks to the collaboration of domain experts.

In this paper, we will focus only on phase 3, 4 and 5 of our approach. Sections 3.1, 3.2 and 3.3 describe these phases.

3.1 Extraction of Topics and Associations from One Document

The goal of this phase is to extract from a document d_i a set of topics and associations between them. As we mentioned in section 3, these associations are defined as onto-logical links ("is a", "part of", etc) and semantic links related to the Topic Map appli-cation domain. For that, we use existing Natural Language Processing techniques and tools to extract concepts and relations from textual documents.

Various approaches are presented in the literature for resolve this problem. Most of them are developed for ontology learning from textual documents. We can classify these approaches according to the techniques used to extract concepts and relations between them: we distinguish statistical techniques and syntactic techniques. The majority of these methods propose tools to implement their techniques.

Statistical methods are based on defining a set of measures to select candidate terms. Among them, the most popular measure is term frequency [16][17][18] used to assign a weight to each term according to its frequency in the corpus, tf-idf, T-test or statistical distribution of terms [19][20][21]. These methods do not propose to extract relations between concepts.

Syntactic methods are based on grammatical functions of words or group of words in a phrase. Some of them explore the hypothesis that grammatical dependencies between terms can be used to define semantic relationships [22][23]. Other works propose to use syntactic patterns [24][25][26] to detect relations between terms. The main advantage of these methods is that they propose not only to extract concepts but also relations between them. However, extracted relations are not semantically named, so, in most cases, domain experts should intervenes to name these relations.

Another class of methods such as [27][21] propose to explore text mining tech-niques, they combine ontology with classification techniques to find similarities be-tween candidate concepts in the document and other concepts present in the ontology. The same idea is used in [16][18] who propose to cluster terms based on term co-occurrence in the corpus. Each cluster should derive possible relations between its concepts. Other methods are based on extracting association rules [28] between terms in the corpus. Each association rule identifies relations between two concepts; pro-duced relations are named by applying a manual labeling process.

Many tools are developed to extract concepts and relations from textual docu-ments, for example, Nomino [29], Lexter [30], Fastr [31], Mantex [32], Likes [33], Acabit [34], Syntex [35], OntoGen [36] and Text2Onto [37]. Syntex is a text analysis tool based on identifying syntactic dependencies between concepts. Text2Onto is an ontology learning tool from textual data. Text2Onto combines machine learning ap-proaches with basic linguistic processing such as tokenization or lemmatizing and shallow parsing in order to identify concepts, relations between them (equivalence relation, hierarchical relation, etc) and concept instances. It is based on the GATE framework [38] for processing texts.

At least, we note that many text mining frameworks have been developed for linguistic processing of textual documents [39][40] such as GATE (General Architecture for text Engineering), a framework used for semantic annotation and information extraction from text corpora.

In our case, we choose GATE platform because it is very flexible with respect to the set of linguistic algorithms used, which means that GATE can be freely configured by replacing existing algorithms or adding new ones according to the domain application and user requirements. GATE proposes also a module called "gazetteer" to recognize named entities using existing dictionaries already included in the general framework. These dictionaries can be enriched with new terms from the application domain. Another benefit of using GATE is the possibility to integrate new external knowledge resources such as domain thesaurus to build topic hierarchies and add new ontological relations to the Topic Map.

3.2 Topic Map Enrichment with New Ontological Links

This phase aims at organizing topics extracted from sources documents by adding new ontological and structural links ("is-a", "part-of", etc). For that, we propose to explore relations between terms present in the thesaurus. The ISO 2788 and ANSI Z39 standards have proposed the guiding principles for building a thesaurus. A thesaurus is a terminological resource in which terms are organized according to restricted relations: equivalence, hierarchical relations and non-taxonomic relations (associative links).

As we know, all the existing Topic Map construction approaches do not propose to use a thesaurus to build a Topic Map. However, many ontology building approaches are based on existing thesaurus as a starting point to create an ontology, such as Hernandez' work [41] who propose to re-use a thesaurus to create and maintain a domain ontology. The authors define a method to extract conceptual schema elements of an ontology from a domain thesaurus and textual documents. The process is based on a set of transformation rules to re-use thesaurus relations. These rules explore "is specific than" (IST), "is generic than" (IGT), "use term instead" (USE) and "used for" (UF) relations to generate ontology concepts, labels associated to each concept and hierarchies of concepts.

Our Topic Map enrichment approach with new ontological links is inspired from Hernandez' approach, since, we propose to re-use thesaurus relations to identify ontological and structural links. In fact, topics are organized in hierarchical structure with "is-a" relation. These relations are directly identified from "is specific than" and "is generic than" explicit relations inherited from the thesaurus. We use also "use term instead" and "used for" relations to add new names to a topic or to group two or more topics in one topic. The method that we propose is defined as an algorithm executed in two steps. The first step concerns the use of (USE) and (UF) relations to group topics. The second step referrers to topics organisation in hierarchical structures.

Let $SYN(Term_i)$ the list of terms composed of $Terme_i$ and all the terms related to $Terme_i$ with USE and UF relations in the thesaurus. Let $/SYN(Terme_i)/$ the preferred term in SYN($Terme_i$). Let $CHILD(/SYN(Terme_i)/)$ the list of terms identified from the thesaurus when parsing all paths starting from $/SYN(Terme_i)/$, these paths contains only "is generic than" links. In the following, we present the algorithm to group topics and affect multiple names to a topic:

For each Topic T do
    If T exists in the thesaurus
        Then construction of SYN(T), /SYN(T) / and CHILD (/SYN(T) /)
            T will have as base name /SYN(T) /and other names
                in SYN(T)
End For
For each couple of topics T1 and T2 do
    If SYN(T1) = SYN(T2)
    Then group T1 and T2 in T3 with base name is /SYN(T1) /et other names
    included in SYN(T1).
    As a result to this merging, all the other characteristics of topics T1 and T2
    (association roles and occurrences) are also merged.
End For

To organize topics, we propose to explore two existing techniques among those that we have already proposed in [42] to build and maintain ontologies. The first one, called "extraction", allows to transform a concepts hierarchy into constraints between these concepts. The second technique, called "normalization", aims at building concepts' hierarchy starting from a set of constraints between concepts. In the Topic Map enrichment process, we apply the first technique on the thesaurus in order to extract constraints between terms represented as topics in our Topic Map based on "is generic than" links, explicit relations inherited from the thesaurus. After identifying these constraints, we apply the normalisation technique on the Topic Map. To summarize, in the following, we present the topics organisation algorithm:

For each topic T1 do
    For each T ∈ CHILD(T1)
        T→T1
        End For
End For
For each couple T1 and T2 present in the Topic Map do
    If CHILD(T1)≠ CHILD(T2) then T1 ↔ T2
    Else if CHILD(T1)∩ CHILD(T2) = T3 then T1, T2→T3
End For
Application of normalization technique

3.3 Topic Map Enrichment with Users Requests

The main goal of our global Topic Map is to support the user in his search and questioning step. It should allow the user navigate across the Topic Map structure in order to find relevant documents related to the selected topics.

We aim at introducing in the Topic Map, knowledge about all the questions frequently asked and related to the sources documents. For that, we define the usage links. This supposes that frequently asked questions are already selected and sorted by document. The selection phase of potential questions is based on analyzing all the possible questioning sources (FAQ, user or expert requests related to the source documents, phone discussions and consultations with people working in the domain, etc).

Each selected question is represented as one topic linked to the associated answers (all the topics referencing the documents allowing to answer the question) using "is an answer to" relation. This question is also linked to all the keywords composing it using "is composed of" relation. This relation gives the possibility to the user to search information when he navigates in the Topic Map and allow automatic search of "similar question" using Salton vector [43] associated to each question (all the terms that composing it).

We suppose that the enrichment process will automatically select all the questions keywords and find the documents to answer associated questions. For that, we use linguistic processing techniques existing in the literature [19][22][23].

To insert a topic, already identified by the enrichment process, in the Topic Map, three cases are possible: (1) No action when the topic already exists in the Topic Map with the same name, (2) add a new name to a topic when this one already exists in the Topic Map but designated with a different name and finally (3) insert a new topic when this one does not exists in the Topic Map. All these possibilities are referred to existing schemas and ontology merging techniques [44][45][46].

4   Topic Map Pruning Process

The Topic Map pruning process is a big issue to be addressed in our work, since a Topic Map is essentially used to organize a content of documents and help users finding relevant information in theses documents. So, it is required to maintain and enrich the Topic Map structure along the time in order to satisfy users' queries and handle possible changes of the document content.

To maintain the Topic Map, we propose to introduce some information, that we have called "meta properties", about a topic relevance according its usage when users explore the Topic Map. This information can be explored to evaluate the quality of a Topic Map.

At a first stage, we propose to define a score (or level) for each topic as a meta property which reflects its importance in the Topic Map. The score is initialized when the Topic Map is created. It can be (a) *very good* when the topic is obtained from three information sources which are documents, thesaurus and requests (b) *good* when the topic is extracted from two information sources or (c) *not very good* when the topic is extracted form one source.

Scores assigned to each topic can be also explored as selecting criteria to visualize the Topic Map. In fact, a topic with a very good score is considered as a main topic so in this case, we will have a default visualization of the Topic Map. The idea is that instead of definitively suppressing a Topic because it is not very used, we prefer just low its score. Indeed, a Topic may be the target of very few queries in one season and coming back to the most frequently asked Topics next season (e.g. many questions concern air conditioners in summer, but only very few in winter). However some Topics - generally concerning case in the news - definitively decrease in importance. So we will add data to capture the type of the Topic's score evolution (season dependant, time dependant, decreasing, increasing, etc.).

In our future works, we will discuss in more detail quality criteria of a Topic Map in order to identify an exhaustive list of meta properties that help managing the Topic

Map in the evolution process. We will also introduce meta-meta data attached to the scores in order to store their evolution's profile and thus automatically update them.

5 Application on the Sustainable Construction Domain

We experiment our approach CITOM on a real corpus from the sustainable construction domain. This corpus contains 105 html documents. We use as input a bilingual thesaurus (French/English) called CTCS (Canadian Thesaurus of Construction Science and Technology) from the domain of construction science [47]. This thesaurus contains 15331 terms organized in hierarchy of 10 levels, each term is described in html file and relations between terms are external cross links contained in this file.

To illustrate the phase 3, 4 and 5 presented in this paper, we consider a fragment of html document D1 (http://www.aceee.org/) related to energy economy. Based on GATE framework, in which we have integrated some of html files from the thesaurus, we identify 44 topics and 16 associations between these topics. We use TM4J tool [48] to visualize the generated Topic Map. Then, we enrich this Topic Map using CTCS thesaurus, consequently new ontological links will be added. For example, we can identify "is-a" hierarchy relation between topics "heating" and "ecological heating".

After that, we select a question "what are the economic heating means?" from a list of FAQ extracted from (http://www.aceee.org/) website and we integrate this question in the Topic Map as shown in figure 2.

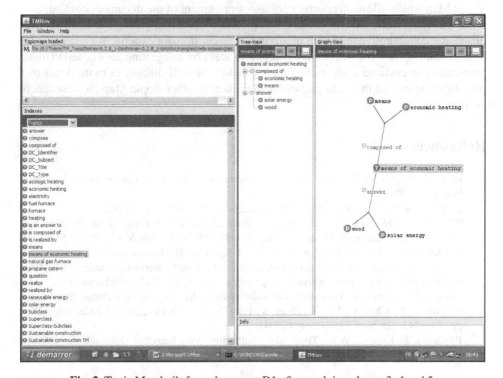

Fig. 2. Topic Map built from document D1 after applying phases 3, 4 and 5

6 Conclusion and Future Work

In this paper, we present CITOM, an incremental approach to build multilingual Topic Map. Our approach is tested on a real corpus from the sustainable construction domain.

CITOM makes advantage in the fact that we take into account the Topic Map usage by defining usage links between potential questions extracted from all the possible questioning sources and associated answers. Each selected question (natural language phrase) is represented as one topic and linked to all the keywords composing it using "is composed of" relation. This relation gives the possibility to the user to search information when he navigates in the Topic Map and allow automatic search of "similar question".

The proposed approach takes into consideration multilingual resources. Thus, when navigating in the Topic Map, a user can access documents written in other languages different from his native language. Also, the main advantage of CITOM is that, with reference to existing answers translation systems, it returns to the user documents associated to concepts that not exists in his language or in his culture. This constitutes an important enrichment to the user.

Moreover, CITOM define a list of meta properties that we have integrated in the Topic Map model. These meta properties are explored in Topic Map pruning and evolution process. Finally, CITOM is based on incremental process to build the global Topic Map which allows supporting possible enrichment of the document content.

In the near future, we will focus on Topic Map validation phase. For that, we will propose a collaborative approach involving different domain experts. We plan also to extend and improve the Topic Map pruning process by integrating new criteria (meta properties) to evaluate a topic relevance. Finally, we will discuss in more detail the possibility to extend the meta properties definition to other Topic Map elements such as associations.

References

1. ISO/IEC :13250. Topic Maps: Information technology-document description and markup languages (2000),
   http://www.y12.doe.gov/sgml/sc34/document/0129.pdf
2. Ellouze, N., Métais, E., Ben Ahmed, M.: State of the Art on Topic Maps Building Approaches. In: Kutsche, R.-D., Milanovic, N. (eds.) MBSDI 2008, Model Based Software and Integration Systems, CCIS 8, pp. 102–112. Springer, Heidelberg (2008)
3. Pepper, S.: Article for the Encyclopedia of Library and Information Sciences (2008),
   http://www.ontopedia.net/pepper/papers/ELIS-TopicMaps.pdf
4. Storey, V.C., Purao, S.: Understanding Relationships: Classifying Verb Phrase Semantics. In: Atzeni, P., Chu, W., Lu, H., Zhou, S., Ling, T.-W. (eds.) ER 2004. LNCS, vol. 3288, pp. 336–347. Springer, Heidelberg (2004)
5. Reynolds, J., Kimber, W.E.: Topic Map Authoring With Reusable Ontologies and Automated Knowledge Mining. In: XML 2002 Proceedings by deepX (2002)

6. Librelotto, G.R., Ramalho, J.C., Henriques, P.R.: TM-Builder: An Ontology Builder based on XML Topic Maps. Clei Electronic Journal 7(2), Paper 4 (2004)
7. Pepper, S.: Topic Map Erotica RDF and Topic Maps "in flagrante" (2002), http://www.ontopia.net/topicmaps/materials/MapMaker_files/frame.htm
8. Pepper, S.: Methods for the Automatic Construction of Topic Maps (2002), http://www.ontopia.net/topicmaps/materials/autogen-pres.pdf
9. LeGrand, B., Soto, M.: Topic Maps et navigation intelligente sur le Web Sémantique, AS CNRS Web Sémantique, CNRS Ivry-sur-Seine (October 2002)
10. Folch, H., Habert, H.: Articulating conceptual spaces using the Topic Map standard. In: Proceedings XML 2002, Baltimore, December 2002, pp. 8–13 (2002)
11. Ahmed, K.: TMShare – Topic Map Fragment Exchange in a Peer-To-Peer Application (2003), http://www.idealliance.org/papers/dx_xmle03/papers/02-03-03/02-03-03.pdf (2003)
12. Lavik, S., Nordeng, T.W., Meloy, J.R.: BrainBank Learning - building personal topic maps as a strategy for learning. In: XML, Washington (2004)
13. Zaher, L.H., Cahier, J.-P., Zacklad, M.: The Agoræ / Hypertopic approach. In: International Workshop IKHS - Indexing and Knowledge in Human Sciences, SdC, Nantes (2006)
14. Dicheva, D., Dichev, C.: TM4L: Creating and Browsing Educational Topic Maps. British Journal of Educational Technology - BJET 37(3), 391–404 (2006)
15. Kasler, L., Venczel, Z., Varga, L.Z.: Framework for Semi Automatically Generating Topic Maps. In: TIR 2006, Proceedings of the 3rd international workshop on text-based information retrieval, Riva del Grada, pp. 24–30 (2006)
16. Agirre, E., Ansa, O., Hovy, E., Martinez, D.: Enriching very large ontologies using the WWW. In: ECAI 2000 workshop on Ontology Learning, Berlin, Germany (2000)
17. Faatz, A., Steinmetz, R.: Ontology enrichment with texts from the WWW. In: The Semantic Web Mining Conference WS 2002 (2002)
18. Parekh, V., Gwo, J.-P., Finin, T.: Mining Domain Specific Texts and Glossaries to Evaluate and Enrich Domain Ontologies. In: International Conference of Information and Knowledge Engineering (2004)
19. Velardi, P., Missikof, M., Fabriani, P.: Using text processing techniques to automatically enrich a domain ontology. In: Proceedings of ACM- FOIS (2001)
20. Xu, F., Kurz, D., Piskorski, J., Schmeier, S.: A domain adaptive approach to automatic acquisition of domain relevant terms and their relations with bootstrapping. In: The 3rd international conference on language resources and evaluation (2002)
21. Neshatian, K., Hejazi, M.R.: Text categorization and classification in terms of multi-attribute concepts for enriching existing ontologies. In: 2nd Workshop on Information Technology and its Disciplines, pp. 43–48 (2004)
22. Bendaoud, R., Rouane Hacene, M., Toussaint, Y., Delecroix, B., Napoli, A.: Construction d'une ontologie à partir d'un corpus de textes avec l'ACF, IC (2007)
23. Roux, C., Proux, D., Rechermann, F., Julliard, L.: An ontology enrichment method for a pragmatic information extraction system gathering data on genetic interactions. In: Proceedings of the ECAI 2000 Workshop on Ontology Learning, OL (2000)
24. Hearst, M.A.: Automatic acquisition of hyponyms from large text corpora, Rapport technique S2K-92-09 (1992)
25. Maedche, A., Staab, S.: Mining ontologies from text. In: Dieng, R., Corby, O. (eds.) EKAW 2000. LNCS (LNAI), vol. 1937, pp. 189–202. Springer, Heidelberg (2000)

26. Stumme, G., Hotho, A., Berendt, B.: Semantic web mining: State of the art and future directions. Web Semantics: Science, Services and Agents on the World Wide Web 4(2), 124–143 (2006)
27. Han, E.-H., Karypis, G.: Centroid based document classification: Analysis and experimental results. In: The 4th European Conference of Principles of Data Mining and Knowledge Discovery, pp. 424–431 (2000)
28. Agrawal, R., Srikant, R.: Mining generalized association rules. Future Generation Computer Systems 13(2-3), 161–180 (1997)
29. Dumas, L., Plante, A., Plante, P.: ALN: Analyseur Linguistique de ALN, vers.1.0. ATO, UQAM (1997)
30. Bourigault, D.: LEXTER, a Natural Language Processing tool for terminology extraction. In: Proceedings of the 7th EURALEX International Congress, Goteborg (1996)
31. Jacquemin, C., Bourigault, D.: Term Extraction and Automatic Indexing. In: Mitkov, R. (ed.) The Oxford Handbook of Computational Linguistics, pp. 599–615. Oxford University Press, Oxford (2003)
32. Frath, P., Oueslati, R., Rousselot, F.: Identification de relations sémantiques par repérage et analyse de cooccurrences de signes linguistiques. In: Charlet, J., Zacklad, M., Kassel, G., Bourigault, D. (eds.) Ingénierie des connaissances, Évolutions récentes et nouveaux défis, Eyrolles, Paris, pp. 291–304 (2000)
33. Rousselot, F., Frath, P., Oueslati, R.: Extracting concepts and relations from Corpora. In: Proceedings of the Workshop on Corpus-oriented Semantic Analysis, European Conference on Artificial Intelligence, ECAI 1996, Budapest (1996)
34. Daille, B.: Identification des adjectifs relationnels en corpus. In: Actes de la Conférence de Traitement Automatique du Langage Naturel (TALN 1999), Cargèse (1999)
35. Bourigault, D., Fabre, C., Frérot, C., Jacques, M.-P., Ozdowska, S.: Syntex, analyseur syntaxique de corpus. In: Actes des 12èmes journées sur le Traitement Automatique des Langues Naturelles, Dourdan, France (2005)
36. Fortuna, B., Grobelnik, M., Mladenic, D.: Semi-automatic data driven ontology construction system. In: Proceedings of the 9th International multiconference Information Society IS 2006, Ljubljana, Slovenia (2006)
37. Cimiano, P., Volker, J.: Text2onto - a framework for ontology learning and data-driven change discovery. In: Montoyo, A., Muñoz, R., Métais, E. (eds.) NLDB 2005. LNCS, vol. 3513, pp. 227–238. Springer, Heidelberg (2005)
38. The GATE platform: http://gate.ac.uk/
39. Ferruci, D., Lally, A.: UIMA: an architecture approach to unstructured information processing in a corporate research environment. Natural Language Engineering 10(3-4), 327–348 (2004)
40. Muller, H.-M., Kenny, E.E., Sternberg, P.W.: Textpresso: an ontology based information retrieval and extraction system for biological literature. PLoS Biology 2(11), 1984–1998 (2004)
41. Hernandez, N., Mothe, J.: D'un thesaurus vers une ontologie de domaine pour l'exploration d'un corpus. In: Actes de la conférence Veille Stratégique Scientifique & Technologique VSST (2006)
42. Lammari, N., Métais, E.: Building and Maintaining Ontologies: a Set of Algorithms. Data and Knowledge Engineering 48(2), 155–176 (2004)
43. Salton, G., Buckley, C.: Term-weighing approaches in automatic text retrieval. Information Processing & Management 24(5), 513–523 (1988)
44. Calvanese, D., Giacomo, G.D., Lenzerini, M.: A framework for ontology integration. In: Proc. of the First Semantic Web Working Symposium (2001)

45. Noy, N.F., Musen, M.A.: Prompt: Algorithm and tool for automated ontology merging and alignment. In: Proceedings of the Seventeenth National Conference on Artificial Intelligence and Twelfth Conference on Innovative Applications of Artificial Intelligence. AAAI Press/MIT Press (2000)

46. Buneman, P., Davidson, S.B., Kosky, A.: Theoretical aspects of schema merging. In: Pirotte, A., Delobel, C., Gottlob, G. (eds.) EDBT 1992. LNCS, vol. 580. Springer, Heidelberg (1992)

47. Canadian Thesaurus of Construction Science and Technology:
    http://irc.nrc-cnrc.gc.ca/thesaurus/

48. TM4J website: http://tm4j.org/

Schema-Matching with Data Dictionaries

Gary Coen and Ping Xue

Boeing Research and Technology
P.O. Box 3707 MC 7L-43
Seattle, WA 98124-2207
gary.a.coen@boeing.com, ping.xue@boeing.com

Abstract. We describe an algorithm for detecting semantically equivalent metadata across namespaces instantiated as database schema, an operation otherwise known as schema-matching. Assuming a metadata description discipline which imposes graph-theoretic constraints on data dictionaries, the algorithm employs analytical techniques used in information retrieval, information theory, and computational linguistics. It exploits the information inherent in textual metadata description and metadata dependency relations in order to match elements across schema boundaries.

Keywords: Schema-matching, database lexicography, metadata management.

1 Problem

For the purpose of discussion, let a *namespace* be a bounded domain of names constructed according to one or more formation rules, each name yielding a unique denotation within the scope of its domain. (Informally, a *name* is a referring term and a *denotation* the item to which it refers.) Namespaces abound in computing. Table names in a relational database and member variables in a UML, Java, or C++ class constitute namespaces. For that matter, the names of all computer languages, living and dead, constitute a namespace too, although not a very interesting one.

The practical use for namespaces in computer science has been to effect modularity. Namespaces are crucial to class encapsulations, generic programming, software modules, compilation units, and a variety of other artifacts of contemporary computing. Any namespace may contain an inventory of unique identifiers, each with discrete semantics. This makes it possible for a particular name to recur in multiple namespaces with a different denotation in each. Whenever a compiler, linker, interpreter, or database management system calls for a function or an instance of a particular element type, that request is satisfied using denotational semantics from an appropriate namespace.

Namespaces organize complexity. A namespace, for instance, may contain one or more namespaces. A column in a relational database table constitutes a namespace within a namespace. The type of a class member variable in a UML model may encapsulate a local namespace of types and functions. Likewise, a Java or C++ class contains a local namespace for its member functions, and each method described therein may contain yet another namespace for its own local variables.

H. Horacek et al. (Eds.): NLDB 2009, LNCS 5723, pp. 62–78, 2010.

A particular name may occur unambiguously in multiple namespaces. In computing environments where multiple namespaces exist, logical distinctions between them are typically maintained by the use of scope identifiers indicating how to resolve the identity of a name across namespaces. Compilers, linkers, interpreters, and database management systems encode the formation rules by which names are composed, and thus resolve such requests appropriately. In this way, the unique denotation of a name is maintained across namespaces.

Frequently, however, it is desirable to translate between namespaces. In multiple system contexts, it is regularly necessary to correlate names with equivalent denotational semantics across namespaces. Given a name and its denotation from one namespace, the necessary operation discovers a name with equivalent denotational semantics in another. Usually this discovery enables data re-use and system interoperability. Unfortunately, the discovery operation resists automation. Today, in fact, it is habitually performed as a time-consuming, error-prone, manual task, one that has become a principle cost driver for computing in multiple system contexts. Since modern computing routinely involves multiple systems, this cost can impede progress.

2 Approach

One might reduce the core problem of translation between namespaces to *schema-matching*, thereby construing the desired solution as an automated operation over the names in two database schemas. (A taxonomy and survey of schema-matching is presented in [1].) While this formulation focuses on databases at the expense of other namespaces, it reduces the general problem to a tractable scope in an area where the beneficial impact of a solution is high. Moreover, the tighter focus on an application domain provides concrete details for the exposition of a fully automated solution. Within such a framework, we present an approach to schema-matching that exploits both the linguistic information inherent in metadata description and its dependency architecture [2]. Since metadata description is a common species of lexicography that proliferates outside database schemas, we trust this approach will serve as a conceptual framework for other aspects of translation between namespaces.

We observe that a database schema is a namespace. Named *metadata* define the design of the data and thus the semantics of each data element. Hence, the data design is a bounded domain of names constructed in a rule-governed way to support the architecture, each name yielding a unique denotation within the database. In this way, a database schema clearly embodies a namespace, and the data dictionary publishes this information for re-use.

2.1 Linguistic, Element-Level, Schema-Based Schema-Matching

Adopting the taxonomy of schema-matching presented in [1], our algorithm can be characterized as linguistic, element-level, and schema-based. It is *element-level* because it discovers candidate matches for individual schema elements, as opposed to complexes of elements. It is *schema-based* because it considers only schema information and ignores instance data. It is *linguistic* because it interprets the information encoded in data dictionary definitions in order to compare metadata elements across schema for equivalent denotational semantics. Like any schema-matching approach of this type, the algorithm requires carefully sculpted data dictionaries, and successful schema-matching is directly related to the integrity of metadata description with respect to the database schemas involved.

We observe that the information content of data depends ultimately on lexicographic knowledge. In other words, interpretation of a data value crucially depends on the metadata type encoding that value. Moreover, the data dictionary is the conventional repository for metadata description. In its highest conception, it records and publishes this information so that data semantics may be shared consistently throughout the architecture. We rely on graph-theoretic methods for managing the shared lexicographic information in metadata description. These methods exploit the data dictionary as a metadata resource characterizing the information structure of instance data. In what follows, we assume that data dictionaries satisfy the lexicographic and namespace constraints of database lexicography [3].

2.2 Lexicographic Constraints

A data dictionary is to a data model as a map is to a terrain. Because the quality of schema-matching is determined by the quality of metadata description, data dictionary information properties are central to our approach. In brief, data dictionary entries should define metadata as precisely and succinctly as possible, and say no more. For example, each entry should consist of (at least) a metadata term and its definition, and each type of entry has specific information requirements. Commonplace properties of metadata include inheritance, aggregation, and existential entailment, and data dictionaries with integrity of description encode this information systematically (*cf.* [3], §3.1).

A lexical graph is a directed acyclic graph with nodes representing dictionary entries as term-definition pairs and edges representing *lexical dependency* relations, where lexical dependency occurs whenever one term's definition uses another term under definition in the dictionary (*v.* [2]). To *overload* a lexical element is to define multiple denotations for it. Hence, overloading occurs when a namespace exhibits one or more instances of lexical ambiguity. Informally, ambiguous lexical elements constitute naming collisions which must be resolved when they occur within a single schema or within multiple schemata considered within the same context.

Finally, a data dictionary suitable for schema-matching should present an underlying lexical graph free of overloading.

2.3 Namespace Constraints

A data dictionary, like any specialized lexicon, has properties that distinguish it from other information resources. An algorithm that interprets the information encoded in a data dictionary must identify these properties and accommodate them. Like any other document, a data dictionary employs standard lexis in normal patterns of exposition according to everyday language use. As a lexicon, however, a data dictionary also defines new meanings: it combines standard lexis in new ways to identify autonomous lexical material and its denotational semantics. The scope of this new lexical material extends throughout the data dictionary, and the network of lexical dependency thus created may be pervasive. In fact, the dependency architecture may influence interpretation of any metadata element.

Typically, a data dictionary presents lexical material in ordinary language correlating elements of a controlled terminology with *definite descriptions*—following Bertrand Russell, a denoting phrase introduced by a definite article or its equivalent—that are true within the boundaries of the associated schema. In other words, a data dictionary augments ordinary language lexis. When its controlled terminology occurs without overloading in the otherwise ordinary language of its definitions, denotations of the controlled terminology contribute alongside ordinary language to define meaning. However, when overloading occurs in a data dictionary definition, it creates a naming conflict between ordinary language and the dictionary's metadata description. A key insight of our approach is that, under these circumstances, the denotational semantics specified in the data dictionary takes precedence over ordinary language semantics. Let this principle be known as *overload precedence*.

Overload precedence complicates data dictionary processing. Even modestly proportioned dictionaries stress attentional and memory resources of human analysts. Each metadata element extends or, potentially, overrides ordinary language use, and any definition may reuse one or more other metadata elements. For interpretation, overload precedence applies equally when the naming conflict involves a collision between single words, word sequences, or their combination. Clearly, an analytical procedure that fails to accommodate this phenomenon will fail to interpret metadata properly.[1]

3 Algorithm

The familiar sequence of data preparation, pattern discovery, and pattern analysis organizes our schema-matching procedure. Alongside elements of the database lexicography discipline, the algorithm employs analytical techniques commonly used in information retrieval, information theory, and computational linguistics. Like a specialized document clustering procedure, it treats each data dictionary entry as an individual document and computes an ordered list of terms to represent its meaning. Initially, it converts metadata describing two database schemas into term lists tagged with their schema of origin. Next, it applies an association measure to the set of term lists to sort it. Finally, the sorted set is evaluated to identify sequential pairs, each from a different schema, with similarity sufficient to suggest equivalence. These pairs are then identified as likely candidates for inter-schema semantic equivalence. A detailed description of the algorithm follows.

3.1 Evaluating Similarity

Conventional information retrieval techniques represent documents as document vectors embedded in an n-dimensional Euclidean space, where n is the total number of

---

[1] For example, failure to account for overload precedence explains in part the inconspicuous results associated with an information retrieval paradigm that identifies the sense of a word's occurrence by calculating the overlap between sense definitions extracted from machine-readable dictionaries and the context of occurrence [4].

index terms appearing in a document collection. Typically, a document vector identifies index terms occurring in a document. (Salton established the paradigm in [5].) Many association measures for document vector similarity (or dissimilarity) have appeared in the related literature. Perhaps the simplest is the basic matching coefficient $|X \cap Y|$, which identifies the number of index terms shared by document vectors X and Y. The higher the number of shared index terms, the higher the presumption of document similarity, and similarity presumably correlates with likelihood of equivalence. (Note that conventional information retrieval algorithms and data structures are typically designed to identify document equivalence, whereas interpretation of document meaning is needed for schema-matching.) Various extensions of the basic matching coefficient attempt to normalize for differences in size between X and Y, often dividing the basic matching coefficient by some related value. Dice's coefficient, Jaccard's coefficient, the cosine coefficient, and the overlap coefficient all qualify as normalized extensions of this type. The goal of this agenda is to control for exaggerated document significance based on size, since larger documents are likely to contain more index terms to share. The disciplinary basis of this paradigm is that all documents should be construed as equals. [6] presents Doyle's influential formulation of this position.

Fundamental problems arise when the documents to be clustered are dictionary entries. Conventional document clustering techniques cannot manage data dictionary entries because metadata elements are inherently unequal. Overload precedence contains the seeds of this inequality, and the network of lexical dependency instantiated in a data dictionary measures out its effects. Recall that conventional information retrieval knows only the words that appear in documents—word occurrence indicates content and equivalence is calculated from word content, not conceptual content. Hence, standard techniques are insufficient for the interpretation of a data dictionary, where definitions are terse and the meaning of a particular metadata element cannot be understood without prior knowledge of the meaning of every other metadata element occurring in its definition.

To interpret data dictionaries, conventional document clustering techniques must be specialized. If the content of a document vector is to represent document meaning, a document vector representing a data dictionary element is obliged to indicate each of the other dictionary elements upon which its meaning depends. If the entire semantic contribution of lexical dependency is represented in its document vector, then the document vector for a data dictionary element should minimally include a generalized union of document vectors for all named metadata occurring in its definition. In order to interpret meaning, this representation should minimally identify conceptual, as opposed to just lexical, content. Accordingly, our schema-matching algorithm specializes conventional document clustering for the purposes of data dictionary analysis so that it produces only document vectors that respect the facts of overload precedence.

3.2 Data Preparation

The data preparation phase of the algorithm takes as input the two data dictionaries involved in schema-matching, each a set of term-definition pairs. In mathematical logic, a term under definition is a *definiendum* and its definition a *definiens*. Formally, it is necessary that a definiendum have a different (*viz.*, wider) scope than a definiens.

([7] provides a treatment formulated in this framework.) For clarity, we adopt this usage henceforth.

Initially, all data dictionary entries are marked to identify schema of origin. The algorithm then rewrites each pair, unambiguously tagging every definiens word contributing to a denotation with an underlying, canonical form representing conceptual content. (In our initial implementation, conceptual content corresponds to a WordNet synset [8].) This step interprets meaning and dispels any lexical ambiguity from the definiens. For each newly rewritten dictionary, it then performs an exhaustive lexical dependency analysis to identify how metadata coupling affects meaning. (For an analysis of meaning and lexical dependency, v. [3], §3.2.1.1.) Thereafter, the algorithm computes a sorted topology of lexical dependency for each dictionary, thus producing a data structure to guide the instantiation of document vectors.

3.3 Pattern Discovery

Once conceptual content and information structure of each dictionary are identified, our algorithm assembles metadata representations. For each dictionary, it lists the unique canonical forms appearing in the set of definiens rewritten during the data preparation phase. These lists account for each multi-word definiendum occurring in the definiens as a single multi-word item, thus respecting overload precedence. For each data dictionary, the result is an inventory of index terms, a set of canonical forms encoding metadata and incidental matter.

Next, the algorithm combines the two lists of index terms and the two sets of rewritten definiens to create a virtual namespace in which the two formerly independent namespaces are local. This combined index term list becomes the inventory of index terms available to the algorithm. Similarly, the combined definiens set becomes the document collection. Document vectors will be constructed from these two data structures.

3.3.1 Term Weighting
Our algorithm employs term weighting techniques to optimize statistical discrimination between documents. Our term weighting is based on single term statistics that control for term certainty, term frequency, term distribution, and document length.

First, considering each index term as a random variable, the algorithm computes an entropy value with respect to its occurrence in the definiens set, thus identifying the certainty of term-document correlation for each index term. The computation customizes Shannon's formula as illustrated in Definition 1:

$$\text{Definition 1: } H(W) \underline{def} - \sum_{w \in V(W)} P(w) \log P(w)$$

Here W is the inventory of index terms, w is a particular term from that inventory, and entropy is a functional of probability distribution functions. Since probabilities lie in the interval between zero and one, their logarithms are always negative, and the minus sign ensures that resultant entropies are always positive. In the language of information theory, $H(w)$ increases with the amount of uncertainty contained in w ([9] explores entropy and information theory), and the algorithm employs this as a measure relating index term to definiens. When $H(w)$ is high, there is uncertainty about the correlation between w and elements of the definiens set; as H(w) decreases, certainty increases until the limit condition in which w correlates with exactly one definiens, the one in which w occurs.

In this schema-matching algorithm, term weighting combines term certainty with term informativeness, another single term statistic. Informativeness, I(w), encapsulates a computation on term frequency, term distribution, and size of the document collection in a function that increases with w's occurrence across elements of the definiens set:[2]

$$\text{Definition 2: } I(w) \underset{=}{\text{def}} \log\left(n - \log\left(\frac{f(w)}{df(w)}\right)\right)$$

Ultimately, the schema-matching algorithm formulates term weighting, W(w), as the product of term certainty and term informativeness. The formula is given in Definition 3:

$$\text{Definition 3: } W(w) \underset{=}{\text{def}} H(w)\, I(w)$$

As in conventional document clustering, term weighting is intended to represent term significance within the document collection. Like inverse document frequency, it compensates for the fact that longer documents would otherwise receive more weight in equivalence comparisons since they are likely to share more index terms. For schema-matching with data dictionaries, W(w) factors term frequency across the definiens set (as opposed to per document, often noted as $tf_{t,d}$) because the relational logic underlying the data dictionary and its corresponding data model conceptualizes multiple occurrences of a term within a definiens as a single, unique denotation, and factoring term occurrences within definiens would override this logic. Furthermore, term frequency per definiens offers little discriminatory power since definiens are characteristically highly abbreviated and regular in length when compared to the documents of generic search algorithms (*cf.* the Okapi BM25 weights discovered mostly through trial and error in TREC research [10]).

In support of schema-matching, term weighting provides the means to induce a weak partial order on a document collection, thus discriminating between its elements. The benefit this provides is illustrated in Table 1, which presents term statistics for a fragment of a data dictionary of 1,000 entries and 20,000 words. Note that the entries are sorted in ascending order according to W(w), illustrating the ordering effect of term weighting, which is lowest when w is highly discriminating. As specified in Definition 3, term weighting is minimal when a term occurs exactly once within a single document of the document collection, and it is maximal when it occurs in every document. Within this interval, term weighting is a function of frequency and distribution, with distribution as the more heavily weighted factor.

---

[2] In Definition 2 and elsewhere, n is the total number of definiens in the dictionary, $f(w)$ is the frequency of term w across the document set, and $df(w)$ identifies the number of definiens in which w occurs. As in the case of term certainty, a lower value for term informativeness indicates a heightened potential for knowledge about the correlation between w and elements of the definiens set.

Table 1. Data Dictionary Term Statistics ($n=10^3$, $|W|=2*10^4$)

f(w)	df(w)	I(w)	P(w)	H(w)	W(w)
1	1	9.965784	0.00005	0.00022	0.002143
2	1	9.964341	0.0001	0.0004	0.003986
2	2	9.965784	0.0001	0.0004	0.003986
3	1	9.963496	0.0002	0.00057	0.005715
3	2	9.964940	0.0002	0.00057	0.005716
3	3	9.965784	0.0002	0.00057	0.005716
4	1	9.962896	0.0002	0.00074	0.007370
4	2	9.964341	0.0002	0.00074	0.007372
4	3	9.965185	0.0002	0.00074	0.007372
4	4	9.965784	0.0002	0.00074	0.007373
5	1	9.962431	0.0003	0.0009	0.008971
5	2	9.963876	0.0003	0.0009	0.008973
5	3	9.964721	0.0003	0.0009	0.008973
5	4	9.965320	0.0003	0.0009	0.008974
5	5	9.965784	0.0003	0.0009	0.008974
10^3	1	9.951335	0.05	0.06505	0.647349
10^3	10^3	9.965784	0.05	0.06505	0.648289

3.3.2 Document Vectors

Term statistics are instrumental in the computation of a document vector format for clustering a document collection. Typically, the target format is a multivariate representation assigned to various kinds of objects in statistical [11] and taxonomic [12] representations. At an abstract level, a document vector is an ordered list of index terms selected to represent a document. Its practical utility is the discriminatory power to partition a document set into meaningful equivalence classes.

3.3.2.1 Defining Document Vector Format. Our schema-matching algorithm establishes a document vector format in two steps. First, the inventory of index terms is filtered to remove ubiquitous elements. This can be accomplished by removing from W, the inventory of index terms, all w for which df(w)=n. Let T denote the set of remaining index terms after this operation is complete. Then the length of the document vector format can be identified as the cardinality of T, |T|.

Once document vector length is established, full specification of the document vector format lacks only a method by which to correlate the elements of *T* with the |T| positions available. To accomplish this, the algorithm resorts to term weighting. Starting with the leftmost vector position and proceeding rightward, the index term from *T* with the lowest term weighting is sequentially correlated with the next available position of the vector until each term is correlated with exactly one position. When complete, index terms with the lowest term weighting are correlated with the leftmost positions, and those with the highest weighting are correlated with the rightmost. This technique invests document representations with the discriminatory power of term weighting described in §3.3.1.

3.3.2.2   Instantiating Document Vectors. Once a format is fixed, the schema-matching algorithm instantiates document vectors to represent the definiens of the document collection. Each of these document vectors depicts meaning in the form of its correlated conceptual content. The set of document vectors produced in this way represents the two document collections which, in turn, contain metadata description for the schemas involved in schema-matching.

Document vectors are instantiated in a straightforward way. In general, new document vectors are initialized with default values (indicating that no index terms have yet been correlated with the document represented). Iterating through the elements of each data dictionary semantic dependency topology S produced in the data preparation phase (v. §3.2), the algorithm executes the following procedure at each step (for a topology of lexical dependency relations, iteration order is from independent to dependent metadata):

(i)	Create and initialize document vector d for current metadata description M in S
(ii)	For each index term t, $t \in T$, in the definiens of M
(iii)	If t is the definiendum of another element of S
(iv)	Mark the position in d correlated with any index term t' upon which the definiens of t depends either directly or indirectly via a transitive closure of dependency relations, in this way relating the definiens of M to index terms upon which t depends
(v)	Else mark the position in d correlated with t (v. §3.3.2.1) to relate the definiens of M to index term t
(vi)	Assign d to represent M
(vii)	Advance to the next metadata item in S

Step (*iv*) of this procedure enforces overload precedence in exactly those instances in which it is necessary for interpreting data dictionary entries. By the time that t appears in the enumeration order on elements of S, the document vector associated with each t' upon which t lexically depends will represent the transitive closure of its lexical dependencies. This ensures that the representation for metadata item M includes all the index terms contained in the definiens of M as well as those implicated in direct reference via chains of lexical dependency.

The strategy behind this approach identifies the semantic contribution to M of lexical dependency t to be the definite description specified in the definiens of t. Assuming a species of direct reference logic, this definiens suffices to fix the referent of t in the database schema. If lexical dependency is directly referential in this sense, then the semantic function of t with respect to M must be to import this referent into M's definiens. This approach approximates Kaplan's treatment of indexicals in ordinary language [13], in which the function of a term that is directly referential (in Russell's sense) is to import its referent into the proposition using it. In Russell's language, the term is a *constituent* of the proposition; following Kaplan, its *content* is the object to which it refers. For the purposes of schema-matching, this is accomplished by assigning to M the representation which results from applying the inclusive OR operator to the document vectors associated with M and t. In this way, the conceptual content of lexical dependent t is a constituent of the representation of M.

3.3.3 Association Measure for Evaluating Similarity

Our schema-matching algorithm employs document vector equivalence as its principle association measure. When two definiens (*i.e.*, documents) share exactly the same index terms, they are presumed maximally similar at the level of conceptual content.[3] Document vector equality indicates high potential for equivalent denotational semantics. Conversely, the lower the inventory of shared index terms, the lower the presumption of equivalent denotational semantics. When considered in this way, the basic matching coefficient can be a crude, but useful, association measure for evaluating metadata description similarity in schema-matching (*cf.* §3.1).

There are three possible outcomes for each application of this association measure. Trivially, when document vectors representing metadata are different, the metadata involved in the comparison are presumed distinct. Otherwise, when document vectors are identical or equivalent, the metadata are presumed to be meaningfully similar. When these metadata originate in the same schema, the comparison may have discovered synonymous metadata, a spurious condition that merits further attention. When these metadata originate in different schema, it is possible the comparison has discovered equivalent metadata across database schema, the goal of schema-matching.

4 Pattern Analysis

Recall that term weighting is minimal when a term occurs exactly once within a single document of the collection and maximal when it occurs in every one. Low term weighting indicates high term certainty and informativeness as well as a potential for a particular index term to occur uniquely in some particular definiens. In our schema-matching algorithm, document vectors embody the discriminatory power of term weighting.

Consider the significance of term weighting when used as the criterion by which document vector elements are correlated with index terms. Since terms with minimal weighting are correlated with leftmost positions in document vectors (*v.* §3.3.2.2), leftmost positional values are the most significant indicators of the potential for a unique relation between a particular index term and the document represented. Assuming document vectors to be ordered sets of binary weights, a positive weighting for a leftmost position indicates the relative likelihood of a correlation between w and some element of the definiens set.

Other things being equal, when the document vectors of metadata items share leftmost positive values, those metadata are more likely to possess equivalent denotational semantics. To a progressively lesser degree, this observation holds true for shared values in a left-to-right iteration of document vector positions. Characteristically, when the document vectors of metadata share all but right-most positive values, those metadata are likely to possess equivalent denotational semantics.

---

[3] This presumption parallels the classic treatment in [4], which correlates each dictionary term under definition with a "signature" list of words appearing in its definiens. Used in word sense disambiguation, this method achieved only modest results, demonstrating a critical sensitivity to the exact wording of each definiens. Our approach addresses this deficiency by incorporating a theory of direct reference semantics implemented with overload precedence in the interpretation of definiens in metadata description.

Clearly, document vectors designed in this fashion assign higher significance to values leftward of a particular position and lower significance to values rightward. This observation suggests an obvious, but convenient, data structure. At a detailed level, document vectors can be realized as bit arrays in which a 0-bit (false) value indicates the absence of the correlated index term from the document and a 1-bit (true) value indicates its presence. This treatment encodes the ordered binary weights of document representations as bit arrays. For example, the sequence 0101011001 could signify a document vector encoding correlations with a modest set of 10 index terms. Obviously, the bit array representation preserves the high-order significance property of the representation when leftmost bits are treated as having higher significance.[4] This insight suggests implementations utilizing user-defined data types that encapsulate bit array representations and provide arithmetic operators for equality, subtraction, and less-than comparisons. Document vectors implemented in this way can be compared for equality, subtracted one from another, and assembled together in a sorted collection ordered by the less-than relation. For simplicity, we assume these operations function as they do when manipulating decimal numbers.

Let *DV* be the user-defined document vector type described in the preceding paragraph. For pattern analysis, our algorithm orders collections of DVs by the less-than relation in order to arrange those with measurably close values in adjacent or proximal contexts. Once sorted in this manner, the collection is scanned to analyze the patterns presented. Adjacent DVs (as well as adjacent groups of DVs) that are equal in value are treated as either spurious synonyms or schema-matches, depending on their schema of origin as described above. This explains the simple cases of pattern analysis.

More complex cases occur when adjacent DVs (or groups of DVs) bear close, but unequal, values. In these cases, our algorithm allows for equivalence (*i.e.*, inexact matches) by calculating a variance value for the collection of DVs and allowing as a possible schema-match any adjacent values for which the difference is less than the standard deviation. To accomplish this, the mean value of the collection of DVs is calculated according to the following formula:

Definition 4: $$\overline{X} \underset{=\!=}{def} \frac{\sum\limits_{dv \in DVs} dv}{n}$$

Next, by subtracting the mean from each DV value, it is possible to identify its distance from the mean, as illustrated in Definition 5. Variance is computed by dividing the square of the total of these values by the number of DVs. Finally, the positive square root of variance for the collection is σ, the standard deviation. Whenever the difference between adjacent DVs (or groups of DVs) with unequal values from

---

[4] This advantageous property is not lost when a bit array is converted to a decimal number (0101011001 binary is 618 decimal, for example), and many programming languages provide bitwise operators to enable processing integers as bit arrays. Nevertheless, document vector format requirements easily exceed the available bits of integer data types–64 is the conventional maximum–rendering this approach untenable when $|T|$ exceeds 64, a likely cardinality in all but the most modest schema-matching contexts. For our initial implementation, we have constructed a wide integer class that provides for arbitrary bit array lengths, the necessary data type conversions, and the necessary arithmetic for document vectors as well as definitions 4 and 5.

different schema of origin is less than σ, our algorithm identifies a potential match, although confidence in this second category of match is less than that for the first.

$$\text{Definition 5:} \quad \sigma^2 \stackrel{def}{=\!=} \frac{\sum_{dv \in DVs} (dv - \overline{X})^2}{n}$$

We are currently focused on refinements of this technique (especially adjustment factors for the divisor n in Definition 5) with which to calibrate the granularity of representation for the identification of candidate schema-matches.

5 Evaluation

We agree with [14] that matching is an inherently subjective operation, and this frustrates evaluation efforts. Unsupervised, fully automated schema-matching algorithms that reason about element-level, structure-level, and schema-level information are rare. CUPID [14] is the most prominent instance of which we are aware. Schema meta-matchers [15] may access a similar variety of information sources, but this is only a side-effect of their rank aggregator architecture--the capability to perform inferential reasoning on the basis of this variety of inputs is not inherent in their logical operations, even though they combine multiple schema-matchers of various sorts.

Amongst linguistic, element-level, schema-based matching frameworks, the algorithm presented here is distinguished insofar as it typically overcomes differences in schema element names to uncover conceptual identities encoded across metadata descriptions. Further, it needs no human supervision to tailor thesauri to address synonymy facts that may vary with application scenario (as does CUPID). Instead, it directly utilizes metadata descriptions in the form of data dictionaries as its singular information resource. In these respects, our algorithm is unique. We have implemented our algorithm as a proof-of-concept demonstrator, and we illustrate system performance in an appendix which presents fragments of two industrial data models, two correlated data dictionaries, and a snapshot of a software user interface presenting matched schema elements from the two models. Inspection of these materials demonstrates how our approach detects semantic equivalence across data models when metadata description is closely aligned as well as when it is widely-divergent.

6 Summary

We have described an algorithm for translating between namespaces instantiated in the form of database schema, an operation otherwise known as schema-matching. Our algorithm can be categorized as element-level inasmuch as it discovers candidate matches for individual schema elements, as opposed to complexes of such elements. Due to its provisions for overload precedence, however, our algorithm obscures the categorical distinction between element- and structure-level matching presented in [1], thus resisting categorization according to the conventional algorithm classification scheme. Our algorithm is schema-based insofar as it considers only schema information and ignores instance data. Furthermore, our algorithm is linguistic because it interprets the information encoded in data dictionary definitions in order to compare metadata elements across schema for equivalent denotational semantics.

Our algorithm employs analytical techniques used in information retrieval, information theory, and computational linguistics. It exploits both the linguistic information inherent in metadata description and its dependency architecture in order to identify metadata with equivalent denotational semantics. When these metadata originate in the same schema, they are likely to be synonymous and thus deserve further scrutiny. When they originate in different schema, they are likely candidates for namespace translation in the sense of schema-matching. In every case, our algorithm determines match candidates associated with higher or lower confidence levels, and users are invited to accept or reject these determinations at their discretion.

References

[1] Rahm, E., Bernstein, P.: A Survey of Approaches to Automatic Schema Matching. The VLDB Journal 10, 334–350 (2001)
[2] Coen, G.: Dictionaries and Lexical Graphs. In: Moreno, A., van de Ried, R.P. (eds.) Applications of Natural Language to Information Systems. Gesellschaft für Informatik, Bonn (2001)
[3] Coen, G.: Database Lexicography. Data and Knowledge Engineering 42, 3 (2002)
[4] Lesk, M.: Automated Sense Disambiguation Using Machine-readable Dictionaries: How to Tell a Pine Cone from an Ice Cream Cone. In: Proceedings of the 5th Annual International Conference on Systems Documentation, Toronto, Ontario, Canada, pp. 24–26 (1986)
[5] Salton, G.: Automatic Information Organization and Retrieval. McGraw-Hill, New York (1968)
[6] Doyle, L.B.: The Microstatistics of Text. Information Storage and Retrieval 1, 189–214 (1963)
[7] Cormack, A.: Definitions: Implications for Syntax, Semantics, and the Language of Thought. Garland, New York (1998)
[8] Fellbaum, C. (ed.): WordNet: An Electronic Lexical Database. MIT Press, Cambridge (1998)
[9] Shannon, C., Weaver, W.: The Mathematical Theory of Communication. University of Illinois Press, Urbana (1949)
[10] Robertson, S.E., et al.: Large test collection experiments on an operational, interactive system: Okapi at TREC. Information Processing and Management 31, 345–360 (1995)
[11] Manly, B.F.J.: Multivariate Statistical Methods: A Primer. Chapman & Hall, London (1986)
[12] Sneath, P.H.A., Sokal, R.R.: Numerical Taxonomy. W.H. Freeman and Company, San Francisco (1973)
[13] Kaplan, D.: Demonstratives. In: Almong, J., et al. (eds.) Themes from Kaplan. Oxford University Press, Oxford (1989)
[14] Madhavan, J., Bernstein, P.A., Rahm, E.: Generic Schema matching with Cupid. In: Proceedings of the 27th VLDB Conference, Rome, Italy (2001)

A Appendix

This appendix exhibits supporting materials in three subsections:

1. Two unified modeling language (UML) models of the aircraft reliability and maintainability domain (§A.1);

2. Two data dictionaries describing the metadata of the two UML sample models (§A.2);

3. A snapshot of a software user interface presenting matched schema elements (§A.3).

After comparing the UML models of §A.1 with the data dictionaries of §A.2, we invite the reader to examine the schema-matching results presented in §A.3. These materials show how our algorithm matches across schemata when metadata exhibit closely aligned descriptions (*cf.* Flight-Year-Month *vs.* Summary-Month) as well as widely-divergent descriptions (*cf.* Maintenance-Record *vs.* Work-Unit).

A.1   Reliability and Maintainability Models

For the reader's convenience, here we present two UML models of the aircraft reliability and maintainability domain. Acknowledging that these logical models reduce the complexity typically encountered in industrial settings, we believe they are sufficiently detailed to illustrate a practical schema-matching algorithm. We invite comparison of these models with the data dictionaries in §A.2, and we submit that these graphical models and their respective data dictionaries describe the same metadata. (The model in Figure 1 describes the system also described by the data dictionary in Table 2 of §A.2; the model in Figure 2 is similarly coordinated with the data dictionary in Table 3.)

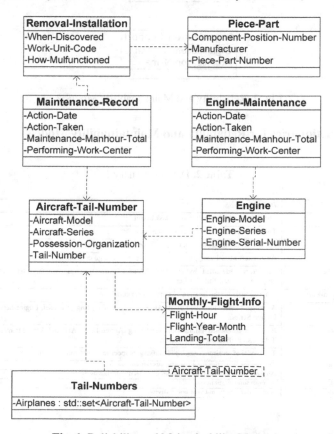

Fig. 1. Reliability and Maintainability Model 1

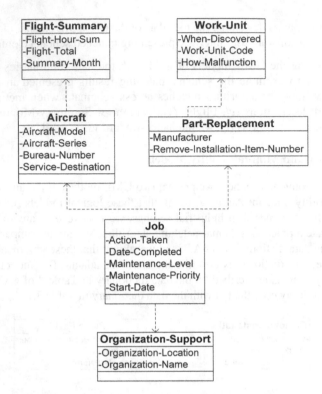

Fig. 2. Reliability and Maintainability Model 2

A.2   Data Dictionaries for Reliability and Maintainability Models

Table 2. Data Dictionary 1

Name	Definition
Action-Date	The date on which a maintenance task begins.
Action-Taken	The action executed to correct a discrepancy.
Aircraft-Model	An aircraft model identifying its specific design and mission.
Aircraft-Series	An Aircraft-Model subtype representing a specific design derivative.
Aircraft-Tail-Number	An aircraft identified by its Aircraft-Model, Aircraft-Series, Monthly-Flight-Info, Possession-Organization, and Tail-Number.
Component-Position-Number	The identifier for the location of a part on an aircraft.
Engine	An aircraft motor identified by its Aircraft-Tail-Number, Engine-Model, Engine-Serial-Number, and Engine-Series.
Engine-Maintenance	Engine maintenance information identifying Action-Date, Action-Taken, Maintenance-Manhour-Total, and Performing-Work-Center.
Engine-Model	The model of an aircraft-engine identifying its specific design and mission.
Engine-Serial-Number	The unique identifier for a specific aircraft-engine.
Engine-Series	An Engine-Model subtype designating a specific design derivative.
Flight-Hour-Total	Aircraft total flight hours.
Flight-Year-Month	A calendar month and year correlativity.

Table 2. (*continued*)

Name	Definition
How-Malfunctioned	The operational circumstances in which a malfunction is ascertained.
Landing-Total	The total number of landings of an aircraft.
Maintenance-Manhour-Total	Total manhours required to complete a specific maintenance task.
Maintenance-Record	A maintenance record identifying Aircraft-Tail-Number, Action-Date, Action-Taken, Maintenance-Manhour-Total, Performing-Work-Center, and Removal-Installation.
Manufacturer	The manufacturer of an aircraft part.
Monthly-Flight-Info	Aircraft operating information identifying the current Flight-Year-Month as well as Flight-Hour-Total and Landing-Total.
Performing-Work-Center	An airplane maintenance service center location.
Piece-Part	Part information identifying Component-Position-Number, Manufacturer, and Piece-Part-Number.
Piece-Part-Number	The Manufacturer provided number for the removed and installed part.
Possession-Organization	The organization to which an aircraft is delegated.
Removal-Installation	Information about a maintenance operation that removed and then installed a Piece-Part, identifying How-Malfunctioned, When-Discovered, and Work-Unit-Code.
Tail-Number	An aircraft's unique identifier.
Tail-Numbers	The inventory of every Aircraft-Tail-Number.
When-Discovered	The operational context in which a requirement for unscheduled maintenance is determined.
Work-Unit-Code	A code identifying the hierarchical decomposition of systems, subsystems, and components within end items.

Table 3. Data Dictionary 2

Name	Definition
Action-Taken	A maintenance action performed to correct a discrepancy.
Aircraft	An airplane described by its Aircraft-Model, Aircraft-Series, Bureau-Number, Flight-Summary, and Service-Designation.
Aircraft-Model	The model of an airplane represented by specific design and mission information.
Aircraft-Series	A subtype of Aircraft-Model corresponding to a particular design derivative.
Bureau-Number	A unique identifier for an airplane.
Date-Completed	The date on which a maintenance task is completed.
Flight-Hour-Sum	Total hours of flight for an airplane.
Flight-Summary	Operating information identifying the current Summary-Month, Flight-Hour-Sum, and Flight-Total for an airplane.
Flight-Total	The number of landings associated with an airplane.
How-Malfunction	The operational context in which a malfunction is determined.
Job	A record of maintenance performed on an Aircraft identifying Action-Taken, Date-Completed, Maintenance-Level, Maintenance-Priority, Organization-Support, Part-Replacement, and Start-Date.
Maintenance-Level	An indication of where a maintenance task is performed.
Maintenance-Priority	An indication of the priority of a maintenance task.
Manufacturer	The producer of an airplane part.
Organization-Location	The location of a maintenance service center.
Organization-Name	The name of an airplane maintenance service center.
Organization-Support	The responsible maintenance service center, its Organization-Location, and Organization-Name.
Part-Replacement	Information about a maintenance operation that removed and then installed a part, identifying Removed-Installed-Item-Number, Manufacturer, and Work-Unit.
Removed-Installed-Item-Number	The part number supplied by the Manufacturer for the part that was removed and installed.
Service-Designation	The organization to which an airplane is assigned.
Start-Date	The date on which a maintenance task starts.
Summary-Month	A correlation between a calendar year and month.
When-Discovered	The operational context in which a requirement for unscheduled maintenance is determined.
Work-Unit	A maintenance record identifying How-Malfunction, When-Discovered, and Work-Unit-Code.
Work-Unit-Code	A code identifying the hierarchical decomposition of systems, subsystems, and components within end items.

A.3   Schema-Matching Human Interface

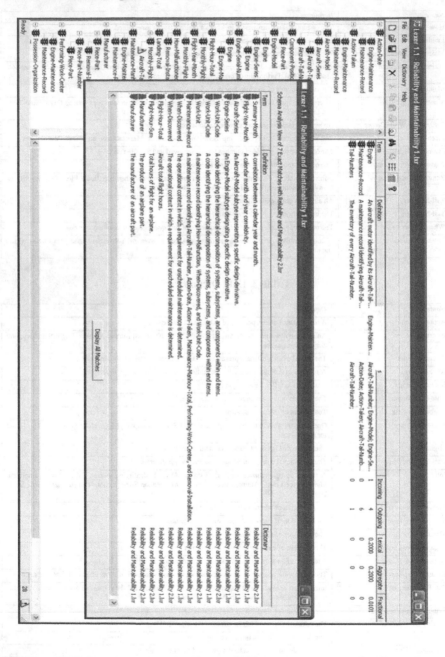

Requirements Analysis: Concept Extraction and Translation of Textual Specifications to Executable Models

Leonid Kof

Fakultät für Informatik, Technische Universität München,
Boltzmannstr. 3, D-85748, Garching bei München, Germany
kof@informatik.tu-muenchen.de

Abstract. Requirements engineering, the first phase of any software development project, is the Achilles' heel of the whole development process, as requirements documents are often inconsistent and incomplete. In industrial requirements documents, natural language is the main presentation means. This results in the fact that the requirements documents are imprecise, incomplete, and inconsistent. A viable way to detect inconsistencies and omissions in documents is to extract system models from them.

In our previous work we developed approaches translating textual scenarios to message sequence charts (MSCs) and textual descriptions of automata to automata themselves. It turned out that these approaches are highly sensitive to proper definition of terms (communicating objects for MSCs, states for automata).

The goal of the presented paper is a systematic comparison of different term extraction heuristics, as a preliminary stage of MSC or automata extraction. The extracted terms were declared to communicating objects (in the case of MSCs) or to states (in the case of automata). The heuristics were compared on the basis of correctness of resulting MSCs and automata. We came to the conclusion that named entity recognition is the best performing technique for term extraction from requirements documents.

1 Requirements Documents Are Inconsistent and Incomplete

At the beginning of every software project, some kind of requirements document is usually written. The majority of these documents are written in natural language, as the survey by Mich et al. shows [1]. This results in the fact that the requirements documents are imprecise, incomplete, and inconsistent, because precision, completeness and consistency are extremely difficult to achieve using mere natural language as the main presentation means.

According to Boehm [2], in software development, the later an error is found, the more expensive its correction. Thus, it is one of the goals of requirements analysis, to find and to correct the defects of requirements documents. A practical way to detect errors in requirements documents is to convert informal specifications to executable models. In this case, errors in documents would lead to inconsistencies or omissions in

H. Horacek et al. (Eds.): NLDB 2009, LNCS 5723, pp. 79–90, 2010.

models, and inconsistencies and omissions are easier to detect in models than in textual documents.

In our previous work [3,4,5,6] we developed approaches extracting behavior specifications (message sequence charts (MSCs) and automata) from requirements documents even in the presence of certain defects. It turned out that the approach to scenario-to-MSC translation is highly sensitive to the proper definition of communicating objects, and the approach producing automata is very sensitive to the proper definition of possible states:

- In the case of text-to-MSC translation, the algorithm tries to identify two communicating objects in every sentence: one before the first verb, and one after the last verb. Communicating objects are just elements of the previously constructed set of objects. Thus, in the case that the set of objects contains wrong terms, wrong communicating objects may be identified in some sentences, which leads to wrong MSCs. Details of the text-to-MSC translation can be found in [3,4].
- In the case of text-to-automaton translation, the algorithm tries to identify a system state after the main verb of every sentence. This allows to translate sentences like "if ⟨some condition⟩, the system goes to ⟨some state⟩" to state transitions. The states are identified as elements of previously constructed set of states. Thus, in the case that the set of states contains wrong terms, wrong states may be identified in some sentences, which leads to wrong state transitions. Details of the text-to-automaton translation can be found in [6].

The goal of the presented work was to compare different term extraction heuristics. The heuristics were compared on the basis of correctness of resulting MSCs and automata: the extracted terms were declared to communicating objects (in the case of MSCs) or to states (in the case of automata). To evaluate the correctness of the automata, we used a manually constructed reference automaton. To evaluate the correctness of the MSCs, we used the evaluation rules developed in our previous work [5]. Surprisingly, it turned out that named entity recognition provided best performance in both cases, despite completely different writing styles.

Terminology: For the remainder of the paper we use the following terminology for MSCs: A *scenario* is a sequence of natural language sentences. An *MSC* consists of a set of *communicating objects*, or *actors*, a sequence of *messages* sent and received by these actors, and a sequence of *assertions* interleaved with the message sequence. Figure 1 illustrates the introduced MSC terminology. For automata, we use the standard definitions of *state* and *state transition* [7].

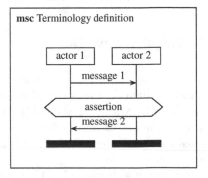

Fig. 1. MSCs: terminology definition

Outline: The remainder of the paper is organized as follows: Section 2 introduces the case studies used to evaluate the term extraction heuristics. Section 3 is the technical core of the paper, it presents the term extraction heuristics and their evaluation. Sections 4 and 5 present the summary of the paper and an overview of related work, respectively.

2   Case Studies

2.1   Automaton: The Steam Boiler

The Steam Boiler Specification was chosen for the automaton-based case study, as it was the standard benchmark for several case studies aiming to compare different formalization methods [8]. This specification describes the steam boiler itself and states the requirements to the control program for the steam boiler. The steam boiler system consists of four pumps to provide the steam boiler with water, one controller for every pump, a device to measure the water level in the steam boiler, and a device to measure the quantity of steam coming out of the steam boiler. The goal of the control program is to maintain the water level between predefined marks, in order to prevent damage of the steam boiler. This water level should be maintained even in case of certain equipment failures. In the case of equipment failures, water levels between certain emergency marks are allowed. Water levels above/below emergency marks cause steam boiler damage.

The control program for the steam boiler should support a number of modes: initialization mode, normal mode, degraded mode, rescue mode, and emergency stop mode. For every mode, the specification describes the required program reactions to different operation situations. An example set of rules, applicable in the normal mode, is shown in Table 1. Table 2 shows the required behavior of the control program (tabular representation of the automaton), manually constructed on the basis of the specification. This manually constructed automaton will be used as the reference for the evaluation of the automatically extracted automata in Section 3.2.

Table 1. The steam boiler, specification excerpt (copied from [9])

Normal mode
1. The normal mode is the standard operating mode in which the program tries to maintain the water level in the steam-boiler between N1 and N2 with all physical units operating correctly.   2. As soon as the water level is below N1 or above N2 the level can be adjusted by the program by switching the pumps on or off.   3. The corresponding decision is taken on the basis of the information which has been received from the physical units.   4. As soon as the program recognizes a failure of the water level measuring unit it goes into rescue mode.   5. Failure of any other physical unit puts the program into degraded mode.   6. If the water level is risking to reach one of the limit values M1 or M2 the program enters the mode emergency stop.   7. This risk is evaluated on the basis of a maximal behaviour of the physical units.   8. A transmission failure puts the program into emergency stop mode.

Table 2. Automaton for steam boiler control, manually constructed

Initial mode	Target mode	Transition condition
initialization	initialization	message steam-boiler-waiting not yet received
initialization	emergency stop	unit for detection of the level of steam is defective
initialization	emergency stop	failure of the water level detection unit
initialization	normal	all the physical units operate correctly
initialization	degraded	any physical unit is defective
initialization	emergency stop	transmission failure
normal	rescue	failure of the water level measuring unit
normal	degraded	failure of any other physical unit
normal	emergency stop	the water level is risking to reach one of the limit values
normal	emergency stop	transmission failure
degraded	normal	defective unit repaired
degraded	rescue	failure of the water level measuring unit
degraded	emergency stop	the water level is risking to reach one of the limit values
degraded	emergency stop	transmission failure
rescue	normal	water level measurement unit repaired
rescue	degraded	water level measurement unit repaired
rescue	emergency stop	the unit which measures the outcome of steam has a failure
rescue	emergency stop	the units which control the pumps have a failure
rescue	emergency stop	the water level risks to reach one of the two limit values
rescue	emergency stop	transmission failure

2.2 MSCs: The Instrument Cluster

The instrument cluster specification describes the optical design of the instrument cluster as a part of the car dashboard, its hardware, and, most importantly, its behavior. The behavior is specified as a set of scenarios, like the below example, taken from [10]:

1. The driver switches on the car (ignition key in position ignition on).
2. The instrument cluster is turned on and stays active.
3. After the trip the driver switches off the ignition.
4. The instrument cluster stays active for 30 seconds and then turns itself off.
5. The driver leaves the car.

For this scenario, several translations to an MSC are possible: For example, the second sentence can be translated both to an assertion and to a message from the instrument cluster to the driver. The same is true for the fourth sentence. Due to the fact that, even for a single scenario, several translations to an MSC are possible, it makes no sense to use a single set of MSCs for evaluation. Instead, we use the following correctness rules for MSCs (cf. [5]):

- General statements that are actually irrelevant for the MSC (e.g., "There is no difference between rising and falling temperature values") should be translated to assertions.
- General statements about the system state (e.g., "The instrument cluster is activated") can be translated both to messages and to assertions.
- For a statement sequence like "X activates Y", "Y is activated", the first statement should be translated to a message, the second one to an assertion.

- If a statement does not have to be translated to an assertion due to one of the above rules, it should be translated to a message.
- If, for any particular actor, it is known that this actor cannot receive messages, as for example some sensors used in automobiles, no messages should be sent to this object.

These rules, applied manually, will be used to evaluate the influence of different term extraction heuristics on the correctness of the extracted MSCs in Section 3.3.

3   Term Extraction Heuristics and Their Influence on Behavior Models

This section is the technical core of the paper. First, in Section 3.1 it presents the term extraction heuristics. Then, it presents the evaluation of the heuristics on the Steam Boiler Specification (Section 3.2) and on the Instrument Cluster Specification (Section 3.3).

3.1   Term Extraction Heuristics

Existing term extraction approaches are based either on named entity recognition or on the analysis of sentence structure, cf. [11,12]. Basic ideas of both approaches are presented below.

Named Entitiy Recognition. Named entity recognition (NER) aims at identification of standard classes of proper names, i.e. people, places, and organizations. For example, the phrase "President Bush visits troops in Iraq" contains two named entities: "President Bush" (person) and "Iraq" (place). In its most simple form, NER procedure just looks up every word in the predefined list of people, places, and organizations and decides whether the given word is a named entity. This procedure would work fine for "Iraq" and other country names, as the number of country names is finite. It makes no sense to apply the lookup procedure to compound names like "President Bush": By means of a lookup table we can identify a finite number of former presidents, but we would be unable to identify future ones.

To identify compound names, the following idea can be applied: We define a set of keywords, each keyword indicating either a person name or a place or an organization. For example, for people, the following keywords make sense: "president", "CEO", "professor", etc. In order to use the keyword approach to named entity recognition, it is sufficient to apply a part-of-speech (POS) tagger to the analyzed text. Then, we can just extract the sequence of nouns (words having tags starting with "NN") following the keyword. For example, for the sentence "President George W. Bush visits troops in Iraq" we would get the tagging "President|NNP George|NNP W.|NNP Bush|NNP visits|VBZ troops|NNS in|IN Iraq|NNP". If we extract the sequence of nouns following the word "president", we obtain the complete named entity, "President George W. Bush".

There already exist tools performing named entity recognition, as for example the C&C tool suite (http://svn.ask.it.usyd.edu.au/trac/candc). However, existing tools are trained on newspapers texts and, thus, are limited to recognition of standard classes of entities. For requirements analysis, we have to recognize other classes of entities, like system components or states, which makes a customized NER procedure necessary. We use the following heuristics: a named entity consists of

- the keyword, followed by any number of substantives (tags beginning with NN), adjectives (tag JJ), or verbs in the past participle form (tag VBD), or
- any number of substantives, adjectives, or verbs in the past participle form, followed by the keyword.

The decision to consider not only substantives, but also adjectives and verbs in the past participle form was motivated by the concrete form of named entities occurring in our case studies. It turned out in the case studies, that these additional terms do not result in wrong communicating objects or wrong states of the automaton.

Extraction Based on Sentence Structure. If we go beyond POS-tagging and use parsing [13], it becomes possible to extract terms with particular grammatical roles. For example, if we extract the grammatical subject from the sentence "The program enters a state in which it waits for the message steam-boiler-waiting to come from the physical units", we get "the program", and if we extract the prepositional object with the preposition "for", we get "the message steam-boiler-waiting".

In the case studies we used the following term extraction heuristics, already proven useful for the Instrument Cluster Specification (cf. [5]):

- subjects of active sentences containing a direct object, also occurring in passive sentences,
- subjects of active sentences containing *no* direct object, also occurring in passive sentences,
- subjects of active sentences containing a direct object, *not* occurring in passive sentences,
- subjects of active sentences containing *no* direct object, *not* occurring in passive sentences,
- subjects of passive sentences,
- direct objects.

This technique should be applied with caution, as the existing parsers are definitely less precise than POS-taggers, so the parser itself could become an error source.

3.2   Evaluation: Term Extraction from the Steam Boiler Specification

The Steam Boiler Specification describes different states and state transitions of the control program. An automaton is the most suitable representation of such a specification. To extract an automaton, it is necessary to know the potential states [6]. In our previous work, we investigated NER as the means of term (=state name) extraction. It turned out that NER works well for the Steam Boiler Specification. In the presented paper, we go further and compare previously applied approaches with term extraction based on sentence structure.

The extracted terms are listed in Table 3. Additionally to the terms extracted by their grammatical roles and by NER, Table 3 lists also the terms explicitly mentioned in the

Table 3. Steam Boiler Specification, extracted terms

explicitly listed modes	initialization mode, normal mode, degraded mode, rescue mode, emergency stop mode
NER, keyword "mode"	mode emergency stop, mode normal, mode rescue, mode degraded, initialization mode, emergency stop mode, normal mode, standard operating mode, rescue mode, degraded mode
subjects of active sentences containing a direct object, also occurring in passive sentences	level, exactly n liters
subjects of active sentences containing *no* direct object, also occurring in passive sentences	water level
subjects of *active* sentences containing a direct object, *not* occurring in passive sentences	program, transmission failure, failure, limit, water level, water level measuring unit, calculation, unit, water level risks, units
subjects of *active* sentences containing *no* direct object, *not* occurring in passive sentences	initialization mode, message, steam, quantity, unit, physical units, physical unit, mode, program, normal mode, ... (list pruned due to space limitations)
subjects of passive sentences	message, level, signal, corresponding decision, risk, units, water level, exactly n liters, calculation, water measuring unit, mode
direct objects	physical units, message, state, steam-boiler, steam, level, detection, emergency stop mode, water, valve, order, pump, ... (list pruned due to space limitations)

specification as mode names. To evaluate the influence of the term extraction heuristics on the behavior extraction, each set of the extracted terms was used as a set of potential states for the automata extraction algorithm [6]. The resulting automata were evaluated according to following criteria:

1. The manually constructed automaton, presented in Table 2, was used as reference.
2. If none of the modes listed in the first line of Table 3 was present in the set of potential states, the corresponding automaton was not evaluated: In this case the automaton can contain correct transitions solely by coincidence.
3. Transition conditions were considered as equivalent, if they had at least two common words.
4. For every transition, it was evaluated whether the transition has the same initial state (called "source" in Table 4) and the same target state as in Table 2.

Evaluation results are presented in Table 4. Each column presents the number of transitions (total, missing, etc.) obtained with the corresponding set of potential states. For sets of potential states containing the word "mode", evaluation was performed twice: for the original set of potential states and for the set of potential states with the word "mode" manually removed from the set. The reason is that "mode" is a constituent of the explicitly listed state names, so the presence of "mode" in the set of potential states can lead to the state called just "mode", and ignoring of the real mode names.

Different heuristics combinations were not considered, as it was known from our previous work [5] that unnecessarily extracted terms reduce the quality of behavior models, and NER (second line of Table 4) resulted in an almost perfect automaton: Out of 20 state transitions in Table 2, 19 were correctly identified, which implies the recall

Table 4. Steam Boiler Specification: evaluation results

			Transitions				
	total	missing	correct condition only	correct condition and target	correct condition and source	correct condition, target, and source	wrong transitions
explicitly listed modes	12	9	—	—	—	11	1
NER, keyword "mode"	20	1	—	—	—	19	1
subjects of active sentences containing a direct object, also occurring in passive sentences	modes completely missing						
subjects of active sentences containing *no* direct object, also occurring in passive sentences	modes completely missing						
subjects of *active* sentences containing a direct object, *not* occurring in passive sentences	modes completely missing						
subjects of *active* sentences containing *no* direct object, *not* occurring in passive sentences	32	5	15	—	—	—	17
subjects of *active* sentences containing *no* direct object, *not* occurring in passive sentences, without "mode"	12	16	4	—	—	—	8
subjects of passive sentences	modes completely missing						
direct objects	29	12	5	3	—	—	21
direct objects, without "mode"	29	11	5	3	1	—	20

of 95%. 19 out of 20 identified transitions were correct, which implies the precision of 95% too. To summarize, for the Steam Boiler Specification, NER was proven the best heuristics to extract potential states of the automaton.

3.3 Evaluation: Term Extraction from the Instrument Cluster Specification

In our previous work we already evaluated the influence of term extraction heuristics based on sentence structure on the quality of the produced MSCs [5]. It turned out that the best heuristics was to declare subjects of active sentences, containing a direct object, *not* occurring in passive sentences, to communicating objects for MSCs. Additionally, this set of communicating objects was augmented with the objects initializing the scenarios: "driver" and "car". Due to good performance that NER has shown on the Steam Boiler Specification, we wanted to investigate how NER can be applied to the Instrument Cluster Specification.

Manual analysis of the Instrument Cluster Specification has shown that following keywords are sensible for NER application to the Instrument Cluster Specification: display, flasher, indicator, indication, light, lights, position, sensor, sensors, warning. Table 5 shows the named entities extracted with these keywords. Some terms present in Table 5 contain not only nouns but also verbs (e.g., "blinks"). This error was introduced by the POS tagger.

On total, the Steam Boiler Specification contains 42 scenarios. 41 out of 42 scenarios were used for evaluation, due to technical difficulties of batch processing one of

Table 5. NER for the Instrument Cluster Specification

Keyword	Extracted terms
display	digital display, dot matrix display, display RPM, warning display, error display, rev meter display, display tolerance, analog display, ... (list pruned due to space limitations)
flasher	warning signal flasher, hazard warning signal flasher, signal flasher
indicator	indicator lights, engine speed indicator, analog indicator, diagram indicator, indicator lights blinks
indication	audible indication, authentic indication, indication (optical display), indication scale, maximum respective minimum possible corresponding indication
light	engine control light, light control unit
lights	instrument cluster lights, warning lights, warning display lights, error display lights, warning symbol lights, ... (list pruned due to space limitations)
position	technical initial position, initial position, technical final position, technical position, degree technical final position, degree technical initial position, pointer position
sensor	outside temperature sensor
sensors	wheel speed sensors, defect wheel speed sensors
warning	hazard warning, engine warning, ice warning, corresponding warning, section warning, warning tone, warning characteristics, ... (list pruned due to space limitations)

Table 6. Instrument Cluster Specification: evaluation results

		correct MSCs		wrong messages	
		total	percentage	total	percentage
all NER-extracted terms		0	0%	17	3,5%
all NER-extracted terms + "driver" + "car"		40	**97,5%**	0	0%
all NER-extracted terms + "driver" + "car" + "instrument cluster"		40	**97,5%**	0	0%
subjects of active sentences, containing a direct object, *not* occurring in passive sentences + "driver" + "car"	without NER	33	80%	20	4.2%
	+ NER, keyword "display"	33	80%	20	4.2%
	+ NER, keyword "flasher"	33	80%	20	4.2%
	+ NER, keyword "indicator"	33	80%	20	4.2%
	+ NER, keyword "light"	33	80%	20	4.2%
	+ NER, keyword "lights"				
	+ NER, keyword "position"				
	+ NER, keyword "indication"	MSCs completely identical to the case without NER application			
	+ NER, keyword "sensor"				
	+ NER, keyword "sensors"				
	+ NER, keyword "warning"				

the scenarios. To evaluate the MSCs resulting from scenarios, the rules introduced in Section 2.2 were used. Additionally to these rules evaluating MSCs as a whole, wrong messages were counted: a message was considered as wrong, only if it was sent to a communicating object not able to process this message type. This implies that an MSC can be wrong without containing a single wrong message, for example if some sentence that should be translated to a message is translated to an assertion. The decision, whether a sentence is translated to a message or to an assertion is influenced by the set of communicating objects too (see [5] for details).

For the first attempt, all terms extracted by NER were joined, this set was declared to the set of communicating objects. However, two important terms, namely "driver" and "car" were missing from this set, which resulted in completely wrong MSCs (cf. first line of Table 6). For the second attempt, the set of NER-extracted terms was augmented with "driver" and "car", and for the third attempt additionally with "instrument cluster". ("Instrument cluster" is an important term in the Instrument Cluster Specification, but it is not identified as a named entity.) This augmentation resulted in many more correct MSCs, which can be seen in the second and third line of Table 6.

To investigate whether the results obtained with the usage of grammatical subjects can be further improved by NER, we augmented the set of communicating objects used before (subjects of active sentences, containing a direct object, *not* occurring in passive sentences + "driver" + "car") with different NER-extracted terms. It turned out that in this case NER barely influences the results: in some cases the resulting MSCs were exactly the same as without NER, and in some cases the MSCs were different, but the percentage of correct MSCs was the same. This can be explained by the fact that the set of grammatical subjects contains unnecessary communicating objects that cause both wrong messages and wrong MSCs. To summarize, for the Instrument Cluster Specification, NER was proven the best heuristics to extract potential communicating objects.

4 Summary

Requirements engineering is an important project phase, influencing all later development phases. Requirements documents are mostly written in natural language, which implies incomplete and inconsistent requirements. Translation of natural language descriptions to executable models is a viable way to deal with incompleteness and inconsistency. The previously developed approaches to text-to-MSC/automaton translation turned out to be highly sensitive to proper term extraction, as a prerequisite for their application. In the presented paper, we systematically compared different term extraction heuristics and came to the conclusion that named entity recognition (NER) provides best performance on two requirements documents, despite of rather different writing styles. However, it is still an open question whether NER provides best results on any requirements document. An answer to this question, obtained in further case studies, would open the way to systematic concept extraction from industrial documents.

5 Related Work

There are three areas where natural language processing is applied to requirements engineering: assessment of document quality, identification and classification of application specific concepts, and analysis of system behavior. Approaches to the assessment of document quality were introduced, for example, by Rupp [14], Fabbrini et al. [15], Kamsties et al. [16], and Chantree et al. [17]. These approaches have in common that they define writing guidelines and measure document quality by measuring the degree to which the document satisfies the guidelines. These approaches have a different focus from our work: their aim is to detect poor phrasing and to improve it, they do not target at behavior analysis.

Another class of approaches, like for example those by Goldin and Berry [18], Abbott [19], or Sawyer et al. [20] analyzes the requirements documents, extracts application specific concepts, and provides an initial model of the application domain. These approaches do not perform any behavior analysis, either.

The approaches analyzing system behavior, as for example those by Vadera and Meziane [21], Gervasi and Zowghi [22], and Avrunin et al. [23] translate requirements documents to executable models by analyzing linguistic patterns. In this sense they are similar to our work. Vadera and Meziane propose a procedure to translate certain linguistic patterns into first order logic and then to the specification language VDM, but they do not provide automation for this procedure. Gervasi and Zowghi go further and introduce a restricted language, a subset of English. They automatically translate textual requirements written in this restricted language to first order logic. The approach by Avrunin et al. is similar to the approach by Gervasi and Zowghi in the sense that it introduces a restricted natural language. The difference lies in the formal representation means: Gervasi and Zowghi stick to first order logic, Avrunin et al. translate natural language to temporal logic. Our work goes further than the above two approaches, as the language is not restricted.

If we consider a more general area of software engineering, it is possible to apply natural language processing to other types of documents too. Witte et al. [12] introduced an approach to software documentation analysis, allowing to link software documentation to code. They extract instances of previously defined concepts, like variables, classes, methods, etc., using named entity recognition with keywords motivated by the previously constructed ontology of programming concepts. In the presented paper, we have to deal with a more challenging situation, as there exists no ontology of modeling concepts, valid for all application domains.

To summarize, to the best of our knowledge, there is no approach to requirements documents analysis, that is able to analyze documents written in non-restricted language, and to extract behavior information from them.

References

1. Mich, L., Franch, M., Novi Inverardi, P.: Market research on requirements analysis using linguistic tools. Requirements Engineering 9, 40–56 (2004)
2. Boehm, B.W.: Software Engineering Economics. Prentice-Hall, Englewood Cliffs (1981)
3. Kof, L.: Scenarios: Identifying missing objects and actions by means of computational linguistics. In: 15th IEEE International Requirements Engineering Conference, New Delhi, India, pp. 121–130. IEEE Computer Society Conference Publishing Services, Los Alamitos (2007)
4. Kof, L.: Treatment of Passive Voice and Conjunctions in Use Case Documents. In: Kedad, Z., Lammari, N., Métais, E., Meziane, F., Rezgui, Y. (eds.) NLDB 2007. LNCS, vol. 4592, pp. 181–192. Springer, Heidelberg (2007)
5. Kof, L.: From Textual Scenarios to Message Sequence Charts: Inclusion of Condition Generation and Actor Extraction. In: 16th IEEE International Requirements Engineering Conference, Barcelona, Spain, pp. 331–332. IEEE Computer Society Conference Publishing Services, Los Alamitos (2008)
6. Kof, L.: Translation of Textual Specifications to Automata by Means of Discourse Context Modeling. In: Glinz, M., Heymans, P. (eds.) REFSQ 2009. LNCS, vol. 5512, pp. 197–211. Springer, Heidelberg (2009)

7. Broy, M.: Informatik. In: Eine grundlegende Einführung, vol. 2. Springer, Heidelberg (1998)
8. Abrial, J.R., Börger, E., Langmaack, H.: Formal Methods for Industrial Applications: Specifying and Programming the Steam Boiler Control. In: Abrial, J.-R., Börger, E., Langmaack, H. (eds.) Dagstuhl Seminar 1995. LNCS, vol. 1165. Springer, Heidelberg (1996)
9. Abrial, J.R., Börger, E., Langmaack, H.: The steam boiler case study: Competition of formal program specification and development methods. In: Abrial, J.-R., Börger, E., Langmaack, H. (eds.) Dagstuhl Seminar 1995. LNCS, vol. 1165. Springer, Heidelberg (1996)
10. Buhr, K., Heumesser, N., Houdek, F., Omasreiter, H., Rothermehl, F., Tavakoli, R., Zink, T.: DaimlerChrysler demonstrator: System specification instrument cluster (2004), http://www.empress-itea.org/deliverables/D5.1_Appendix_B_v1.0_Public_Version.pdf (accessed 11.01.2007)
11. Kof, L.: Natural Language Processing: Mature Enough for Requirements Documents Analysis? In: Montoyo, A., Muñoz, R., Métais, E. (eds.) NLDB 2005. LNCS, vol. 3513, pp. 91–102. Springer, Heidelberg (2005)
12. Witte, R., Li, Q., Zhang, Y., Rilling, J.: Ontological Text Mining of Software Documents. In: Kedad, Z., Lammari, N., Métais, E., Meziane, F., Rezgui, Y. (eds.) NLDB 2007. LNCS, vol. 4592, pp. 168–180. Springer, Heidelberg (2007)
13. Clark, S., Curran, J.R.: Wide-coverage efficient statistical parsing with ccg and log-linear models. Comput. Linguist. 33, 493–552 (2007)
14. Rupp, C.: Requirements-Engineering und-Management. In: Professionelle, iterative Anforderungsanalyse für die Praxis, 2nd edn. Hanser–Verlag (2002) ISBN 3-446-21960-9
15. Fabbrini, F., Fusani, M., Gnesi, S., Lami, G.: The linguistic approach to the natural language requirements quality: benefit of the use of an automatic tool. In: 26th Annual NASA Goddard Software Engineering Workshop, Greenbelt, Maryland, pp. 97–105. IEEE Computer Society, Los Alamitos (2001)
16. Kamsties, E., Berry, D.M., Paech, B.: Detecting ambiguities in requirements documents using inspections. In: Workshop on Inspections in Software Engineering, Paris, France, pp. 68–80 (2001)
17. Chantree, F., Nuseibeh, B., de Roeck, A., Willis, A.: Identifying nocuous ambiguities in natural language requirements. In: RE 2006: Proceedings of the 14th IEEE International Requirements Engineering Conference (RE 2006), pp. 56–65. IEEE Computer Society, Washington (2006)
18. Goldin, L., Berry, D.M.: AbstFinder, a prototype natural language text abstraction finder for use in requirements elicitation. Automated Software Eng. 4, 375–412 (1997)
19. Abbott, R.J.: Program design by informal English descriptions. Communications of the ACM 26, 882–894 (1983)
20. Sawyer, P., Rayson, P., Cosh, K.: Shallow knowledge as an aid to deep understanding in early phase requirements engineering. IEEE Trans. Softw. Eng. 31, 969–981 (2005)
21. Vadera, S., Meziane, F.: From English to formal specifications. The Computer Journal 37, 753–763 (1994)
22. Gervasi, V., Zowghi, D.: Reasoning about inconsistencies in natural language requirements. ACM Trans. Softw. Eng. Methodol. 14, 277–330 (2005)
23. Smith, R.L., Avrunin, G.S., Clarke, L.A., Osterweil, L.J.: Propel: an approach supporting property elucidation. In: ICSE 2002: Proceedings of the 24th International Conference on Software Engineering, pp. 11–21. ACM, New York (2002)

Score-Based Approach for Anaphora Resolution in Drug-Drug Interactions Documents[*]

Isabel Segura-Bedmar, Mario Crespo, and Cesar de Pablo-Sánchez

Computer Science Department, Universidad Carlos III de Madrid, Leganés,
28911 Madrid, Spain
{isegura,mcmiguel,cdepablo}@inf.uc3m.es

Abstract. Drug-drug interactions are frequently reported in biomedical literature and Information Extraction (IE) techniques have been devised as a useful instrument for managing this knowledge. Nevertheless, IE at the sentence level has a limited effect because there are frequent references to previous entities in the discourse, a phenomenon known as 'anaphora'. The problem of resolving pronominal and nominal anaphora to improve a system that detects drug interactions is addressed in this paper. To our knowledge, this is the first research article that tackles this issue. A corpus and a system for the evaluation of drug anaphora resolution have been developed and an analysis of the phenomena is also included. The system uses a domain-specific syntactic and semantic parser, UMLS Metamap Transfer (MMTx) [1], to select anaphoric expressions and candidate references. It is shown that a combination of the domain-specific syntax and semantic information with generic heuristics can be leveraged to produce good results comparable to other related domains. Furthermore, the analysis of the errors suggests that the use of additional semantic knowledge is needed to improve results and deal with this linguistic phenomenon in this particular domain.

Keywords: Information Extraction, Anaphora Resolution, Drug-Drug Interactions.

1 Introduction

In recent years there has been a huge development in the field of Biomedicine and the volume of documents related to it. One clear example of this growth can be found in the pharmaceutical industry. During the whole drug development life cycle, tens of thousands of documents are generated to be subsequently analyzed by experts. An essential stage of this life cycle is *pharmacovigilance*, which aims at the detection, assessment, understanding and prevention of adverse effects or any other drug-related problems [2]. Several published drug safety issues have showed that adverse effects of drugs may be detected too late, when millions of patients have already been

---

[*] This research paper is supported by the Regional Government of Madrid under the Research Network MAVIR (S-0505/TIC-0267) and by the Spanish Ministry of Science and Innovation under the project BRAVO (TIN2007-67407-C03-01).

H. Horacek et al. (Eds.): NLDB 2009, LNCS 5723, pp. 91–102, 2010.
© Springer-Verlag Berlin Heidelberg 2010

exposed. This fact poses a serious problem for patient safety, motivating a growing interest in improving the early detection of adverse effects, as shown in recent research like ALERT[1].

A drug-drug interaction occurs when one drug influences the level or activity of another drug [3]. The detection of drug interactions, such as a type of adverse drug event in clinical care, is an important research area in patient safety since these interactions can become very dangerous, have a high incidence among certain population groups and increase health care costs [4]. Although there are different resources supporting health care professionals in the detection of drug interactions, this kind of resource is rarely complete. Drug interactions are frequently reported in journals of clinical pharmacology, making medical literature the most effective source for the detection of drug interactions [5]. However, the increasing volume of the literature overwhelms health care professionals trying to keep an up-to-date collection of all reported drug-drug interactions. The development of automatic methods for collecting, maintaining and interpreting this information is crucial to achieving a real improvement in their early detection. Natural Language Processing can provide an interesting way to reduce the time spent by health care professionals on reviewing the literature. Nevertheless, only a few approaches have tackled this issue [6].

Our current research aims at developing an automatic system for extracting drug interactions from pharmacological texts. This is a difficult task whose complexity increases when one or more drugs involved in an interaction are expressed with an anaphoric expression, as shown in the following excerpts taken from the DrugBank database [7]:

Table 1. Examples of anaphoric expressions taken from DrugBank

- Since both **acitretin**$_1$ and **tetracyclines**$_1$ can cause increased intracranial pressure, **their**$_1$ **combined use** is contraindicated.
- Cimetidine is reported to reduce hepatic metabolism of **certain tricyclic antidepressants**$_3$, thereby delaying elimination and increasing steady-state concentrations of **these drugs**$_3$.

In this paper, an approach for anaphora resolution of both pronominal and definite nominal phrases is presented as part of a future system of drug-drug interaction extraction. Unlike other approaches devoted to anaphora resolution in biomedical texts [8, 9, 10], this research is, to our knowledge, the first work that specifically focuses on anaphoric expressions involved in drug interactions and uses a dedicated parser to deal with the syntactic complexity of the biomedical language: MMTx, a program developed by the National Library of Medicine (NLM) to provide a linguistic analysis of biomedical texts and a semantic mapping between the text terms and concepts from the UMLS Metathesaurus[2]. Antecedent selection is based on a scoring procedure in which a salience measure is obtained by weighting the distance and morphological similarity of those candidates identified according to a set of

---

[1] Early detection of adverse drug events by integrative mining of clinical records and biomedical knowledge : http://www.alert-project.org/
[2] http://www.nlm.nih.gov/research/umls/

semantic and morpho-syntactic restrictions. Results show that this approach gets results similar to other approaches for anaphora resolution in this area.

The paper is organized as follows: Section 2 reviews the state of the art in anaphora resolution in the biomedical domain. Section 3 gives a detailed description of the building of the corpus. Section 4 describes this proposal for anaphora resolution. Experimental results and evaluation of the system are presented in Section 5. Finally, in Section 6, the results are discussed and some conclusions are offered.

2   Related Work

Biomedical text mining and information extraction have received increased attention in recent years as a way to reduce time and costs in reviewing health-care literature [11, 12, 13]. Regarding anaphora resolution as a way to improve performance in biomedical information extraction, Castaño et al. [9] present a method for resolving anaphoric expressions for candidates taken from MedLine[3] articles and abstracts. By defining a different range of resolution scope for each type of anaphoric expression, it uses different linguistic, morphologic, syntactic and semantic features such as number or semantic type agreement (UMLS typing-based system), longest common subsequence for similarity among candidate antecedents and coercion-type matching (most suitable agent / patient linguistic role according to the verb) from the most frequent bio-relevant verbs in Medline. Each possible antecedent of a certain anaphora was given a different cumulative score according to the significance of its linguistic features and the one with the best salience measure was chosen. General results are 73.8% F-score over a corpus of 46 MedLine abstracts which were annotated by a domain expert.

Lin and Liang [10] also apply this scoring technique but they restrict the types of nominal anaphoric expressions, enrich the syntactic features and apply coercion-type matching as before, using GENIA[4] corpus. General results are 92% F-score in pronominal anaphora and 78% in nominal anaphora in 32 Medline abstracts (MedStract) [14]. This approach is improved in [15] by using new resources like WordNet or PubMed for finding semantic relationships among concepts not found in UMLS. They extend the MedStract corpus with 100 Medline abstracts obtaining 87.43% F-score for pronominal anaphora and 80.61% for nominal anaphora.

Anaphoric expressions are resolved in [16] by the use of rich domain resources such as the FlyBase [17] database for the recognition of biomedical entities and the Sequence Ontology [18] for semantic tagging. Nominal phrases are identified by the use of the domain-independent parser RASP [19]. The system was evaluated against two hand-annotated full papers. Both documents contain 302 sentences, in which 314 anaphoric expressions were found. The anaphora resolution system reaches 58.8% precision and 57.3% recall.

---

[3] http://medlineplus.gov/

[4] A subset of the substances and the biological locations involved in reactions of proteins were annotated, based on a data model (GENIA ontology) of the biological domain, in XML format (GPML). Version 3.0x consists of 2000 abstracts. http://www-tsujii.is.s.u-tokyo.ac.jp/~genia/topics/Corpus/

Anaphora resolution applied to specific fields in biomedicine can be found in [8], which presents an anaphora resolution system integrated in a larger protein interaction extraction study, called BioAR. It identifies antecedents of pronouns by applying patterns for parallelism and centering theory [22]. Nominal phrase anaphoras are identified according to the most salient score, using similar features as in [9]. Additionally, it also restores the missing arguments of keywords like 'interaction', 'association', etc., which are also referred to interactions by observing omitted arguments in the previous context. Experimental results are 75% precision and 56.3% recall in pronoun resolution and 75% precision and 52.2% in definite noun phrase resolution from 120 unseen biological interactions extracted by BioIE.

Likewise, in [23] the impact of anaphora resolution on the result of a protein interaction extraction system is analyzed by using the Guitar system [24] over the 20 full texts and abstracts of the Medstract corpus and three articles taken from the Journal of Biological Chemistry. From the 402 protein-protein interactions in the corpus, only 20 were conveyed by an anaphoric expression. Results show 70% recall in anaphora resolution in abstracts and 52.65% in full texts. No data about precision are available. Results suggest small improvements in protein extraction.

3 Building a Corpus to Support the Anaphora Reference Resolution for Drug-Drug Interactions

There is no existing corpus devoted to anaphoric expressions resolution in pharmacological texts about drug interactions, so a corpus was built for research and evaluation. DrugBank is an annotated database with approximately 4900 drug entries. Each entry contains more than 100 data fields that gather detailed chemical, pharmaceutical and pharmacological drug information. In particular, there is a field called 'interactions' whose link provides a long description about drug interactions in unstructured text. Each of these documents has on average 40 sentences and 716 words and it was decided to use these documents as source of plain textual information on drug interactions to build a corpus. Thus, 49 drugs were randomly selected and the "interactions" file associated with them was extracted by using an automatic robot developed with the free tool openKapow[5].

3.1 Preprocessing the Corpus

Each of the 49 documents was preprocessed by DrugNer [25]. This system is based on a set of nomenclature rules recommended by the World Health Organization (WHO) International Nonproprietary Names (INNs) Program[6] to identify and classify pharmaceutical substances. In addition, DrugNer uses the MMTx program to analyze the text syntactically by splitting it into components including sentences, phrases and tokens. In addition, for each phrase, MMTx selects the concepts of the UMLS Metathesaurus that best fit a certain phrase. Table 2 shows the linguistic features provided by MMTx to depict phrases:

---

[5] http://alpha.openkapow.com/
[6] http://www.who.int/medicines/services/inn/en/

Table 2. Some features extracted by MMTx and used by this approach

Feature	Description
Sentence Text	Text of each sentence
Phrase Text	Text of each phrase
Phrase Type	Type (NP, PP, VB, V/be, ADJ, ADV, CONJ, UNK)
Semantic Types	Semantic types of UMLS concepts
CUI	Concept Unique Identifiers in UMLS Metathesaurus
Word Token	Word of each token
Head Token	Boolean value indicating the head noun of the phrase

Table 3 summarizes some statistics of the corpus created for evaluation. Phrases whose type MMTx was not able to determine were assigned type UNK. The third column represents the number of phrases assigned to the semantic type pharmacological substance (phsu), antibiotic (antb) or clinical drug (clnd) by MMTx. These types are especially interesting since they represent drugs.

Table 3. Some statistics of the corpus

Type of Phrase	N° Phrases	N° Drugs
Noun (NP)	4935	406
Prepositional (PP)	2157	119
Verbal (VP)	4317	3
Adjectival (ADJ)	89	1
Adverbial (ADV)	605	0
Conjunctions (CONJ)	2535	0
Unknown (UNK)	2535	14
Total Phrases:	18035	689
Total Sentences:	1975	

3.2 Manual Annotation of the Corpus

The corpus was annotated manually by a linguist with the assistance of a pharmaceutical expert. This annotation was made on the output of DrugNer in XML format.

Anaphora is a linguistic procedure to refer to biological entities that usually come up in the recent discourse (antecedents). Its resolution is essential to understanding the meaning of a certain expression. There are two kinds of anaphors that are prevalent in biomedical literature:

1. *Pronominal anaphora*. In this case, an entity is referred to by a pronoun. The set of more prevalent pronouns was identified in the drug interactions corpus: personal *it, they,* reflexive *itself, themselves,* relative *which, that,* distributive, *each, either, neither* and indefinite *all, some, many, one*. As this approach focuses

on drug interactions those pronouns that could not refer to drug entities such as *I, me, you, your, who,* etc., were ruled out.

2. *Nominal (phrase) anaphora.* This is the case of an entity being referred to by a nominal phrase. This approach focuses on the domain-relevant nominal phrases, that is, those that refer to drugs or drug properties in pharmacological documents. These phrases consist of the definite article *the*, possessives *its, their,* demonstratives *this, these, those,* distributives *both, such, each, either, neither* and indefinites *other, another, all, a* followed by a generic term for drugs (such as *antibiotic, medicine, medication,* etc) or a drug property or effect, e.g., '*the drugs*', '*these anticoagulants*', '*its pharmacological effects*', '*their anticoagulant properties*'.

The linguist annotated all anaphora in the corpus and their corresponding antecedents in the XML format so such linguistic relations could be retrieved automatically. The corpus contains a total of 331 anaphoric expressions. Table 4 and Table 5 show the distribution of the pronominal and nominal anaphors in the corpus.

Table 4. Distribution of pronominal anaphors in the corpus

Pronominal Anaphors	N°
Personal (it, they)	23
Reflexives (itself, themselves)	1
Relatives (which, that)	120
Distributives (both, each, either, neither)	8
Demonstratives (these, this, those, that)	12
Indefinites (all, some, many, one)	8
Total Phrases:	165

Table 5. Distribution of nominal anaphors in the corpus

Nominal Anaphors	N°
Definite (the)	37
Possessives (its, theirs)	52
Distributives (both, each, either, neither)	11
Demonstratives (these, this, those, that)	58
Indefinites (other, another, all)	8
Total Phrases:	166

4   Anaphora Reference Resolution for Drug Interactions

The anaphora resolution issue can be split into two different phases: identification of anaphoric expressions and selection of their antecedents.

4.1   Identification of Anaphoric Expressions

The identification of anaphoric expressions is carried out through several steps of selective filtering. A first filter will restrict the type of the phrase by selecting those with type NP (noun phrase), PP (prepositional phrase) or UNK (unknown phrase) as possible candidates. Moreover, a detailed observation of the corpus revealed that MMTx misidentified phrases with 'both', 'either', 'neither', or 'each', annotated as CONJ instead of NP. Thus, these kinds of phrases were also selected as candidates.

4.1.1 Pronominal Candidates for Anaphoric Expressions

Regarding pronominal anaphora, the module selects those elements referred to in Table 4. Singular and plural pronouns in first and second grammatical person were filtered out because they usually refer to other entities (usually patients or health care professionals) rather than pharmaceutical substances. Moreover, the pleonastic-it expressions are excluded by using the rules proposed in [10]. These rules were extended to recognize negation and modal verbs as possible arguments in these kinds of expressions (see Table 6).

Table 6. Rules to recognize pleonastic-it expressions

	Rules	Examples
1.	IT [MODALVERB [NOT]?]? BE [NOT]? [AJD|ADV| VP]* [THAT|WHETHER]	**It should be recognized that** a positive test may be due to the drug.
2.	IT [MODALVERB [NOT]?]? BE [NOT]? ADJ [FOR NP] TO VP	If **it is not possible to discontinue** the diuretic, the starting dose of trandolapril should be reduced.
3.	IT [MODALVERB [NOT]]? [SEEM|APPEAR|MEAN|FOLLOW] [THAT]*	**It does not appear that** the SSRIs reduce the effectiveness of a mood stabilizer in these populations

4.1.2 Nominal Candidates for Anaphoric Expressions

In the case of nominal phrase anaphora, the module selects those phrases with determiners or articles in Table 5. However, it must be born in mind that anaphora is a linguistic device for referring to previous entities in the discourse and this reference is carried out generically. This was the reason why a semantic restriction based on the semantic type of phrases was used to rule out those phrases that were not classified as pharmacological substance (phsu), antibiotic (antb) or clinical drug (clnd). Besides, nominal phrase anaphor candidates must be attached to a generic term for drugs ('*this medicine*', '*the medication*' or '*both drugs*') or to a drug family ("*the oral anticoagulant*", "*these analgesics*", etc.), while the candidates consisting of specific terms for drugs are disregarded ('*the serum digoxin concentration*', '*the warfarin drug*'). To achieve this restriction, the module uses the concept-unique identifier (CUI) provided by MMTx to distinguish between phrases linked to abstract or concrete drugs. Therefore, only phrases attached to the concept '*pharmacological substance*' (CUI=*C1254351*), their direct hyponyms and their hyponym descendants will be selected. Those hyponyms included in the *Medical Entities Dictionary* representing specific terms for drugs were ruled out.

Candidate anaphors consisting of a possessive article will have a different semantic restriction. These phrases are usually linked with several UMLS concepts, called MultiMap, one of them representing an abstract drug and the other representing a property or an effect of the drug with the semantic type 'Qualitative Concept', e.g., *'its pharmacological effect'*, *'their anticoagulant properties'*. These phrases will be selected if such properties are present.

For distinguishing nominal phrases and pronouns consisting of units 'both', 'either', 'neither' from correlative expressions, a regular expression was developed as exemplified in Table 9.

Table 7. Regular expression for detecting the correlative expressions

Rules	
4.	[BOTH|EITHER|NEITHER] [NP|PP|UNK] [AND|OR|NOR] [NP|PP|UNK]

Once a nominal candidate has been selected, it is necessary to determine its grammatical number. Unfortunately, MMTx does not provide this information, so every phrase's head noun was matched against a set of lexical patterns to decide its number.

Table 8. Lexical Patterns for deciding the number of the grammatical form

	Rules	Grammatical Form
5.	[A-Z]+(S |ES|OES|XES|SHES|CHES|SES|ZES)	Plural
6.	[A-Z]+(U|S)S	Exception for Singular.

4.2 Identification and Selection of the Antecedents

In this stage, antecedents of each of the anaphoric expressions selected are found in the text. Corpus observation showed that most antecedents usually occur in the previous context of their referring expressions, so only phrases in the range of two sentences were considered. According to this scope, the module will select those phrases whose syntactic type is NP, PP or UNK as possible candidates and semantically classified as *phsu*, *antb* or *clnd*. These semantic restrictions will not be applied for pronominal anaphora resolution since antecedents are not semantically determined by their pronominal anaphoric expressions.

From the resulting list number agreement between anaphora and its candidate antecedent is checked out. The same lexical patterns (Table 8) as for the analysis of anaphoric expressions were applied in the determination of the number of the antecedent.

Finally, a regular expression was applied in sentences to detect coordinative structures. Such structures will be taken as possible antecedents in plural grammatical number. Table 11 shows the rules used:

Table 9. Rule for detecting coordinative structures

	Rules	Example
7.	([NP|PP|UNK],)* [NP|PP|UNK] [AND|OR|NOR] [NP|PP|UNK]	While all the selective serotonin reuptake inhibitors (SSRIs) e.g., **fluoxetine$_1$**, **sertraline$_1$** and **paroxetine$_1$** inhibit P450 2D6, **they$_1$** may vary in the extent of inhibition.

Finally, the module assigns a salience measure to each antecedent candidate according to distance. The closer a certain candidate and anaphora are, the more probable it is that the candidate will be selected as the antecedent. The formula used was the following:

$$Distance(Cand_i) = N * \frac{d_{max} - d_i}{d_{max} - d_{min}}$$

Fig. 1. Formula for calculating the distance score of each candidate, where N is the maximum weight assigned to the distance factor (N=3), *dmax* is the distance between the most faraway candidate and anaphora according to the number of phrase elements between them, *dmin* is the distance between the closest candidate and a certain anaphora, and *di* is the distance between the anaphora and the candidate to be evaluated

In addition to the distance factor, the longest common subsequence shared between the anaphora and the antecedent was also considered. The function for weight assignation according to common morphological subsequence is expressed as follows:

$$Morpho(Cand_i) = N * (1 - \frac{min - LCS(anaphor, cand_i)}{min})$$

Fig. 2. Formula for calculating the morphological score of each candidate where N is the weight assigned to this factor (N=1.5 in experiments), *min* is determined by selecting the smaller length from between the anaphora and the candidate and *LCS* is the shared length between the anaphora and the candidate

Finally, the sum of both scores is assigned to the candidate. The candidates are ordered by score and only those exceeding the threshold (1.5) were selected; if candidates did not go beyond this value, the anaphora was unresolved.

5 Results and Discussion

As there is no previous work on anaphora resolution for drug-drug interaction expressions, it was decided to develop an ad-hoc baseline strategy for anaphora resolution that simply selects the closest nominal phrase. Anaphoric expressions considered are those referred to in Tables 4 and 5. Results obtained for the baseline system are shown in the following tables.

Results of the anaphora resolver were compared with those provided by the manually annotated corpus. From the 330 anaphoric expressions obtained for the types analyzed in the corpus, 265 were detected by the system and 222 were successfully resolved, that is, attached to the correct antecedent. For testing the performance of the system, the F-score measure[7] with β=1, also called *balanced*

---

[7] F-score=(1+ β²) * (precision*recall) / ((β² * precision)+recall).

F-score, was used. This is a weighted harmonic mean of precision and recall. Precision is the ratio between the anaphors successfully resolved by the approach and the anaphors proposed by the approach. Recall is the ratio between the anaphors successfully resolved by the approach and the number of anaphors occurring in the corpus. Results are the following:

Table 10. Global results for anaphora resolution

	Baseline			Approach			
Total	Precision	Recall	F-score	Precision	Recall	F-score	Inc[8]
330	0.49	0.40	0.44	0.77	0.62	0.69	0.57

Table 11. Results for pronominal anaphora resolution

		Baseline			Approach			
Type	Total	P	R	F	P	R	F	Inc
Personal	23	0.26	0.26	0.26	0.52	0.52	0.52	1.00
Reflexive	1	1.0	1.0	1.0	1.0	1.0	1.0	0
Relative	120	0.83	0.81	0.82	1.0	0.92	0.96	0.17
Distributive	8	0.33	0.12	0.18	0.85	0.75	0.8	3.44
Demonstrative	11	0	0	0	0.33	0.09	0.14	∞
Indefinite	8	0.25	0.12	0.16	0.8	0.5	0.61	2.81
Global results	164	0.67	0.65	0.66	0.9	0.82	0.85	0.29

Table 12. Baseline and approach results for nominal anaphora resolution

		Baseline			Approach			
Type	Total	P	R	F	P	R	F	Inc
Definite	37	0	0	0	0.63	0.37	0.47	∞
Possessive	52	0.53	0.42	0.47	0.67	0.67	0.67	0.43
Distributive	11	0.20	0.27	0.23	0.60	0.54	0.57	1.48
Demonstrative	58	0.03	0.01	0.02	0.53	0.25	0.34	16.00
Indefinite	8	0	0	0	0.33	0.12	0.34	∞
Global results	166	0.23	0.15	0.18	0.61	0.42	0.50	1.78

Our approach obtains a 57% relative improvement over the baseline model in overall results. The difference is even more pronounced for the case of nominal anaphora resolution with an increase of 178%. Clauses in the corpus are characterized by frequent coordinative and subordinate structures, along with numerous prepositional phrases which explain the difficulty of the task and the results of the baseline. From the results it is clear that linguistic information is needed in order to

---

[8] Inc= (F-approach − F-baseline) / F-baseline.

deal with anaphora in this kind of document. The contribution of semantic resources like MMTx becomes evident when comparing the approach against the baseline.

Pronominal anaphora resolution performs better than its counterpart not only in precision but also in recall. Likewise, the good performance in the resolution of relative pronoun antecedents must be emphasized, explained by the fact that these units are mostly located directly to the right of the anaphoric expressions. In addition, as pointed out in [26], pronominal anaphora is easier to resolve than the nominal one because the latter requires an encyclopedic knowledge source.

6 Conclusion

Our approach obtains results similar to other systems referred to in the biomedical domain [9] [10] [15] [8], but it is, to our knowledge, the first research that tackles this issue for the case of drug-drug interaction documents. It shares with these articles the use of a set of linguistic features and a semantic resource. However, it is believed that features weighted in previous approaches [9, 10, 15] such as number constraints must always be satisfied. In addition, other approaches in this area make use of domain-independent parsers that do not adequately deal with the syntactic complexity of the biomedical language. Good performance is explained in part by the analysis provided by this specific domain parser (MMTx).

Future work will focus on the development of linguistic rules based on the Centering Theory [27]. Additionally, there are plans to extend current approach with the use of the pharmacological family of drugs (information also provided by DrugNer) in the recognition of expressions such as *'this anticoagulant'*, whose antecedent must be a member of the class 'anticoagulants'. It is expected that such extensions improve results of the current approach.

Finally, it can be emphasized that although the characteristics of the corpus are similar to other approaches, it is essential to increase its size in order to make reliable conclusions. The style of the language used in this corpus is mostly devoted to descriptions of drug interactions, so the evaluation of the approach about texts from other resources like MedLine will be considered to estimate its performance over a less specific domain.

References

1. Aronson, A.R.: Effective mapping of biomedical text to the UMLS metathesaurus: the Metamap program. In: Proceedings of AMIA Symp., pp. 17–21 (2001)
2. WHO. The Importance of Pharmacovigilance: Safety Monitoring of Medicinal Products. World Health Organization (2002)
3. Stockley, I.: Stockley's Drug Interactions. Pharmaceutical Press (2007)
4. Jankel, C., McMillan, J., Martin, B.: Effects of drug interactions on outcomes of patients receiving warfarin or theophylline. Am. J. Hosp. Pharm. 51, 661–666 (1994)
5. Aronson, J.K.: Communicating information about drug interactions. British Journal of Clinical Pharmacology 63(6), 637–639 (2007)
6. Duda, S., Aliferis, C., Miller, R., Statnikov, A., Johnson, K.: Extracting Drug-Drug Interaction Articles from MEDLINE to Improve the Content of Drug Databases. In: AMIA Annual Symposium Proceedings (2005)

7. Wishart, D.S., Knox, C., Guo, A.C., Cheng, D., Shrivastava, S., Tzur, D., Gautam, B., Hassanali, M.: Drugbank: a knowledgebase for drugs, drug actions and drug targets. Nucl. Acids Res. (2007), doi:10.1093/nar/gkm958
8. Kim, J.J., Park, J.C.: BioAR: Anaphora Resolution for Relating Protein Names to Proteome Database Entries. In: Proceedings of ACL, pp. 79–86 (2004)
9. Castaño, J., Zhang, J., Pustejovsky, J.: Anaphora resolution in biomedical literature. In: Int'l Symp. Reference Resolution in NLP, Alicante, Spain (2002)
10. Lin, Y.H., Liang, T.: Pronominal and sortal anaphora resolution for biomedical literature. In: Proceedings of ROCLING XVI: Conference on Computational Linguistics and Speech Processing (2004)
11. Temkin, J., Gilder, M.: Extraction of protein interaction information from unstructured text using a context-free grammar. Bioinformatics 19(16), 2046–2053 (2003)
12. Fundel, K., Kuffner, R., Zimmer, R.: RelEx-Relation extraction using dependency parse trees. Bioinformatics 23(3), 365 (2007)
13. Kolarik, C., Hofmann-Apitius, M., Zimmermann, M., Fluck, J.: Identification of new drug classiffcation terms in textual resources. Bioinformatics 23(13), i264 (2007)
14. Pustejovsky, J., Castaño, J., Saurí, R., Rumshisky, A., Zhang, J., Luo, W.: Medstract: Creating Large-scale Information Servers for Biomedical Libraries. In: Proceedings of ACL 2002 Workshop on Natural Language Processing in the Biomedical Domain, Philadelphia (2002)
15. Liang, T., Lin, Y.: Anaphora Resolution for Biomedical Literature by Exploiting Multiple Resources. In: Proceedings of IJCNLP, pp. 742–753 (2005)
16. Gasperin, C.: Semi-supervised anaphora resolution in biomedical texts. In: Proceedings of BioNLP in HLT-NAACL, New York, pp. 96–103 (2006)
17. FlyBase, http://www.flybase.org
18. Eilbeck, K., Lewis, S.E., Mungall, C.J., Yandell, M., Stein, L., Durbin, R., et al.: The Sequence Ontology: a tool for the unification of genome annotations. Genome Biol. (2005)
19. Briscoe, T., Carroll, J.: Robust accurate statistical annotation of general text (2002)
20. Kim, J.J., Park, J.C.: BioIE: retargetable information extraction and ontological annotation of biological interactions from the literature. J. Bioinformatics and Computational Biology 2(3), 551–568 (2004)
21. Bairoch, A., Apweiler, R.: The swiss-prot protein sequence database and its supplement TrEMBL in 2000. Nucl. Acids Res. 28(1), 45–48 (2000)
22. Grosz, B.J., Joshi, A.K., Weinstein, S.: Centering: A framework for modelling the local coherence of discourse. Computational Linguistics 21(2), 203–225 (1995)
23. Sanchez, O., Poesio, M., Kabadjov, M., Tesar, R.: What kind of problems do protein interactions raise for anaphora resolution? - A preliminary analysis. In: SMBM, Jena, Germany (2006)
24. Poesio, M., Kabadjov, M.A.: A general-purpose, off-the-shelf anaphora resolution module: implementation and preliminary evaluation. In: Proceedings of LREC, Lisbon, Portugal (2004)
25. Segura-Bedmar, I., Martínez, P., Segura-Bedmar, M.: Drug Name Recognition and classification in biomedical texts. Drug Discovery Today 13(17), 816–823 (2008)
26. Poprat, M., Hahn, U.: Quantitative Data on Referring Expressions in Biomedical Abstracts. In: Proceedings of the Workshop on BioNLP 2007, pp. 193–194 (2007)
27. Grosz, B.J., Joshi, A.K., Weinstein, S.: Centering: A framework for modelling the local coherence of discourse. In: Computational Linguistics, pp. 203–225 (1995)

Relation Extraction for
Monitoring Economic Networks

Martin Had, Felix Jungermann, and Katharina Morik

Technical University of Dortmund
Department of Computer Science - Artificial Intelligence Group
Baroper Strasse 301, 44227 Dortmund, Germany

Abstract. *Relation extraction* from texts is a research topic since the message understanding conferences. Most investigations dealt with English texts. However, the heuristics found for these do not perform well when applied to a language with free word order, as is, e.g., German. In this paper, we present a German annotated corpus for *relation extraction*. We have implemented the state of the art methods of *relation extraction* using kernel methods and evaluate them on this corpus. The poor results led to a feature set which focusses on all words of the sentence and a tree kernel which includes words, in addition to the syntactic structure. The *relation extraction* is applied to monitoring a graph of economic company-directors network.

1   Introduction

Social networks have raised scientific attention, the goals ranging from enhancing recommender systems [4,15,5] to gaining scientific insights [6,22]. Where the taggings, mailings, co-authorship, or citations in communities have well been investigated, the economic relationships between companies and their networking have less been studied.

Today's search engines are not prepared to answer questions like "show me all companies that have merged with Volkswagen". In order to get that information anyway, it would be necessary to do an extensive search and consider several sources. This is time consuming and tedious. This is why question answering approaches require automatic *relation extraction*.

Moreover, it is important to represent the extracted information in a compact and easily to access manner. Especially concerning *relation extraction*, the extracted entities and relations can be represented using an (un-)directed graph.

In this paper, we present an approach to monitoring economic information in the world wide web using a graph-based representation. We will show that it is possible to extract additional information using *relation extraction* techniques, which have not yet successfully been used on German texts, because German language features problems, which other languages – especially English – do not face. A comparison of our feature set and enhanced tree kernel with state of the art methods illustrates the importance of a balanced use of semantic and syntactic information. First, we describe the state of the art in *relation extraction*

H. Horacek et al. (Eds.): NLDB 2009, LNCS 5723, pp. 103–114, 2010.

using kernel methods, then we present our application, before we introduce or enhancement of the method and the experimental results.

2   Kernel Methods for Relation Extraction

The ACE RDC task [11] defines a relation as a valid combination of two entities that are mentioned in the same sentence and have a connection to each other. Relations may be symmetric or asymmetric. The schema of i relations in a sentence s is defined as follows:

Definition 1. Relation candidates *in a sentence:*

$$R_i(Sentence\ s) = <Type_m \in relationtypes,$$

$$(Argument_1 \in entities_s, Argument_2 \in entities_s) >$$

where $entities_s$ is the set of entities contained in the current sentence, and *relationtypes* is the set of possible relations in the corpus.

Structured information of a sentence e.g. is the syntactic parse tree (an example can be seen in Figure 1), where each node follows a grammar production rule. By splitting up a tree in subtrees (see Figure 2) it is possible to calculate the similarity of two trees by counting their common subtrees. The set of subtrees of a parse tree consists of every substructure that can be built by applying the grammatical rule set of the original tree.

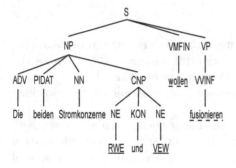

Fig. 1. A parse tree for a German sentence containing a merger-relation

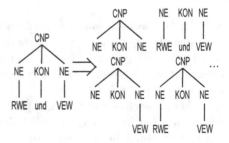

Fig. 2. Some subtrees of a tree

2.1   Linear Kernels

First experiments on *relation extraction* have been done by just using feature-based methods. That made it necessary to manually create a large set of 'flat' features describing the relation and comparing the similarities of these feature-vectors in order to find the best discriminating classification function. The most efficient way to compare feature-vectors is based on kernel functions which can be embedded in various machine-learning algorithms like support vector machines or clustering methods. A linear kernel on feature-vectors, x and z, is defined as their inner product:

Definition 2. *A* linear kernel*:*

$$K(x, z) = \sum_n \phi_n(x)\phi_n(z) \tag{1}$$

where $\phi_n(x)$ is the n-th feature of x.

2.2   Convolution Kernels

Converting syntactic structures into feature-vectors is tedious [21] [23]. This overhead is avoided when using a kernel function, which operates on any discrete structure [7]. Because of the formulation as a kernel, the calculation of the inner product requires the enumeration of substructures only implicitly.

Definition 3. *Suppose $x \in X$ is a composite structure and $\boldsymbol{x} = x_1, ..., x_p$ are its parts, where each $x_i \in X_i$ for $i = 1, ..., p$ and all X_i are countable sets. The relation $R(\boldsymbol{x}, x)$ is true, iff $x_1, ..., x_p$ are all parts of x. As a special case, X being the set of all p-degree ordered, rooted trees and $X_1 = ... = X_p = X$, the relation R can be used iteratively to define more complex structures in X.*

Given $x, z \in X$ and $\boldsymbol{x} = x_1, ..., x_p$, $\boldsymbol{z} = z_1, ..., z_p$ and a kernel K_i for X_i measuring the similarity $K_i(x_i, z_i)$, then the similarity $K(x, z)$ is defined as the following generalized convolution

$$K(x, z) = \sum_{\{\boldsymbol{x}|R(\boldsymbol{x},x)\}} \sum_{\{\boldsymbol{z}|R(\boldsymbol{z},z)\}} \prod_{i=1}^{p} K_i(x_i, z_i) \tag{2}$$

[7]p.5f

Convolution kernels characterize the similarity of parse trees by the similarity of their subtrees [3]. Within the kernel calculation, all subtrees of the trees are compared. They are (implicitly) represented as a vector:

$$\Phi(T) = (subtree_1(T), ..., subtree_m(T)) \tag{3}$$

where $subtree_i$ means the number of occurrences of the i-th subtree in T. The number of common subtrees is summed up. The worst case runtime is $O(|N_1| \times |N_2|)$, being N_t the set of nodes of a tree T_t.

Definition 4. *The* tree kernel *computes a scalar product:*

$$K(T_1, T_2) = <\mathbf{h}(T_1), \mathbf{h}(T_2)> \tag{4}$$

$$h_i(T_1) = \sum_{n_1 \in N_1} I_i(n_1) \tag{5}$$

where the indicator function I_i is defined for the nodes n_1 in N_1 of T_1 and n_2 in N_2 for T_2 as 1, iff the $i-$th subtree is rooted in node n. Hence,

$$K(T_1, T_2) = \sum_{n_1 \in N_1} \sum_{n_2 \in N_2} \sum_i I_i(n_1)I_i(n_2) \tag{6}$$

$$= \sum_{n_1 \in N_1} \sum_{n_2 \in N_2} \Delta(n_1, n_2) \tag{7}$$

The calculation of Δ is done recursively by following three simple rules:

- If the grammar production rules of n_1 and n_2 are different: $\Delta(n_1, n_2) = 0$
- If the production rules in n_1 and n_2 are equal and n_1 and n_2 are pre-terminals (last node before a leaf): $\Delta(n_1, n_2) = \lambda$
- If the production rules in n_1 are n_2 equal and n_1 and n_2 are non pre-terminals:

$$\Delta(n_1, n_2) = \lambda \prod_{j=1}^{nc(n_1)} (1 + \Delta(ch(n_1, j), ch(n_2, j))) \tag{8}$$

$nc(n_1) = $ number of children of node n_1

$ch(n_1, i) = i$th child of node n_1

$\lambda = $ parameter to downweight the contribution of large tree

fragments exponentially with their size.

[13] designed an algorithm for the above calculation that has linear runtime on average due to a clever preprocessing step. Nodes that don't need to be considered by the kernel are filtered out by sorting and comparing the production rules of both trees in advance ("Fast Tree Kernel" FTK).

[24] extended the FTK kernel to become context sensitive by looking back at the path above the ancestors of the root node of each subtree. The left side of the production rule is taking into account m-1 steps towards the root. The kernel calculation itself sums up the calculations for each set of production rules created for $1 \ldots m$. In the special case m=1 the kernel result is the same as with the non context-sensitive kernel.

$$K_C(T[1], T[2]) = \sum_{i=1}^{m} \sum_{n_1^i[1] \in N_i^1[1], n_1^i[2] \in N_i^1[2]} \Delta(n_1^i[1], n_1^i[2]) \tag{9}$$

- m the number of ancestor nodes to consider.
- $n_1^i[j]$ is a node of tree j with a production rule over i ancestors. $n_1[j]$ is the root node of the context free subtree the ancestor node of $n_k[j]$ is $n_{k+1}[j]$.
- $N_1^i[j]$ is the set of all nodes with their production rules over i ancestor.

2.3   Composite Kernels

[20] showed that better results can be achieved with a combination of linear and tree kernels. [7] showed that the class of kernels is closed under product and addition. This implies that combining two kernels is possible and results in a new kernel which is called a *composite kernel*, defined as follows:

$$K(x, z) = K_1(x, z) \circ K_2(x, z) \tag{10}$$

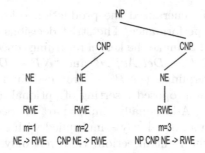

Fig. 3. Production rules of subtree root node (NE) for different m values

In the case of *relation extraction* the composite kernel combines a linear and a convolution tree kernel. [20] propose a linear combination (11) and a polynomial expansion (12).

$$K(x,z) = \alpha \cdot \hat{K}_L(x,z) + (1-\alpha) \cdot \hat{K}_T(x,z) \tag{11}$$

$$K(x,z) = \alpha \cdot \hat{K}_L^P(x,z) + (1-\alpha) \cdot \hat{K}_T(x,z) \tag{12}$$

where x and z are *relation candidates* that consist of flat features and structured information, each kernel is given the right input for its kind. $K_L^P(x,z)$ is the polynomial expansion of $K_L(x,z)$ with degree $d = 2$ i.e. $K_L^P(x,z) = (K_L(x,z) + 1)^2$. By setting the α value the influence of each kernel type can be adjusted. Both kernels are normalized in kernel space before the combination:

$$K'(T_1,T_2) = \frac{K(T_1,T_2)}{\sqrt{K(T_1,T_2)K(T_1,T_2)}} \tag{13}$$

2.4   State-of-the-Art Composite Kernels for Relation Extraction

In addition to just using the composite kernel on the full parse tree of a sentence, [20] examined several ways of pruning the parse tree in order to get differently shaped subtrees on which the treekernel performs as well or better as on the full tree. They showed that the shortest path-enclosed tree (SPT) which is the minimal subtree containing the two entities of a *relation candidate* performs best for the ACE 2003 RDC corpus.

But [24] showed that the ACE corpus contains relations for which the SPT is not sufficient. These relations are indicated by their related verb. Figure 1 shows a relation of our corpus which is indicated by its related verb, too. But in contrast to the ACE corpus which just contains a few relations of this type, our corpus has many. The type of relation and the specialty of the German language are responsible for this fact.

The strict and binary decisions of the tree kernel are the main disadvantage of this method. [24] tried to overcome this problem by embedding syntactic features into the parse tree directly above the leaf-nodes. Moreover, syntactic structure is already covered by the tree kernel, adding it in terms of features does not help generalization. [19] generalized the production rules of the parse tree in order to achieve better performance. The strict decisions of a convolution tree kernel (remind Section 2.2) make the kernel returning 'unequal' confronted with two production rules "$NP \rightarrow Det\ Adj\ N$" and "$NP \rightarrow Det\ N$" although they might contain similar terminals ("$NP \rightarrow$ a red car" and "$NP \rightarrow$ a car"). To avoid such behavior they proposed inserting of optional nodes into production rules to generalize them. Additionally similar part of speech tags in the parse tree can be processed in an equal way – multiplied with a penalty term.

This is a step into the right direction. However, only syntactic variance is handled. Since words carry most of the semantic information, moving them into the tree kernel could well help to generalize in a more semantic way.

Related kernels for Relation Extraction. There are several related approaches for *relation extraction* differing from the ones already presented. [17] presented a general kernel function for trees and its subtrees. [18] used a kernel function on shallow parse trees. Bunescu and Mooney used a kernel function on the shortest path between two entities in a dependency tree [1]. Additionally they used the context of entities for *relation extraction* [2].

3   Monitoring the Merger Event in an Economic Network

The enhancement of the state of the art in *relation extraction* which is described in Section 4 became necessary when we developed the economic network based on German sources. We did not want to manually build and update its database. Extracting relations from documents directly allows to automatically accomplish the data about companies and their board members with relationships between them. Hence, the network can be monitored and is always up-to-date.

3.1   Building-Up the Economic Network

Building up the economic network starts with extracting companies and their board of directors. The extraction of the named entities "company" and "board member" is quite simple, because there exist several web archives of companies which are semi-structured. Hence, companies and their representatives can easily be extracted using simple regular expressions. The initial stock of data is stored in a SQL database. It consists of about 2.000 different big companies from throughout the world. Basic information includes only address and industry but most entries provide a lot more details about members of the board of directorate, share ownership, shareholding and some key performance indicators. Many of these companies (here: 1,354) are connected to one other company at least, by sharing a member of the directorate. The best connected company even has 37 different outgoing directorate connections. These numbers support the assumption, that the graph built from these relations can reveal significant structures in the business world. From the SQL data base, a network $G = (V, E)$ is built containing entities ($v \in V$) and relations ($e \in E$) between entities. Its visualization is performed using the JUNG-Framework [14]. The human-computer

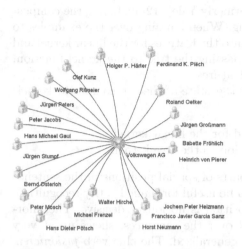

Fig. 4. Selecting Volkswagen AG (VW) from all companies, the involved responsible persons are displayed

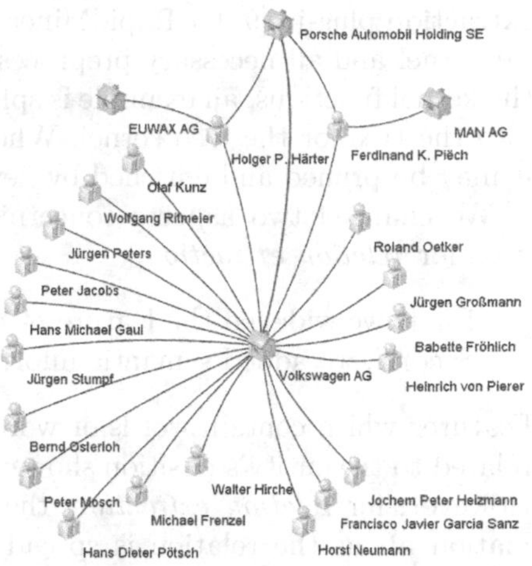

Fig. 5. Two directors of the board of VW are directors of Porsche, as well. The merger-relation holds between VW and Porsche (indicated by a thicker line between the companies).

interface allows users to select a company and move to the involved persons, from which the user may move to all the companies in which they play a role – thus browsing through the basic social network graph of economy. Figure 4 shows an example. Since the archives do not change their structure whenever the content changes, the database is easily updated.

3.2 Extracting the Merger-Relation from Web Documents

For monitoring the web of economy, the merger-relation is most interesting. To get a preselection of relevant documents the the web is crawled for information about the 30 DAX indexed German companies. Given a list of known company names, the texts of the resulting websites are tagged in the IOB-scheme indicating "company"-entities. Only those sentences containing at least two company entities are selected for further processing. It is then the task of *relation extraction* to identify the true merger-relations between two companies. Of course, simple co-occurrence is not sufficient for this task. Note, that a sentence with three company names can include none, one, or two merger-relations. Hence, we applied our method described in Section 4. Details on the experiments are given in Section 5. Figure 5 shows an example of a found merger-relation.

4   Relation Extraction with an Enhanced Composite Kernel

We have implemented the state-of-the-art kernel method in Java, extending the kernel functions of SVM^{light} [8]. We also have developed an information extraction plug-in [9] for RapidMiner (formerly Yale) [12] including the composite kernel and all necessary preprocessing. When handing over the examples to the kernel functions, an example is split into the features for the linear kernel and into the tree for the tree kernel. When passing the tree to the kernel function, it may be pruned and enriched by new features.

We changed two aspects concerning the state-of-the-art composite kernels used for *relation extraction*:

- First, we widened the featureset used for the linear kernel.
- Second, we added semantic information to the tree kernel.

Features which contain words or word-parts of special positions in the sentence related to the entity's position showed to be useful for named entity recognition. However, for *relation extraction*, the position is no longer decisive. The information about the relation is spread all over the sentences, shows up at very different places, and can, hence, not be generalized. The clue verb *fusionieren*, for instance, may occur at various positions of the word sequence (see Figure 1). Especially for distinguishing between positive and negative *relation candidates*, the contextual information is not restricted to the words between the entities or to some words in front or behind the entities, as assumed by the feature set in [23]). In order to capture the influence of words that can act as an indicator for a relation we extract the word vector (containing just the word stems) of the complete sentence and add it to the linear features. In a separate experiment setting we use the features presented by [23], for comparison.

The second of our enhancements concerns the tree kernel. Figure 1 shows a parse tree of our corpus containing two entities (underlined solid) and the merger-indicating verbs (underlined dashed). It is easy to see that well-known subtrees for better *relation extraction* like shortest path-enclosing trees (SPT) will not work well in this context. But using the whole and unaltered parse tree will not work as well. The reason is, if a sentence contains positive and negative *relation candidates* the identical parse tree would be used for both *relation candidate*-types.

Using the context-sensitive parse tree of [24] is promising. But this approach needs well-trained parsers which are still not available in an appropriate version for the German language. We therefore generalized the parse tree by adding syntactical information directly into the tree. First of all we marked the entities of the current *relation candidate* in the corresponding parse tree. In addition, we added semantic information into the parse tree by introducing extra nodes containing the word stems of the sentence at different depths.

Figure 6 shows four different types of parse trees used in our experimental settings. The first one (1)) is the original parse tree. In parse tree 2) we have replaced all terminals by their stems. Parse tree 3) is the tree after inserting the stem at depth 0. The depth is the depth of the stem in relation to the depth of the pre-terminal symbol. 4) is the tree after inserting the stem at depth 1. The word 'VW' has no stem, so nothing is inserted.

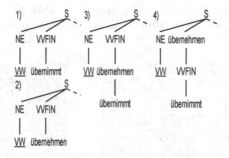

Fig. 6. Word stems at different depth-levels in the parse tree

5   Experiments

Our companies corpus consists of 1698 sentences containing 3602 *relation candidates*. 2930 of these *relation candidates* are negative ones (being no merger-relation), and 672 *relation candidates* are true merger-relations. Only 98 of these 1689 sentences contain multiple *relation candidates* with different labels. Compared to other *relation extraction* datasets this distribution is very skewed and leads to the behavior described in the following Section. The ACE04 corpus for instance contains 2981 sentences out of which 1654 sentences contain at least a true relation and a negative candidate.

We produced several training sets with different attribute sets to compare our enhancements with the state-of-the-art composite kernel methods for *relation extraction*. All the training sets consist of all *relation candidates*, i.e., pairs of entities found in one sentence. Each example consists of a relation label, e.g. *merger* or *nomerger*, the syntactic tree of the sentence in which the arguments occur, and several features which are now described in detail.

The *baseline-featureset* contains the word-features proposed by [23]. These features are mainly based on words of the *relation candidate* entities or words nearby in the sentence. For its use by the tree kernel, the feature set also contains the parse tree of the sentence the *relation candidate* is extracted from.

The *word-vector-featureset* contains just the word-vector of the sentence from which the *relation candidate* is extracted, and the parse tree.

The *big-word-vector-featureset* contains just the word-vector, the parse tree and the *baseline-featureset*.

The *stem-x-tree-featuresets* are equal to the *word-vector-featureset* but the parse tree contains the word stems inserted at depth-level x or as a replacement of the original terminal symbol.

The parse tree is given by running the Stanford parser [10] trained on the NEGRA corpus [16]. We applied 10-fold cross validation using the composite kernel with a parameter setting of $C = 2.4$, $m = 3$ and $\alpha = 0.6$ (see Section 2.2).

Performance. Table 1 shows the performance of the state of the art method and the two versions of our new method. Table 2 shows the standard deviation of the performance measures in 10-fold cross validation.

Table 1. Performance of *relation extraction* on the companies corpus using 10-fold cross validation

Table 2. Standard deviation of the performance of *relation extraction*

Featureset	Precision	Recall	F-meas.
baseline	33,47%	52,27%	38,64%
word-vector	36,41%	69,93%	45,45%
big-word-vector	36,83%	74,86%	**48,73%**
stem-replace-tree	31,46%	**76,03%**	44,08%
stem-0-tree	37,94%	47,90%	41,79%
stem-1-tree	**44,33%**	53,42%	47,51%
stem-2-tree	36,28%	62,91%	45,64%

Precision	Recall	F-meas.
3,88%	21,99%	8,88%
12,16%	11,64%	5,12%
5,15%	9,55%	**3,12%**
4,63%	**8,99%**	4,32%
6,29%	14,55%	9,07%
7,58%	9,89%	4.40%
3,95%	10,89%	4.62%

As can be seen, recall increases significantly using word vectors in the linear kernel and word stems in the tree kernel while at the same time the deviation decreases. Precision is best when semantic information in the tree is used at level 1. The best F-measure achieved by the big-word-vector is to be explained by the very few sentences containing a positive and a negative candidate of a relation. If sentences include either a positive or a negative example of a relation, the *relation extraction* is downgraded to sentence classification, where word vectors are a well suited representation. Hence, for *relation extraction*, the enhanced trees remain important.

6 Conclusions and Future Work

We proposed an economic network that is built up extracting semi-structured websites containing financial stock information.

The network – consisting of entities and relations between them – should be kept up to date automatically. Therefore we presented an enhancement to state-of-the-art *relation extraction* methods. Our enhancements take into account the problems German language faces in contrast to the well-examined English language.

To evaluate our method we extracted a German document corpus of the economic domain. We tagged all the firms in our corpus and extracted all possible *relation candidates*. We tested state-of-the-art *relation extraction* methods on our *relation extraction* corpus and compared the results with the results achieved by our enhancements. Our enhanced composite kernel method achieves significantly better performance compared to the baseline. Although using just the linear kernel performs best, the usage of the composite kernel will be needed if the relations become more frequent and the number of relation-types becomes bigger.

Future work will implement better measures for the evaluation, so that sentence classification effects in *relation extraction* can properly be detected. Our approach should be evaluated on English benchmark datasets. Additionally our approach to add semantic information in the parse trees could be replaced by using dependency trees. But unfortunatelly the used library (stanford parser) just offers trained dependency parsers for English and Chinese. Using dependency trees therefore might be tested on English datasets.

References

1. Buncescu, R.C., Mooney, R.J.: A shortest path dependency kernel for relation extraction. In: HLT 2005: Proceedings of the conference on Human Language Technology and Empirical Methods in Natural Language Processing, Vancouver, British Columbia, Canada, pp. 724–731. Association for Computational Linguistics (2005)
2. Bunescu, R.C., Mooney, R.J.: Subsequence kernels for relation extraction. In: Weiss, Y., Schölkopf, B., Platt, J. (eds.) Advances in Neural Information Processing Systems, vol. 18, pp. 171–178. MIT Press, Cambridge (2006)
3. Collins, M., Duffy, N.: Convolution kernels for natural language. In: Advances in Neural Information Processing Systems, vol. 14, pp. 625–632. MIT Press, Cambridge (2001)
4. Debnath, S., Ganguly, N., Mitra, P.: Feature weighting in content based recommendation system using social network analysis. In: WWW 2008. ACM Press, New York (2008)
5. Domingos, P., Richardson, M.: Mining the network value of customers. In: Procs. KDD, pp. 57–66. ACM Press, New York (2001)
6. Golder, S.A., Wilkinson, D.M., Huberman, B.A.: Rhythms of social interaction: Messaging within a massive online network. In: Procs. 3rd Intl. Conf. on Communities and Technologies (2007)
7. Haussler, D.: Convolution kernels on discrete structures. Technical report, University of California in Santa Cruz, Computer Science Dept. (1999)
8. Joachims, T.: Text categorization with support vector machines: Learning with many relevant features. In: Procs. of European Conference on Machine Learning, pp. 137–142. Springer, Heidelberg (1989)
9. Jungermann, F.: Information extraction with rapidminer. In: Hoeppner, W. (ed.) Proceedings of the GSCL Symposium 'Sprachtechnologie und eHumanities', pp. 50–61. Universität Duisburg-Essen, Abteilung für Informatik und Angewandte Kognitionswissenschaft Fakultät für Ingenieurwissenschaften (2009)
10. Klein, D., Manning, C.D.: Fast extract inference with a factored model for natural language parsing. In: Proceedings of Advances in Neural Information Processing Systems (2002)
11. Linguistic Data Consortium. In: The ACE 2004 Evaluation Plan (2004)
12. Mierswa, I., Wurst, M., Klinkenberg, R., Scholz, M., Euler, T.: Yale: Rapid prototyping for complex data mining tasks. In: Procs. 12th ACM SIGKDD Int. Conf. Knowledge Discovery and Data Mining, KDD (2006)
13. Moschitti, A.: Efficient convolution kernels for dependency and constituent syntactic trees. In: Fürnkranz, J., Scheffer, T., Spiliopoulou, M. (eds.) ECML 2006. LNCS (LNAI), vol. 4212, pp. 318–329. Springer, Heidelberg (2006)
14. O'Madadhain, J., Fisher, D., White, S., Boey, Y.-B.: The JUNG (java universal network/graph) framework. Technical Report Technical Report UCI-ICS 03-17, School of Information and Computer Science University of California, Irvine, CA 92697-3425 (2003)
15. Palau, J., Montaner, M., López, B., de la Rosa, J.L.: Collaboration analysis in recommender systems using social networks. In: Procs. Cooperative Information Agents VIII, pp. 137–151. Springer, Heidelberg (2004)
16. Skut, W., Krenn, B., Brants, T., Uszkoreit, H.: An annotation scheme for free word order languages. In: Proceedings of the Fifth Conference on Applied Natural Language Processing ANLP 1997, Washington, DC (1997)

17. Vishwanathan, S.V.N., Smola, A.J.: Fast kernels for string and tree matching. In: NIPS, pp. 569–576 (2002)
18. Zelenko, D., Aone, C., Richardella, A.: Kernel methods for relation extraction. Journal of Machine Learning Research 3, 1083–1106 (2003)
19. Zhang, M., Che, W., Aw, A.T., Tan, C.L., Zhou, G., Liu, T., Li, S.: A grammar-driven convolution tree kernel for semantic role classification. In: Procs. 4th Annual Meeting of ACL, pp. 200–207 (2007)
20. Zhang, M., Zhang, J., Su, J., Zhou, G.: A composite kernel to extract relations between entities with both flat and structured features. In: Procs. 44th Annual Meeting of ACL, pp. 825–832 (2006)
21. Zhao, S., Grishman, R.: Extracting relations with integrated information using kernel methods. In: ACL 2005: Proceedings of the 43rd Annual Meeting on Association for Computational Linguistics, Morristown, NJ, USA, pp. 419–426. Association for Computational Linguistics (2005)
22. Zhou, D., Orshanskiy, S.A., Zha, H., Giles, C.L.: Co-ranking authors and documents in a heterogeneous network. In: 7th IEEE ICDM (2007)
23. Zhou, G., Su, J., Zhang, M., Zhang, J.: Exploring various knowledge in relation extraction. In: ACL, pp. 427–434 (2005)
24. Zhou, G., Zhang, M., Ji, D.H., Zhu, Q.: Tree kernel-based relation extraction with context-sensitive structured parse tree information. In: Proceedings of the 2007 Joint Conference on Empirical Methods in Natural Language Processing and Computational Natural Language Learning, Association for Computational Linguistics (2007)

Real-Word Typo Detection

Dmitri Asonov

Moscow, Russia
asonov@fastpl.com

Abstract. Context-sensitive spelling correction (CSSC) is a widely accepted and long studied formalization of the problem of finding and fixing contextually incorrect words. We argue that CSSC has its limitations as a model, and propose a weakened CSSC model (RWTD) to partially counter these limitations. We weaken the CSSC model by canceling its word-correction role. Thus, RWTD is focused solely on finding words that require correction. Once this is done, the actual correction process is performed by a human or a CSSC solution.

We propose a preliminary solution for RWTD model that differs from related CSSC work in several ways. The solution does not rely on a set of confusion lists and detects not only a limited set of confusion typos, but almost any class of typos. The solution offers a flexible trade-off between the time a human is willing to spend on the task and the quality of the proofreading. It does not require POS tagging and may be applied seamlessly to different languages. Experiment running times prove to be acceptable for real-world applications.

We report Brown corpus real-word typos that were exposed by implementing our solution. We also discuss experiments in applying the solution to other real-world test texts and demonstrate improved false positive and hit rates.

1   Introduction

Finding and fixing contextually incorrect words is an important task in publishing, optical character recognition (OCR), end user word processing, translation, speech recognition and other industries. We refer to contextually incorrect words as real-word typos or typos for short. We refer to typo detection as proofreading.

Typos can still be found in newspapers, magazines, and books that are published today. Emails, Wikipedia articles, and other electronic documents produced by end user word processing systems often contain typos, sometimes regrettable and embarrassing ones[1].

Documents produced by OCR are prone to typos as are manually and automatically translated documents which also have a higher frequency of contextually incorrect words and irregular word usage than other sorts of documents. The Internet is full of typo collections from printed and electronic media.

---

[1] In just one of many such examples witnessed personally , one high-ranking executive sent another an email that began: "thank you for your massage".

H. Horacek et al. (Eds.): NLDB 2009, LNCS 5723, pp. 115–129, 2010.
© Springer-Verlag Berlin Heidelberg 2010

Aside from causing potential embarrassment and being difficult to read, typos create unwanted added costs within the publishing, translation, and OCR industries. These expenses include hiring proofreaders to parse the documents before and/or after editors and style correctors have finished with them. Additional expenses are incurred when publishing errata and letters of apology. Typos also cause unwanted delays in time critical businesses, such as newspapers, where a precious time slot must be allocated for proofreading rather than editing. One might assume that typos are less of a problem for online news sources than for the printed press. Unfortunately, typos are as troublesome, if not more so, in the online press due to the fact that once an article is published online it is immediately copied by other news channels and blogs, making it impossible to correct the typos.

For a human, proofreading is a time-consuming task because the entire document or book must be read even if there are only a handful of typos present in the text. In addition to being time-consuming, occasionally proofreading can be difficult for a human: the term "stealth typo" speaks for itself. The stealth typo may be roughly defined as a typo that is especially hard for a human to detect.

For a computer, proofreading is also a difficult task. By definition, typos escape detection by all dictionary and many grammar checks, so a more advanced technology is required to deal with them. Context-sensitive spelling correction (CSSC) is a widely accepted and long studied formalization of the problem of computer-aided typo detection and correction [1,2,3]. We believe that CSSC as a problem definition has two major shortcomings. Firstly, it focuses on typo correction in addition to typo detection. As a result, CSSC solutions inherently focus on and are designed for finding only those typos that can be corrected by a computer. This in turn limits the number of typo classes that can be detected. Secondly, CSSC implies and relies upon the concept of confusion sets and this further narrows the range of typo classes that can be found.

This paper is structured as follows. In Section 2 we define and discuss a weakened CSSC problem which focuses solely on typo detection and does not rely on confusion sets. In Section 3 we provide a preliminary solution to this problem. We present experimental results in Section 4. We finish with an overview of related work and our conclusion.

2   RWTD Model

Context-sensitive spelling correction is the task of detecting and fixing spelling errors that result in valid words, as in *I'd like a peace of cake* where *peace* was typed when *piece* was intended [1]. The real-word typo detection (RWTD) problem is defined here as the task of detecting typos that result in valid words. Once typos are found, their correction is left to a human or, potentially and partially, to a CSSC solution.

Our motivation for defining the RWTD problem is threefold. Firstly, typo detection makes up a major part of the time it takes a human to find and correct typos. Thus, effectively dealing with the "detection" aspect solves most of the overall task. Secondly, RWTD is a weakened version of CSSC. When down-scaling any problem, solutions for a weakened version of the problem may potentially yield better performance. Finally, RWTD problem definition does not rely on confusion sets, explicitly or implicitly. Thus, RWTD solutions are not inherently limited to detecting only confusion typos.

A thorough review and discussion of various CSSC metrics can be found in [4]. Precision, recall, and F-measure are commonly used in the field to measure performance, and are taken from the information retrieval research area. False positive and hit rates are also used, albeit much less frequently.

We suggest the following requirements for measuring the performance of a RWTD method:

1. To take into account the time sensitivity of many of the proofreading applications the RWTD metrics shall include the program run time and program space complexities, RT and S respectively.
2. To measure the proofreading quality the RWTD metrics Q shall indicate how the system makes a human proofreader's work easier compared with the best and worse case scenarios. In our model, we take the worst case scenario to be a human proofreader reading the entire text. The best case scenario is a human proofreader reading only those phrases that must be corrected.

For the sake of simplicity, we will report RT and S in words per second and megabytes respectively. Providing the respective complexities does not appear necessary here. As far as proofreading quality is concerned, we believe that the false positive rate (FP) *on checked text* [3] and the hit rate (HR) are most relevant for RWTD setting among existing measures. This is because these two rates together with the coverage directly reflect the time saved by the human proofreader and the percentage of typos detected.

As with any typo correction quality metrics, FP and HR are only meaningful if used against the same input texts. Furthermore, such metrics are posterior by definition. In other words, given a fresh new input text, it is impossible to calculate FP and HR without a human proofreading the entire text first. This shortcoming is extremely inconvenient when it comes to giving the proofreader a choice in determining a time slot size to cover a given typo percentage (or vice versa). To counter this shortcoming one can suggest first putting the new text into one of the categories for which approximate metrics values are already known. For example, the text can be categorized by an author or news reporter. An alternative approach to finding the required trade-off is specific to our solution and is discussed in Sec. 4.5.

3   A RWTD Method

The task of detecting *any* real-word typo is a difficult one. Correction-tied algorithms cope better in this regard because they only have to assume a relatively narrow class of typos in order to offer a legitimate automated correction. Once the class of typos is assumed, detecting typos and even measuring the success of an algorithm becomes much more defined.

Dropping the correction requirement, however, opens up many new avenues for exploration. A detection method does not have to follow the assumptions and restrictions made to adhere to the automated correction requirement.

The expected performance of a detection-only method shall therefore, on the one hand, lag behind the success of the correction methods that are focused on detecting specific classes of typos. On the other hand, a detection-only method aimed at finding any and all typos does have a chance of outperforming the correction-tied methods.

We hypothesize that once we build a collocation database on a large enough training corpora, we might be able to draw conclusions about the probability of a specific collocation of words containing a real-word typo. We would then sort all of the collocations in the input text in the order of this probability decreasing. Given the time a human is willing to spend proofreading the input text, we would provide him/her with the top-k pairs (and corresponding sentences, if needed) to check, and to correct if necessary. Alternatively, we could set the threshold and generate the top-k based on this threshold and not on the time available to the proofreader.

We intend to use raw (untagged) corpora to facilitate the construction of our collocation database which is to be based on diverse sources and for use with different languages. We also speculate that building collocations without POS discrimination may improve the quality of our algorithm. In this study we implement bigram collocations only.

3.1 Typo Likelihood Definition

Our algorithm shall determine the approximate probability of a pair of words $< w_1 \ w_2 >$ containing a real-word typo. The algorithm has the following input parameters:

1. Number of appearances of w_1 and w_2 in training corpora, denoted as C_{w_1} and C_{w_2}
2. Number of collocations of $< w_1 \ w_2 >$ in training corpora, denoted as $C_{w_1 w_2}$

Note that for the purpose of sorting the pairs, the algorithm does not have to output a probability. The output only has to be suitable for sorting the pairs in the order of their decreasing typo probability. We argue that the following formula provides such an output, albeit approximate:

$$TL_{w_1 w_2} \stackrel{def}{=} C_{w_1} \cdot C_{w_2} / C_{w_1 w_2} \tag{1}$$

We expect $TL_{w_1 w_2}$ to be larger the higher the likelihood is of the pair $< w_1 \ w_2 >$ containing a typo. We will later evaluate the performance of this measure of typo probability in the experimental section.

Informally, the higher the frequencies of the words, and the lower the frequency of these two words collocating, the greater the likelihood that we face a typo or a bad word choice. Formally, our TL measure can logically be deduced from the mutual information concept [5]. We offer somewhat alternative reasoning for the TL definition in the next section.

Special attention has to be drawn to the cases where the number of colloca-tions $C_{w_1 w_2}$ is zero and Eq. 1 becomes invalid. One approach is to assign $C_{w_1 w_2}$ a number between 0 and 1. Another approach is to presume $C_{w_1 w_2} = 1$ and to handle such pairs separately, i.e., such pairs are put aside and sorted and top-k_0 collocations are marked as "detected" in addition to the top-k collocations where $C_{w_1 w_2} > 0$.

3.2 Alternative Reasoning for TL Definition

We offer another way of looking at the definition of TL (Eq. 1) to see that it linearly depends on the probability of two words coinciding based on word frequencies divided by the probability of two words coinciding based on word pair frequencies. Let us denote the total word count in the corpora as $\mathcal{W}$, and the total bigram count as $\mathcal{B}$. Now the probability P_a of the $< w1\ w2 >$ pair based on the C_{w1}, C_{w2}, and CA is calculated as:

$$P(w_1) = C_{w_1}/\mathcal{W} \tag{2}$$
$$P(w_2) = C_{w_2}/\mathcal{W} \tag{3}$$
$$P_a = P_{<w_1 w_2>} = P(w_1) \cdot P(w_2) = C_{w_1} \cdot C_{w_2}/\mathcal{W}^2 \tag{4}$$

Probability P_b of the $< w1\ w2 >$ pair based on its count $C_{<w_1 w_2>}$ in the corpora and $\mathcal{B}$ is:

$$P_b = P_{<w_1 w_2>} = C_{<w_1 w_2>}/\mathcal{B} \tag{5}$$

Our definition of TL basically states that the less P_b is than P_a, the more we suspect $< w1\ w2 >$ of containing a typo:

$$TL = P_a/P_b = \frac{C_{w_1} \cdot C_{w_2} \cdot \mathcal{B}}{C_{<w_1 w_2>} \cdot \mathcal{W}^2} \tag{6}$$

Because $\mathcal{W}$ and $\mathcal{B}$ are constants, the $\mathcal{B}/\mathcal{W}^2$ part of TL is constant for all words and pairs, and so can be omitted, thus making Eq. 6 become Eq. 1. We provide this alternative explanation of Eq. 1 because it allows for a more transparent expansion of the approach for triples and quadruples.

3.3 Potential Disadvantages

There are at least two disadvantages of this method that can be observed prior to experimentation.

1. Pairs consisting of very infrequent words will not be in the top-k if they constitute a real-word typo. There are two arguments to remedy this flaw. Firstly, the probability of two words of very low frequency collocating is lower than the three alternatives: only one of the words being very infrequent , and the two words being relatively frequent. Secondly, we believe this problem diminishes as the collocation database grows.
2. We have not yet prototyped trigram and fourgram control. Thus, the real-word typos that do not introduce irregularity in any two consecutive words will not be detected. However, our experiments indicate that such typos are in the minority.

3.4 Implementation

The collocation database may be created as a flat file matrix $M_{N x N}$, where N is the size of the dictionary of the training corpora. Each cell $< i, j >$ in the matrix stores a single integer value – the number of collocations of a pair w_i w_j in the training corpora. The matrix structure enables easy address resolution for any i, j, wherein the position of the required collocation value in the file is calculated as $(N \cdot i + j) \cdot sizeof(int)$. However, the hard disk space required is quadratic in the size of the dictionary. In one of our experiments, the dictionary of our training corpora (of 600 million words) contains 4.5 million words. This implies 20 Terabytes of disk space required for storing the collocation database assuming a four byte integer size.

The SQL database approach can reduce the disk space requirements tremendously. This is because storing the zero value tuples corresponding to zero collocations is unnecessary, and zero values are undoubtedly in the majority. Without delving into discussion about the possible table structures of an SQL database, we assume that for each pair of words at least one SQL query will be executed. Based on this assumption, we speculate that one SQL query per word constitutes an unacceptably slow processing speed for most of the applications outlined above.

In our implementation we combined the simplicity and low overhead of the flat file approach and the flexible indexing of a database approach. We do not store zero collocation values to save on disk space, and we have an index data structure that points to a sorted list of all the collocations of a specific word within a flat file. Because the list is sorted, finding a specific collocation value is logarithmic in the number of distinct collocations of a given word. Employing the collected knowledge about the distribution pattern of the collocations for each word and using memory file mapping further decreases the number and duration of disk accesses.

4 Experiments

We conducted experiments on real-world English and Russian texts. Exactly the same code was used to produce collocation databases and detection results for both English and Russian. English test texts are of two types: Brown corpus that dates back to 1961, and top quality news articles, most published in the second half of 2008. Russian test texts are two relatively large novels dated 1992 and 2008. We outline some preliminary statistics in the following section and some of our reporting presumptions in [6] in Sec. A.

Collocation data. The texts for gathering the collocation data were collected in an automatic manner from the Internet. We tried to have the corpora compiled mostly of books in txt and html format. Project Gutenberg (English texts) and Lib.ru (Russian texts) are two examples of extremely large book collections available on the Internet. The decision to focus on online books rather than random web pages was influenced by time and bandwidth limitations. Downloading bundled collections of books takes much less time and requires much less open connections compared to page crawling. An additional secondary reason for using books was acquiring better text quality compared to that of an average web page. Our hope was that the lower typo rate of the collocation data will improve the quality of experimental results. While it remains to be proven, we now speculate that the quality of the text does not appear to have a significant influence on the efficiency of the method because the relatively common typos contained even in our high-quality collocation data experimentally scored as high as some unique typos.

The English corpora consists of 461 million (m) words, and the dictionary size is 2.31m words. The collocation database was collected from 387m word pairs, resulting in 29.8m distinct word pairs. The Russian corpora consists of 598m words, and the dictionary size is 4.57m words. The collocation database was collected from 436m word pairs, resulting in 80.9m distinct bigrams.

4.1   High-Quality News

There are two reasons behind our choice of prime news as a test corpus. The first is that we are interested in understanding the performance of our algorithm for high-quality, low-level typo rate texts. The second reason is that we are non-native speakers, so proofreading a large volume of foreign text is a problem. However, we frequently read English news sources, so reading approximately 60–70 news articles in English for this experiment was not a task that required an enormous amount of extra time. Simply running our prototype on hundreds of news articles and selecting those with typos detected would certainly have saved a lot of time. However, this would not have been a clean-room experiment as only those typos that could be detected by the method would have found their way into the experimental results. Therefore, we started reading well before the prototype was ready.

From mid August 2008 to mid January 2009 we read top news articles on topics ranging from economics to politics to IT, finding most of them on the Google News front page. We collected all the news articles in which we found typos. As a result, we were left with 9 news articles (Table A in [6]) with 10 typos.[2]

These 9 news articles contain 9753 words, not counting punctuation marks. There are 8224 word pairs, of which 6664 (81%) have a non-zero count in the collocation database, 1368 (17%) have zero count in the collocation database, and 192 (2%) have one or two words not found in the dictionary.

Initially, we ran experiments on each of these news articles separately to demonstrate the TL scores and positions gained by each of the typos (Table B

---

[2] Unfortunately, we did not count the number of articles in which we found no typos. However, we believe that for approximately every 6–8 top news articles there was one with a typo. At the same time, we believe the typo rate may be twice as high as this as we may have failed to detect every second typo. We believe that further research into the actual typo rate of the top quality magazines and newspapers would make an interesting future work.

Table 1. HR and FP for different thresholds and coverages for 9 news as one text

(b) Coverage 98%		
$TL \cdot 10^{-9}$	HR	FP
0.9, 0.5	90%	460+151 (7.6%)
0.9, 5.58	80%	460+13 (5.9%)
2.4, 0.5	70%	167+151 (4.0%)
2.4, 5.58	60%	167+13 (2.2%)
5.6, 0.5	60%	62+151 (2.7%)
5.6, 5.58	50%	62+13 (0.9%)
16, 0.5	50%	16+151 (2.1%)
16, 5.58	40%	16+13 (0.4%)
114, 0.5	40%	151 (1.9%)
114, 5.58	30%	13 (0.0%)

(a) Coverage 81%		
$TL \cdot 10^{-9}$	HR	FP
0.7	100%	667 (10.0%)
0.9	88%	460 (6.9%)
2.4	63%	167 (2.5%)
5.6	57%	62 (0.9%)
16	38%	16 (0.2%)
114	25%	0 (0.0%)

in [6]). However, presenting the hit rate in this experiment is not a straight-forward task due to the fact that most texts contain only one typo, resulting in almost imperceptible (0% or 100%) HR. To get more useful statistics, we combined 9 news articles to form one test text containing 10 typos and ran the prototype on it.

Table 1a demonstrates the HR and FP for the case where only bigrams presented in the collocation database are covered. Table 1b shows the results for the case where bigrams not present in the collocation database are covered as well, with the presumption of $C = 1$. In Table 1a each row corresponds to a specific TL threshold, while in Table 1b rows display two TL thresholds: one for the top-k list and another for the top-k_0 list. The rows are ordered by thresholds.

4.2   Brown Corpus

HR and FP are calculated and reported for the news articles and novels, and are not calculated for Brown corpus. In order to calculate these values for Brown corpus, the corpus needs to be thoroughly proofread by a human, and proofreading such a large volume of text is unfortunately impossible for the authors of this study. Nevertheless, the real-word typos that we found using our method provide an argument against using Brown as the golden standard in typo-detection experiments. Another reason for reporting *some* typos in this study is to observe the wide range of typo classes detected by the method (Table D in [6]).

While we selectively report only 30 typos, we approximate the total number of real-word typos in Brown corpus to be over 100 at the very least. This actually corresponds to the typo rate order of magnitude that we estimated for prime news articles.

4.3   Russian Novels

We read two novels in Russian to find all the typos in the text. The choice of these novels was based purely on our own personal reading habits, and so the selection can be assumed to be random. Neither of these novels is in the

Table 2. HR and FP for different thresholds and coverages for the two novels

TL·10⁻⁹	\(\text{novel 1}\)		\(\text{novel 2}\)	
	HR	FP	HR	FP
C>0	(a) coverage 74%		(b) coverage 87%	
1.9	100%	2467 (5.4%)	83%	1566 (2.7%)
10	71%	342 (0.8%)	75%	136 (0.2%)
31	43%	89 (0.2%)	75%	14 (0.02%)
C≥0	(c) coverage 97%		(d) coverage 99%	
1.9,1.5	92%	2467+531 (5.0%)	83%	1566+305 (2.8%)
1.9,10	75%	2467+89 (4.3%)	83%	1566+33 (2.4%)
10,1.5	75%	342+531 (1.5%)	67%	136+305 (0.7%)
10,10	58%	342+89 (0.7%)	67%	136+33 (0.3%)
31,1.5	58%	89+531 (1.0%)	67%	14+305 (0.5%)
31,10	42%	89+89 (0.3%)	67%	14+33 (0.1%)

collocation database. The first text appears to have been produced by a typist, while the second text appears to be an OCR-ed version of the published book (unfortunately, we do not know for sure).

The first novel contains 12 typos (Table E in [6]), 94117 words, and 61836 word pairs of which 45502 (74%) have a non-zero count in the collocation database, 14341 (23%) have a zero count in the collocation database, and 1993 (3%) have one or two words not found in the dictionary.

The second novel contains 19 typos (Table F in [6]), 96875 words, and 65929 word pairs of which 57532 (87%) have a non-zero count in the collocation database, 8208 (12%) have a zero count in the collocation database, and 189 (less than 1%) have one or two words not found in the dictionary.[3] HP and FP rates are presented in Table 2.

4.4 Analysis of Typos in the Experimental Texts

The confusion class of typos are obviously in the minority. However, it can be argued that the professional writers of the texts used in the experiments rarely confuse words. The outliers of typos in the second novel are quite insightful regarding the potential weaknesses of the solution. Potentially, some may be corrected by implementing trigrams and by increasing the volume of collocation data, but some cannot.

Typo number 20 of novel 2 (Table F in [6]) went completely undetected by the method. Although this was unexpected, it can be explained by the fact that bigrams separated by punctuation marks were not considered as such. At least two approaches can be further investigated to counter such rare cases. A trigam implementation may help deal with the punctuation mark in between

---

[3] The considerable difference in zero counts between the two novels can be explained by the fact that the first novel was written in 2008, while the second one dates back to at least 1992. The 1992 novel is thus using relatively conservative language resulting in relatively low zero count and not-in-dictionary hits.

the two words. Alternatively, the two words can still be checked by removing the punctuation mark altogether. The assumption is that the resulting bigram will produce a higher TL if there is a non-punctuation typo present, and a lower TL if only a punctuation mark is of importance in the given case.

A rough comparison with Microsoft Proofing Tools [7] can be found in Sec. D in [6]. A performance comparison with related research work is discussed in Sec. 5.3 and 5.4.

4.5 Fixing Trade-Off Based on TL Distribution

We report the distribution of TL values for the tested texts in Figure 1. The reason for reporting these distributions is twofold. Firstly, the graph visually demonstrates how TL values for collocation cases $C > 0$, $C = 0$ behave similarly in that only tiny intervals exhibit high TL values. Secondly, the distributions are given to support the argument that in addition to fixed TL thresholds and the time limit available to the human proofreader there is a third approach to determine the best suspect list cut-off point. Namely, the cut-off can be calculated based on the TL distribution for the given input text. For example, the user may decide to consider only the steep upper parts of the curves.

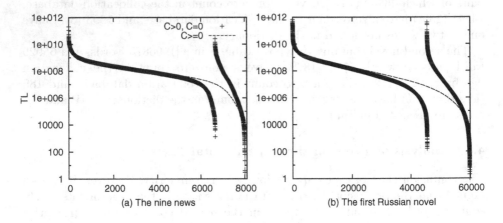

Fig. 1. The word pairs sorted by TL values

4.6 Varying Database Size

Extensive time-consuming investigation is required to determine the exact effect the size of a collocation database has on the quality of results. However, we feel obliged to report several preliminary observations based on one isolated experimental result gained from varying the collocation database size. Tables 3a and 3b report the TLs, FP and HR rates for the nine news articles as an input text and a collocation database that is approximately 50 times smaller than the one previously described. Two hypotheses can be drawn by comparing these tables with Tables 1a and 1b. Firstly, increasing the database size seems to

Table 3. HR and FP for different thresholds and coverages for 9 news as one text (reduced database size)

(a) Coverage 62%		
TL·10^{-6}	HR	FP
18	100%	257 (5.0%)
23	86%	205 (4.0%)
42	71%	95 (1.9%)
53	57%	75 (1.5%)
61	43%	65 (1.3%)
85	29%	39 (0.8%)

(b) Coverage 93%		
TL·10^{-6}	HR	FP
18, 3.9	90%	257+355 (8.0%)
18, 500	80%	257+10 (3.5%)
42, 3.9	70%	95+355 (5.9%)
42, 500	60%	95+10 (1.4%)
85, 3.9	40%	39+355 (5.1%)
85, 500	30%	39+10 (0.6%)

improve the FP rates for small hit rates, while slightly worsening FP rates for large hit rates. Secondly, and perhaps surprisingly, overall performance does not seem to suffer as a result of a database size reduction. Further experimentation is required before we may confirm or dismiss these hypotheses.

4.7   Execution Time and Space

The experiments were run on a workstation equipped with an Intel Core 2 Duo E7300 (2.66 GHz) processor, 4 Gb of memory, 1 Tb 7200 RPM 32 Mb cache SATA-II Seagate hard drive, Microsoft XP Professional OS, and Java SDK v.1.6.

We measured the time it took to load the dictionary and the collocation data into the memory of the computer (a one off process). This time is approximately 50 seconds (s) for English and 60s for Russian. The disk space required for the dictionary and collocation data is 740Mb for Russian and 330Mb for English. We measured the time required to retrieve and output word pairs and associated word and collocation counts, and to calculate TL. For Brown corpus it took 46.6s to process and output all 925401 bigrams, for the 9 news articles as one text the time was less than 0.3s. These times correspond to 0.04-0.05 milliseconds per bigram. The Java code was run with a 700Mb heap size limit for both English and Russian. An additional 400–800Mb were reserved by OS and JVM to support collocation file memory mapping. A version of the prototype without memory mapping and no additional memory requirement processes the bigrams approximately 30–50 times slower. The difference is less than that between RAM and hard disk access times due to the fact that OS and JVM are caching some disk reads anyway. It appears impossible to force JVM and OS to turn off internal read caching.

5   Related Work

Most of the related research is based on confusion sets. Whenever a phrase belonging to one of the confusion sets is found in the text it is suspected of being a typo. Phrases from the related confusion set are automatically sorted in the order of the likelihood of a phrase being a typo in this specific context. Then the phrase from the confusion set that has the lowest likelihood of being a typo is selected. If the selected phrase is not the one that is currently being parsed, a typo is declared to be found and is automatically corrected by replacing the current phrase with the selected phrase. The major differences between various approaches lie in the mechanisms used to sort the phrases in a confusion set based on the context, as well as in the mechanisms used for training based on a large corpus.

The second category of related work, of which there is very little, is connected with methods that do not rely on confusion sets. Both categories seem to consider the typo problem purely in the context of subsequent automatic correction. Thus, the name commonly used for this field of research is context-sensitive spelling *correction*.

Due to space limitations, we have compiled this section in the following concise manner. After making reference to papers with good overviews and bibliographies [8,2], we will (i) focus on the general potential shortcomings of related algorithms without delving into any particular study; (ii) discuss only two recent studies that, in our opinion, are the most relevant to our study. Other examples of related studies include but are not limited to [9,10].

5.1   General Shortcomings

Confusion sets are the most commonly used instrument in the field [2]. The only apparent advantage of this instrument is that it is easy to compare these studies because many use the same 18–22 confusion word sets. However, there are many disadvantages. First and foremost, confusion typos are a relatively narrow class of typos, and many other classes get past the algorithms undetected. Even if we stick with the confusion set setting, the scalability of many algorithms will be unclear if the number of confusion sets and the number of members in each grows by one or two orders to demonstrate the practical applicability of the algorithm in question. Furthermore, increasing the size of confusion sets is, in itself, not a straightforward task.

We would also like to point to other common shortcomings, independent of those of confusion sets. Many studies take Brown or another corpus as the golden standard for experiments. As we saw above, such an approach will never allow us to achieve (a true) 100% HR rate. Furthermore, such approaches to experimentation may cause researchers to focus too much on fine-tuning their algorithm to perform best on the "golden" standard. Another common approach is to insert artificially compiled typos into the text which is a well-recognized pitfall in the field which lacks an easy solution. Bearing in mind the above, we were pleased that the nature of our method excluded the possibility of us working with either technique. We could not use Brown as a golden standard because our algorithm would have detected the real-word typos in it, and we did not have to insert typos because our method detected typos even in high-quality low-typo-rate texts.

5.2   Most Related Research

Bolshakov et al research [5,11] comes closest to the described approach, employing word pair probability estimation as one of its instruments. Moreover, our colleagues have foreseen the need for creating a local collocation database from large corpora, and then deriving word pair probabilities from it. For the sake of brevity, we will refrain from a full description of these studies here and will instead focus solely upon what makes our approach different.

In [11] our colleagues have focused entirely on detecting and correcting malapropisms, that is, real words that sound similar to the correct word but differ in spelling. In [5] the focus is on erroneous collocations involving a "noun and its own modifier". Other classes of real-word typos are neither accounted for nor experimented with. The word pairs with a zero collocation count are considered errors by definition, which differs from our somewhat lenient treatment. Another difference is that the outlined approaches require or presume the gathering of some prior information about the language, such as a list of nouns and their modifiers, whereas our approach requires no information about a language whatsoever, except perhaps knowledge of its punctuation marks. In our opinion, these related works would gain a lot by departing from fixed thresholds and providing more flexibility for trade-offs, where different false positive rates are offered for respective hit rates by varying the thresholds of the algorithms.

Fossati et al [12,3] suggest that a confusion set for any word might be compiled by applying an editing function (letter insertion, deletion, and substitution) to the word and storing the resulting real-words in the set. After producing relatively large confusion sets in such a manner, the results of experiments with the so called mixed trigram approach are reported. While this work is not closely related to ours algorithmically, it is discussed here due to its relatively broad coverage of different classes of typos and the availability of FP and HR statistics. Both these facts make it easier to make approximate performance comparisons below.

5.3   A Comparison Based on False Positive and Hit Rates

The proposed method brings the false positive rates on checked text down from 19–25% in [3] to 2–6%, while keeping the coverage and hit rate percentages in the eighties – nineties. Sub-percent false rates are achieved by decreasing the hit rate to 50–60%. However, the method tackles an almost unlimited range of typos and has been tested on real-world texts with real-world typos. This fact makes it difficult to draw FP rate comparisons with methods detecting limited classes of typos artificially inserted into texts.

5.4   Precision, Recall, and F-Measure

We think this metrics falls short of being effective in professional proofreading environment. Namely, it lacks the trade-off between the time spent proofreading and the percentage of typos detected.

Furthermore, this metrics depends heavily on the typo rate. In other words, increasing the typo rate (by artificially inducing typos, for example) boosts the output of the metrics, and vice versa. We will demonstrate this point below.

We will take one cell from Table 2c to calculate P, R, F-measure as an example[4]: $R = HR = 75\%$, $FP = 342 + 521 = 873$, $P = 9/(9 + 873) = 0.01$, $F = 0.02$. The tuple {P, R, F}={0.75,0.01,0.02} is not a good result compared to the related work ([12] being one example).

Let us assume that we populate the existing typos within the text from 12 to 1200. This results in 1 typo per 80 words, which is close to the typo rates tested in related work. Populating typos will not change HR, but will increase precision and F-measure: $P = 900/(900 + 873) = 0.5$, $F = 0.6$. Now the tuple {0.75,0.5,0.6} is very competitive.

Since the typo rate we consider in this work differs greatly from the related work it appears that comparing the results in the PRF space does not make much sense.

6   Conclusion and Future Work

We proposed a new approach to real-word context-sensitive spelling correction that relies on neither confusion sets nor part of speech tagging. The approach does not offer automatic correction, but rather focuses solely on typo detection. Once detected, the typos can be corrected either by a human proofreader or by one of the numerous solutions offered in related research.

This study offers fresh scope for future work. For instance, will using POS tagging on top of the given approach improve results, or does the method partially owe its success to neglecting POS discrimination? It is natural to try to extend the approach to other languages. Our results are somewhat better in Russian than in English and the Russian collocation database is larger than English one. It would therefore be logical to investigate to what extent increasing the size of the databases would improve results. We have demonstrated the applicability of the method to top quality news articles as well as OCR-ed and typed books. Thus, yet another area for future research would be to validate and optimize this approach for other applications, such as speech recognition, manual and automatic translation.

Acknowledgments. Alexander Gelbukh and Pavel Velikhov, as well as anonymous reviewers provided many insights into how to improve the presentation.

References

1. Golding, A.R., Roth, D.: A winnow-based approach to context-sensitive spelling correction. Machine Learning 34(1) (February 1999)
2. Wilcox-O'Hearn, L.A., Hirst, G., Budanitsky, A.: Real-word spelling correction with trigrams: A reconsideration of the mays, damerau, and mercer model. In: Gelbukh, A. (ed.) CICLing 2008. LNCS, vol. 4919, pp. 605–616. Springer, Heidelberg (2008)

---

[4] Similarly, PRF can be calculated for all our experiments based on the data already presented.

3. Fossati, D., Eugenio, B.D.: A mixed trigrams approach for context sensitive spell checking. In: Gelbukh, A. (ed.) CICLing 2007. LNCS, vol. 4394, pp. 623–633. Springer, Heidelberg (2007)
4. Reynaert, M.: All, and only, the errors: more complete and consistent spelling and ocr-error correction evaluation. In: Proceedings of the Sixth International Language Resources and Evaluation (LREC 2008), Marrakech, Morocco (2008)
5. Bolshakov, I.A., Bolshakova, E.I., Kotlyarov, A.P., Gelbukh, A.F.: Various criteria of collocation cohesion in internet: Comparison of resolving power. In: Computational Linguistics and Intelligent Text Processing, Haifa, Israel (2008)
6. Asonov, D.: Real-word typo detection: Supplementary material (2009), http://www.fastpl.com/pubs/nldb09supm.pdf
7. Hirst, G.: An evaluation of the contextual spelling checker of microsoft office word 2007 (2008)
8. Mitton, R.: Spellchecking by computer. Journal of the Simplified Spelling Society 20(1) (1996)
9. Lapata, M., Keller, F.: Web-based models for natural language processing. TSLP 2(1), 1–31 (2005)
10. Morris, R., Cherry, L.L.: Computer detection of typographical errors. IEEE Transactions on Professional Communication 18(1) (1975)
11. Bolshakova, E., Bolshakov, I., Kotlyarov, A.: Experiments in detection and correction of russian malapropisms by means of the web. International Journal Information Theories and Applications 12 (2006)
12. Fossati, D., Eugenio, B.D.: I saw tree trees in the park: How to correct real-word spelling mistakes. In: Proceedings of the Sixth International Language Resources and Evaluation (LREC 2008), Marrakech, Morocco (2008)

The Impact of Semantic and Morphosyntactic Ambiguity on Automatic Humour Recognition*

Antonio Reyes, Davide Buscaldi, and Paolo Rosso

Natural Language Engineering Lab - ELiRF
Departa mento de Sistemas Informáticos y Computación
Universidad Politécnica de Valencia, Spain
{areyes,dbuscaldi,prosso}@dsic.upv.es

Abstract. Humour is one of the most amazing characteristics that defines us as human beings and social entities. Its study supposes a deep insight into several areas such as linguistics, psychology or philosophy. From the Natural Language Processing (NLP) perspective, recent researches have shown that humour can be automatically generated and recognized with some success. In this work we present a study carried out on a collection of English texts in order to investigate whether or not semantic and morphosyntactic ambiguities may be employed as features in the automatic humour recognition task. The results we have obtained show that it is possible to discriminate humorous from non humorous sentences through features like perplexity or sense dispersion.

1   Introduction

Humour is one of the most amazing characteristics that defines us as human beings and social entities. Most people reacts to an amusing stimulus with humour, regardless of their cultural differences, beliefs or social status. However, this stimulus which makes people laugh is an *ethereal* object that can hardly be generalized for everybody. Cognitive aspects as well as cultural knowledge, for instance, are some of the huge and complex aspects that must be analysed in order to get some explanations about how humour works and what makes us laugh. Despite the huge effort that the study of humour implies, many disciplines such as philosophy, psychology or linguistics have attempted to know its basic features in order to explain it through different methodologies. A good recent example of this kind of efforts is Attardo's work (1994, 2001), which tries to characterize humour by means of linguistic patterns. On the basis of linguistic layers, specially semantic and pragmatic ones, Attardo tries to explain verbal

* We would like to thank Rada Mihalcea and Carlo Strapparava for kindly sharing their one-liners corpus. The MiDEs (CICYT TIN2006-15265-C06) and TeLMoSis (UPV PAID083294) research projects have partially funded this work. The National Council for Science and Technology (CONACyT - Mexico) has funded the research work of Antonio Reyes.

H. Horacek et al. (Eds.): NLDB 2009, LNCS 5723, pp. 130–141, 2010.

humour[1] as a phenomenon that supposes some elements or *knowledge resources* such as language, narrative strategies, target, situation, logical mechanism and opposition, which combined may, with some probability, produce humour[2].

From a psychological perspective, the research performed by Willibald Ruch (2001) has shown how humour appreciation is linked to the personality and how this feature has an important weight which directly affects the kind of stimulus that produces humour. On the technological viewpoint, humour has gained increasing importance in the recent years. Researchers working in areas like computational linguistics, artificial intelligence or NLP have tried to analyse what kind of features are related to humour and how to understand and simulate them in order to automatically generate and recognize it. These tasks imply a serious challenge: it is not matter of knowing the linguistic information about words, but going beyond it and being able to identify, through their senses and internal and external referents, the mood of a person, her/his way of thinking or, as in our case, if s/he is expressing humour.

Some research works about the computational processing of humour have shown that it may automatically be generated and recognized with success: Binsted (1996); Stock and Strapparava (2005); Mihalcea and Strapparava (2006a, 2006b); Mihalcea (2007); Mihalcea and Pulman (2007); Sjöbergh and Araki (2007); Buscaldi and Rosso (2007). Therefore, on the basis of some assumptions and results derived from these works, we analyse how semantic and morphosyntactic information could provide valuable features to recognize humour.

The rest of the paper is organized as follows. Section 2 describes the state of the art of computational humour. Section 3 underlines the importance of the ambiguity in humour and presents our assumptions on the role of morphosyntactic and semantic ambiguity and the aim of our research. Section 4 describes the data sets and all the experiments we carried out. In Sect. 5 we discuss the results obtained and the implications they suggest. Finally, in Sect. 6, we draw some conclusions and address further work.

2   Computational Humour

As noted by Buscaldi and Rosso (2007), "the nature of humour is elusive, it is expressed in many different forms and styles" and is, in almost all senses, subjective. This assertion is not surprising, after all, humour is a characteristic as complex as the human behaviour. Therefore, properties such as the previous ones turn humour in a fuzzy attribute that can hardly be addressed by computational systems. However, recent approaches from the NLP area have shown that it is possible to find some patterns which define data as humorous. Through machine learning techniques, Mihalcea and Strapparava (2006a, 2006b); Mihalcea (2007);

---

[1] Verbal humour refers to the kind of humour that is generated by linguistic strategies, i.e., by language.

[2] On the subject, cf. Attardo's research work about the General Theory of Verbal Humour, and that one performed by Raskin (1985) about the Semantic Script Theory of Humour.

Mihalcea and Pulman (2007); Sjöbergh and Araki (2007) and Buscaldi and Rosso (2007), have demonstrated how some features have been employed to differentiate between humorous and non humorous sentences in order to automatically generate and recognize humour. Both in Automatic Humour Recognition (AHR) and Automatic Humour Generation (AHG) subareas, the results obtained show that it is feasible to implement computational models to solve the problem of the automatic humour processing. For instance, in AHG the research has focused on the study of specific features that could be handled by a computer in order to simulate them and generate a humorous output. A sample about the generated outputs is the sentence (a) (Binsted, 1996):

(a) What do you use to talk to an elephant? An elly-phone.

In this sentence we can see how structural features, codified through linguistic information, are used to automatically generate a humorous effect. Analysing the sentence (a), it can be noted that *elly-phone* has a phonological similarity with telephone. Moreover, *elly-phone* is related, phonologically and "semantically", to the word which gives its right meaning: elephant. This kind of funny question answering structure, named *punning riddles*, uses information like the one pointed out above in order to produce an amusing sentence. Binsted (1996) and Binsted and Ritchie (1997, 2001) found out that features such as the one above may be simulated by rules to automatically generate sentences like (b):

(b) What do you call a depressed engine? A low-comotive.

Another sample about the features that can be identified to represent humorous data appears in the research carried out by Stock and Strapparava (2005). They have shown how to generate humour combining words into existing templates exploiting incongruity theories. An example about the results of their investigation is the new funny sense for the acronym MIT (Massachusetts Institute of Technology) which appears in (c):

(c) Mythical Institute of Theology

In AHR the research works described in Mihalcea and Strapparava (2006a, 2006b); Mihalcea and Pulman (2007); Mihalcea (2007); Sjöbergh and Araki (2007) and Buscaldi and Rosso (2007), have focused on the analysis of some particular humorous structures: *one-liners* (OLs), in order to find out what features could define this kind of humour. The sentence (d) is an example of an OL extracted from Mihalcea and Strapparava (2006b):

(d) Infants don't enjoy infancy like adults do adultery.

Despite the fact that such OLs are short sentences with a very simple syntactic structure (Mihalcea, 2007; Mihalcea and Strapparava, 2006a, 2006b), its analysis supposes to learn more complex features in order to automatically recognize whether an input is humorous or not. As can be noted, the sentence (d) contains phonological information which helps to produce a humorous effect, but this is not everything. There is also a pun that plays an oppositional role between the meaning of the words. Together, they produce an amusing result. In their research work, Mihalcea and Strapparava described how they identified features that define a corpus of OLs and how, once learned, they are used to differentiate between OLs and proverbs or news. For instance, such features are: alliteration, antonymy and adult slang. Moreover, they found some semantic spaces that are potentials triggers of humour like: human centric vocabulary (e), negative orientation (f) or professional communities (g) (Mihalcea and Strapparava, 2006b):

(e) Of all the things **I** lost, **I** miss **my** mind the most.

(f) Money **can't** buy your friends, but you do get a better class of **enemy**.

(g) It was so cold last winter that I saw a **lawyer** with his hands in his own pockets.

According to Mihalcea and Strapparava's results (2006a, 2006b), irony, incongruity, idiomatic expressions, common sense knowledge and ambiguity, are some of the paths where it is possible to find new features for characterizing the humour produced by the OLs. This has been verified by the work of Sjöbergh and Araki (2007), whose aim was to recognize humour through the identification of syntactic and semantic features. Some of the ones they have detected are: similarity, style, idiomatic expressions and ambiguity. The research carried out by Buscaldi and Rosso (2007) has also pointed in that direction. Their goal was to study whether or not humorous quotes may be separated from non humorous ones through a set of features such as n-grams, sentence length or bag of words. They carried out some experiments on a corpus of Italian quotations[3] and their results showed how, employing features such as bag of words or sentence length, it was possible to discriminate humorous from non humorous sentences.

3   Ambiguity

Most of the works related to humour stress the importance of ambiguity in the production of the humorous effect. However, ambiguity has been one of the most difficult topics in NLP research. The results obtained are important for tasks such as POS tagging or parsing. But the question is: how to translate this knowledge for computational humour purposes, if humorous ambiguity covers a wide range of phenomena that, most times, goes beyond the word or the sentence? For instance, let us consider the sentence (h):

(h) Jesus saves, and at today's prices, that's a miracle!

There is no ambiguity at phonological, morphological or syntactic level. Humour is produced in the semantic and pragmatic level. The ambiguity relies on a sense dispersion combined with cultural facts which cause that the **figure** of

---

[3] This corpus is available for the AHR research community and can be downloaded from http://www.dsic.upv.es/grupos/nle/downloads.html.

the sentence: *Jesus saves*, becomes the **ground**[4], shifting the profiled sense of the sentence. This new profiling of roles generates ambiguity and, consequently, an humorous result. In other words, this sentence has two readings: the first one is related to the figure and logical sense about preserving someone from harm or loss. The second one shifts this reading from the logical sense related to a religious meaning to a ground sense related to economy. This sense is promoved as figure and the meaning of the entire sentence becomes funny. This kind of strategies, according to Mihalcea and Strapparava (2006b), leads surprise and create the humorous effect. Thus, our aim in this work is to study the ambiguity of a corpus of OLs in order to find out how, from different linguistic layers, we can tackle this problem for the AHR purposes.

3.1 Structural Ambiguity

As noted above, according to the results obtained by Mihalcea and Strapparava (2006b), the OLs, besides having a simple syntactic structure, use some predictable features such as rhetoric devices in order to get humorous effects. On the basis of these results, we aim at investigating how valuable information could be extracted from a measure such as perplexity which, according to Jurafsky and Martin (2007), "given two probabilistic models, the better model is the one that has tighter fit to the test data". Therefore, because the humour exploits a characteristic such as ambiguity for generating the humorous effect, our hypothesis is that OLs maximize this characteristic profiling some kind of structural ambiguity, which may be codified by a dispersion in the number of combinations among the words which constitute them. This behaviour functions as a trigger of potential humorous situations and should appear in a lower degree with negative examples. Thus, measuring the OLs perplexity with respect to a Language Model (LM), we expect to find some clues which indicate that there is ambiguity in a structural level. In Sect. 4.2 we show the results of the experiments we carried out in order to investigate whether perplexity may be considered a discriminative feature for AHR or not.

3.2 Morphosyntactic Ambiguity

In several research works it has been pointed out that ambiguity covers all linguistic levels: from morphological up to discursive level (Mihalcea and Strapparava, 2006b; Sjöbergh and Araki, 2007). Therefore, we also considered important the study of the impact that morphosyntactic ambiguity may have on AHR. On this subject, we think that the number of POS tags which a word can have within a sentence, may be an evidence about the different morphosyntactic functions that this word could play: thus, the hypothesis is that the morphosyntactic ambiguity could be useful for identifying potential humorous situations. Moreover, according to Sjöbergh and Araki (2007), it is known that "ambiguity is not only caused by word senses", phonological similarity or by their morphosyntactic

---

[4] cf. Langacker (1991) for a detailed explanation about the concepts employed in the Cognitive Grammar.

tags. Therefore, to supplement the morphosyntactic analysis, we aim at measuring, in terms of a syntactic parsing, how complex a sentence is. In Sect. 4.3, we investigate the issue about morphosyntactic tags and, in Sect. 4.4, we study whether or not information about sentence complexity could be useful in AHR.

3.3 Semantic Ambiguity

Ambiguity is closely related to the different meanings that a word, phrase or sentence may produce. As we already mentioned, in sentence (h) the trigger that enables the humorous sense is linked to semantic and pragmatic referents. This represents a great challenge for the NLP research. On this subject, both semantic and pragmatic levels constitute an important source of triggers of ambiguity and, thus, we aim at studying how to obtain valuable information from these layers. In Sect. 4.5 we study whether a simple feature such as the number of senses contained in the OLs, with respect to negative examples, could be important in order to represent the ambiguity of the data and to characterize humorous information. We also aim at investigating a more complex feature from the semantic layer: in Sect. 4.6 we analyse the sense dispersion. That is, we think that a factor that may cause humour is related to the ground sense of a word. If the ground sense is profiled, then the logical sense is broken and humour is produced. Therefore, we defined a measure based on the hypernym distance between synsets, calculated with respect to the WordNet ontology (Miller, 1995), in order to estimate the sense dispersion.

4 Experiments

We performed five kinds of experiments with respect to the study of the ambiguity in order to find out which features are the most important to identify potential humorous situations. The aspects we analysed were: Perplexity (PPL); POS tags; Sentence Complexity (SC); Mean of Senses and Sense Dispersion.

In the following subsections we describe the corpus we used for our experiments and the results obtained in each of them.

4.1 Corpus

Our data is divided in two sets. The positive set is the corpus of English OLs collected by Mihalcea and Strapparava. This corpus has 16,000 OLs automatically extracted from the web (Mihalcea and Strapparava, 2006a, 2006b). The negative examples were extracted from the English version of the *Leipzig Corpus* (LC) (Quasthoff, Richter and Biemann, 2006). This corpus is composed by one million of sentences extracted from the Associated Press, Financial Times, Wall Street Journal and OTS News. This corpus was randomly undersampled to 16,000 sentences.

4.2 Perplexity

In order to estimate the PPL we divided our positive and negative examples in two subsets: the training subset with 12,000 sentences and the test subset with 4,000. Using the SRILM Toolkit (Stolcke, 2002) we generated two LMs, one for each sample (positive and negative). Every LM was composed by trigrams and, to address the "poor estimates that are due to variability" (Jurafsky and Martin, 2007), we used interpolation and Kneser-Ney discount as smoothing. We obtained 2,628 trigrams for the positive sample and 2,053 for the negative one. The process performed to estimate the PPL was to compare the positive test subset against the negative LM and, the negative test subset, against the positive LM. The PPL we obtained appear in Table 1.

Table 1. Perplexity for OLs and LC

	Positive LM	Negative LM
PPL	1071,64	886,46
OOVs	35,934	66,910

As can be noted in this table, the sample whose PPL is higher is the positive one. This means that, given two different schemas of distribution, the structures which have a broader range of combinations are the OLs. This measure shows how, according to our initial assumption, the structure in the OLs are less predictable and, probabilistically, more ambiguous. Moreover, if we take into account the Out Of the Vocabulary words (OOVs), we will see how the number of OOVs is higher in the negative examples: almost the double than the positive ones. Therefore, based on this difference, we would expect a higher PPL in the LC, situation that does not occur. We think that this is a hint that indicates how the OLs are intrinsically more difficult to be classified.

4.3 POS Tagging

In order to investigate whether or not morphosyntactic ambiguity could provide useful information to recognize a humorous sentence, we carried out an experiment to know if the number of POS tags that a word could have within a sentence could be a feature to characterize the morphosyntactic ambiguity of our sets. The process we performed was to label both positive and negative sets using the TreeTagger software (Schmid, 1995). The only requirement was to label the sets with a threshold of probability of (0,3) in order to get all the possible tags that a word could have. The number of tokens for both sets are: 171,327 for the positive set and 228,522 for the negative one. The total tags that could be assigned to every token is shown in Table 2.

We can see in this table how, considering the number of tokens per set, the OLs have a wider range of possibilities to be tagged. This information strengthens the previous behaviour that the OLs follow a pattern of ambiguity in order

Table 2. Number of tags for OLs and LC

Tags	Positive Set	Negative Set
2	13,926	13,450
3	1,679	1,427
4	142	65
5	35	4
6	1	0

to produce humour. Thus, based on the information of this table, we can consider that, given a word w_i from the OLs set, there is more ambiguity in the morphosyntactic functions that a word can play (due to the fact that w_i may be noun, verb, adjective, etc., with higher probability than a word w_j from the LC). Therefore, this behaviour may be a feature that generates ambiguity and, consequently, a potential humorous situation.

4.4   Sentence Complexity

With this experiment we tried to know whether or not a measure such as SC which, according to Basili and Zanzotto (2002), "captures aspects like average number of syntactic dependencies", could be a feature to characterize the syntactic ambiguity. The underlying aim of the experiment was to know how complex the syntactic structures are in the OLs. The experiment was ran on all positive and negative sets. The software we used to generate the syntactic trees was the Chaos Parser of Basili and Zanzotto (2002). The SC was calculated according to the following formula:

$$SC = \frac{\sum V_L, N_L}{\sum Cl} \tag{1}$$

where V_L and N_L are the number of verbal and nominal links respectively, divided by the number of clauses (Cl).

The SC we obtained was (0,99) for the positive set and (2,02) for the negative one. This result indicates that there is not enough ambiguity in the syntactic structure of the OLs. Therefore, this information is not useful for the purpose of discriminating the positive from the negative samples.

4.5   Mean of Senses

The last two experiments are focused on the semantic layer. We aim at finding out how semantic information could provide good patterns for identifying the triggers of the ambiguous situations. As noted in Subsect. 3.3, it is important to find more complex features in order to characterize humorous data. The first experiment measures the number of senses for both positive and negative sets for the following morphosyntactic categories: noun (N), adjective (ADJ) and adverb (ADV). Then, we searched in WordNet (Miller, 1995) all the senses for each one

Table 3. Number of words per category

Category	Positive Set	Negative Set
N	3,466	3,300
ADJ	926	1,172
ADV	265	349

of these words. The total number of words per category is shown in Table 3. The mean of senses per word, according to its category, appears in Table 4.

As can be noted in Table 4, the words of the OLs, except for the ADJ category, seem to be more ambiguous due to the range of senses that they have. However, from a statistical viewpoint, the difference is not so significant. Thus, to be sure about this behaviour, we carried out another experiment whose goal was to analyse the dispersion of these senses calculating the distance among the senses of a word and their hypernyms.

Table 4. Mean of senses per category

Category	Positive Set	Negative Set
N	1,409	1,224
ADJ	1,169	1,169
ADV	1,104	1,051

4.6   Sense Dispersion

In order to know how much dispersed the senses of both positive and negative sets are, we measured the length among the synsets of each noun and their first common hypernym. For instance, the noun *killer* has four synsets[5]. Taking only into account the synsets s_0 and s_1 we obtain as first common hypernym *physical entity*. The number of nodes to reach this hypernym is 6 and 2, respectively. Thus, the dispersion of *killer* is the sum of those distances divided by 2. Now, considering all its synsets, we obtain six possible combinations whose distance among them and their first common hypernym generates a dispersion of 6,83. The formula we employed in order to obtain this distance appears in (2):

$$\delta(w_s) = \frac{1}{P(|S|, 2)} \sum_{s_i, s_j \in S} d(s_i, s_j) \tag{2}$$

where S is the set of synsets $(s_0, ..., s_n)$ for the word w; $P(n,k)$ is the number of permutations of n objects in k slots; and $d(s_i, s_j)$ is the length of the hypernym path between synsets (s_i, s_j) according to WordNet. The underlying assumption of this experiment is that a humorous effect can be produced by words that have very different meanings. For instance, let us examine the following header: *The*

---

[5] cf. WordNet v. 3.0.

assembly passed and sent to the senate a bill requiring dog owners in New York City to clean up after their dogs, in penalty of $100 fine. The bill also applies to Buffalo. The ambiguity of "Buffalo", that can be a city or a bison, is key to produce the humorous effect. However, if Buffalo's ambiguity was only related to two different types of bison, or two cities, the ambiguous effect would disappear. Therefore, we think that the degree of ambiguity is key to the humorous effect. A word with senses that differ significantly between them is more likely to be used to create humour than a word with senses that differ slightly. The dispersion measure we introduced is a way to quantify the difference between the senses of a word. The formula in (3) shows how the total dispersion per word is calculated:

$$\delta_{TOT} = \sum_{w_s \in W} \delta(w_s) \qquad (3)$$

where W is the set of nouns in the collection N. Finally, we estimated the average dispersion per word as: $\overline{\delta}_W = \frac{\delta_{TOT}}{|W|}$; whereas the one per sentence as: $\overline{\delta}_N = \frac{\delta_{TOT}}{|N_s|}$; being N_s the set of sentences of the collection. The results related to these average dispersions appear in Table 5.

Table 5. Dispersion per word

	Positive Set	Negative Set
$\overline{\delta}_W$	7,63	6,94
$\overline{\delta}_N$	1,85	2,25

In the following Sect. we analyse all the results obtained.

5   Discussion

According to Mihalcea and Strapparava (2006a, 2006b) and Sjöberg and Araki (2007), ambiguity is one of the sources which potentially produces humour. Therefore, we performed some experiments whose final goal was to analyse whether it could be possible to find some features in order to characterize the humour of the OLs taking into account linguistic information. On the basis of these premises, the results of each experiment suggest the following inferences.

The perplexity reported in Table 1 indicates that words of the OLs have a wider range of combination. That is, in terms of language models, it is more probable to predict the word $w+1$ given a word w which belongs to the negative examples (i.e., non OLs). This information points out a minor predictability in the positive set. Thus, on the basis of this fact, we think that this kind of dispersion is a focus of ambiguity that may produce humour.

The results of the POS tagging experiment have shown how the words of the positive set have a bigger probability to play different morphosyntactic functions. In Table 2 it can be noted how this dispersion, that we assume as ambiguity, is greater when a tag was assigned to a positive word than when it was assigned to a negative one. On the basis of this behaviour, we may infer that, given an isolated word w_i, the probability to be assigned to various categories can break a logical meaning producing ambiguity and, therefore, a humorous effects.

Our third experiment aimed at verifying whether the syntactic structure could provide useful information about the impact of the syntax on the humorous process. The results we obtained suggest that the sentences of LC are syntactically more complex than the OLs. This behaviour supports what Mihalcea and Strapparava (2006a, 2006b) have pointed out about the simple syntactic information of the OLs, and rejects our hypothesis about the SC as a feature to characterize the syntactic ambiguity. Therefore, on the basis of this result, we can consider that the ambiguity does not appear from a syntactic viewpoint because the OLs are well formed structures that exploit other kind of strategies, (for instance, semantic and pragmatic information) in order to produce humour.

The impact of the semantic information as a trigger to generate humour seems to be more relevant. According to the experiments of Subsect. 4.5 and 4.6, there are more elements that strengthen the assumption about that humour is produced employing semantic information that generates ambiguity. As can be noted in Table 4, the mean of senses for the OLs is a clue about the greater ambiguity in the ground meanings that the words of the positive set profile. This assumption is supported by the information given in Table 5, where it is shown, by one side, how the sense dispersion per word marks the difference between both sets of data and, by the other, the sense dispersion per sentence confirms the results obtained for SC. If SC is lower in the OLs corpus, it means that the ambiguity that produces the humorous meaning is not at syntactic level but at semantic and/or pragmatic level. On the basis of these results, we may infer that the use of the semantic strategies is relevant in the generation of hollows of ambiguity that contribute to produce the humorous meaning.

6   Conclusions and Further Work

In this paper we investigated one of the most important sources that generates humorous situations: ambiguity. We analysed, from different linguistic layers, whether the information that we obtained studying the ambiguity could be taken into account as a set of features to automatically recognize humour or not. Some of the results obtained confirmed our initial assumptions about the usefulness of this kind of information to characterize humorous examples, especially with respect to measures such as perplexity, mean of senses and sense dispersion. As final remark, we can say that the OLs contain elements that could potentially be considered as triggers of humour, i.e., the OLs are ambiguous but, according to our results, not from a syntactic viewpoint. This behaviour could be due to the fact that the OLs are syntactically well formed structures which exploit ambiguous referents related to words and not to the whole sentence. Summarizing, although the preliminary findings are interesting, they must be further tested with a classifier. We also plan, as further work, to integrate these features in a mixture model in order to identify their relevance in the AHR task.

References

Attardo, S.: Linguistic Theories of Humor. Mouton de Gruyter, Berlin (1994)

Attardo, S.: Humorous Texts: A semantic and pragmatic analysis. Mouton De Gruyter, Berlin (2001)

Basili, R., Zanzotto, F.: Parsing Engineering and Empirical Robustness. Journal of Natural Language Engineering 8(3), 97–120 (2002)

Binsted, K.: Machine humour: An implemented model of puns. PhD thesis. University of Edinburgh, Edinburgh, Scotland (1996)

Binsted, K., Ritchie, G.: Computational rules for punning riddles. Humor. Walter de Gruyter Co. 10, 25–75 (1997)

Binsted, K., Ritchie, G.: Towards a model of story puns. Humor 14(3), 275–292 (2001)

Buscaldi, D., Rosso, P.: Some experiments in Humour Recognition using the Italian Wikiquote collection. In: Masulli, F., Mitra, S., Pasi, G. (eds.) WILF 2007. LNCS (LNAI), vol. 4578, pp. 464–468. Springer, Heidelberg (2007)

Jurafsky, D., Martin, J.: Speech and Language Processing: An introduction to natural language processing, computational linguistics, and speech recognition, Draft of June 25 (2007)

Langacker, R.: Concept, Image and Symbol. In: The Cognitive Basis of Grammar. Mounton de Gruyter, Berlin (1991)

Mihalcea, R.: Multidisciplinary Facets of Research on Humour. In: Masulli, F., Mitra, S., Pasi, G. (eds.) WILF 2007. LNCS (LNAI), vol. 4578, pp. 412–421. Springer, Heidelberg (2007)

Mihalcea, R., Strapparava, C.: Technologies that make you smile: Adding humour to text-based applications. IEEE Intelligent Systems 21(5), 33–39 (2006)

Mihalcea, R., Strapparava, C.: Learning to Laugh (Automatically): Computational Models for Humor Recognition. Journal of Computational Intelligence 22(2), 126–142 (2006)

Mihalcea, R., Pulman, S.: Characterizing Humour: An Exploration of Features in Humorous Texts. In: Gelbukh, A. (ed.) CICLing 2007. LNCS, vol. 4394, pp. 337–347. Springer, Heidelberg (2007)

Miller, G.: Wordnet: A lexical database. Communications of the ACM 38(11), 39–41 (1995)

Quasthoff, U., Richter, M., Biemann, C.: Corpus Portal for Search in Monolingual Corpora. In: Proceedings of the fifth international conference on Language Resources and Evaluation, LREC, pp. 1799–1802 (2006)

Raskin, V.: Semantic Mechanisms of Humor. D. Reidel, Dordrecht (1985)

Ruch, W.: The Perception of Humor. In: Kaszniak, A. (ed.) Emotion, Qualia, and Consciousness, Tokyo, pp. 410–425 (2001)

Schmid, H.: Improvements in Part-of-Speech Tagging with an Application to German. In: Proceedings of the ACL SIGDAT Workshop (1995)

Sjöbergh, J., Araki, K.: Recognizing Humor without Recognizing Meaning. In: Masulli, F., Mitra, S., Pasi, G. (eds.) WILF 2007. LNCS (LNAI), vol. 4578, pp. 469–476. Springer, Heidelberg (2007)

Stock, O., Strapparava, C.: Hahacronym: A computational humor system. In: Demo Proc. of the 43rd annual meeting of the Association of Computational Linguistics (ACL 2005), pp. 113–116 (2005)

Stolcke, A.: SRILM - An Extensible Language Modeling Toolkit. In: Proc. Intl. Conf. Spoken Language Processing, Denver, Colorado (2002)

Semantic Approaches to Fine and Coarse-Grained Feature-Based Opinion Mining

Alexandra Balahur and Andrés Montoyo

University of Alicante, Department of Software and Computing Systems
Apartado de Correos, 99
03080 Alicante, Spain
{abalahur,montoyo}@dlsi.ua.es

Abstract. Feature-based opinion mining from product reviews is a difficult task, both due to the high semantic variability of opinion expression, as well as because of the diversity of characteristics and sub-characteristics describing the products and the multitude of opinion words used to depict them. Further on, this task supposes not only the discovery of directly expressed opinions, but also the extraction of phrases that indirectly or implicitly value objects and their characteristics, by means of emotions or attitudes. Last, but not least, evaluation of results is difficult, because there is no standard corpus available that is annotated at such a fine-grained level and no annotation scheme defined for this purpose. This article presents our contributions to this task, given by the definition and application of an opinion annotation scheme, the testing of different methodologies to detect phrases related to different characteristics and the employment of Textual Entailment recognition for opinion mining. Finally, we test our approaches both on the built corpus, as well as on an ad-hoc built collection of reviews that we evaluate on the basis of the stars given. We prove that our approaches are appropriate and give high precision results.

Keywords: opinion mining, emotion detection, polarity classification, Textual Entailment Recognition.

1 Introduction

Recent years have marked the beginning and expansion of the Social Web, in which people freely express and respond to opinion on a whole variety of topics. The analysis of this phenomenon showed that free opinion expression in blogs, forums and e-commerce sites led to profound changes, both at the level of individual users, as well as for companies, whose business intelligence strategies have been modified to accommodate this new "global" opinionated vision of markets and competitors.

While the growing volume of opinion information available allows for better and more informed buying decisions of the users and strategic measures of the companies, the quantity of data to be analyzed imposed the development of specialized Natural Language Processing systems that automatically extract, classify and summarize the opinions available on different topics.

H. Horacek et al. (Eds.): NLDB 2009, LNCS 5723, pp. 142–153, 2010.

Research in this field, of opinion mining (sentiment analysis), has shown opinion mining is a difficult problem and addressed it from different perspectives and at different levels, depending a series of factors. These factors include: *level of interest* (overall/specific), *querying formula* ("Nokia E65"/"Why do people buy Nokia E65?"), *type of text* (review on forum/blog/dialogue/press article), and *manner of expression of opinion* - directly (using opinion statements, e.g. "I think this product is wonderful!"/"This is a bright initiative"), indirectly (using affect vocabulary, e.g. "I love the pictures this camera takes!"/"Personally, I am shocked one can propose such a law!") or implicitly (using adjectives and evaluative expressions, e.g. "It's light as a feather and fits right into my pocket!").

Further on, different "objects" on which opinion is expressed require the use of distinct strategies for sentiment analysis. For example, when the object of the review is a movie, one can be interested in the overall sentiment expressed, as a recommendation to see the movie or not. If the object of the review is a washing machine, one will be interested in four or five different aspects – such as durability, quality, easy use, price and size and most people will not be interested in the other details. In the case of a laptop or camera, nonetheless, more fine-grained information will be requested from a user at the time of mining for opinions. The most difficult problem of all consists in determining, at the time of mining a review for opinions, the characteristic assessed. There is a whole variety of linguistic possibilities to express, for example, the fact that a washing machine is small, starting with directly valuing statements and ending with the description of installing it the corner of a studio kitchenette.

On the other hand, at the time of implementing the different opinion mining strategies, we are confronted with different issues: lack of annotated corpora, lack of an annotation scheme that is able to detect the different means of opinion expression, lack of methods to automatically map the discussed characteristics of a product onto a taxonomy-like structure and finally, a method to evaluate the results obtained by an opinion mining system.

In the remaining sections, we will first describe related work in the field of opinion mining and emotion detection and present the motivation and contribution of our work. Further on, we will present an annotation scheme we defined for the labeling of review corpora, the corpus of opinions we built and annotated and the interesting aspects we discovered in the annotation process, as well as the problems we encountered at the time of annotating. We will show the manner in which we overcame the difficulty of annotating the different classes of opinion expression and how we used this information at the time of automatic feature discovery in opinion reviews. Next, we evaluate our approach using our fine-grained annotated corpus. Last, but not least, we will present a method to employ textual entailment recognition for the detection and classification of opinions on characteristics that users previously assess using the "stars" facility and subsequently evaluate against the stars information.

2 Related Work

In recent years, there has been growing interest in the field of emotion detection as an NLP subtask. Different authors have addressed the necessities and possible methodologies from the linguistic, theoretical and practical points of view. Thus, the first step involved resided in building lexical resources of affect, such as WordNet Affect [1], SentiWordNet [2], Micro-WNOP [3], or "emotion triggers"[4]. Further, an initial distinction between objectivity and subjectivity was made. [5] centered the idea of subjectivity around that of private states, and set the benchmark for subjectivity analysis as the recognition of opinion-oriented language in order to distinguish it from objective language and giving a method to annotate a corpus depending on these two aspects – MPQA [6]. Subsequently, authors show that this initial discrimination is crucial for the sentiment task, as part of Opinion Information Retrieval (last three editions of the TREC Blog tracks competitions, the TAC 2008 competition), Information Extraction [7] and Question Answering [8] systems. Once this discrimination is done, or in the case of texts containing only or mostly subjective language (such as e-reviews), opinion mining becomes a polarity classification task.

Previous work in customer review classification includes document level sentiment classification using unsupervised methods [9], machine learning techniques [10], scoring of features [11], using PMI, syntactic relations and other attributes with SVM [12], sentiment classification considering rating scales [10], supervised and unsupervised methods [13] and semi supervised learning [14]. Research in classification at a document level included sentiment classification of reviews [15], sentiment classification on customer feedback data [16], comparative experiments [17]. Other research has been conducted in analysing sentiment at a sentence level using bootstrapping techniques [18], considering gradable adjectives [19], semi supervised learning with the initial training set identified by some strong patterns and then applying NB or self-training [20], finding strength of opinions [21], summing up orientations of opinion words in a sentence (or within some word window) [22, 23], determining the semantic orientation of words and phrases[24], identifying opinion holders [25], comparative sentence and relation extraction and feature-based opinion mining and summarization [9]. Finally, fine grained, feature-based opinion summarization is defined in [26] and researched in [9] or [10].

3   Motivation and Contribution

The work we present herein is motivated by a series of factors. The first one is the lack of a fine-grained opinion annotation scheme for reviews that takes into consideration the characteristics of this type of writing, both at the opinion versus fact level, as well as, within the opinion category, among the different methods to express it. To our knowledge, no similar approach has been taken so far. Our first contribution is thus the definition of such an annotation scheme that encloses all the elements that should be considered at the time of opinion mining. Further on, to evaluate systems implementing the paradigm of feature-based opinion mining, there are two available corpora, which are small, only concentrate on electronic products and are annotated at a sentence level, using a simple scheme under the form of [feature name, polarity, value]. Our second contribution is annotating a corpus of 100 reviews in English of different product categories (ranging from electro domestics to restaurants and books), using the created annotation scheme. We make this corpus available to further research. Thirdly, we evaluate and validate our corpus annotations for fact versus opinion, using n-gram similarity to the annotated phrases and further classify opinionated sentences according to the polarity expressed, using n-gram similarity to the annotated sentences. Finally, our contribution resides in proposing a method to use textual entailment in the opinion mining task. As far as we are aware of, such research has not been done so far. From the categories on which stars are given, we generate short positive and negative phrases (e.g. Ease of Use – "This is easy to use" versus "This is not easy to use.") To these short hypotheses that we want to verify, we add examples of positive and negative phrases from the annotated corpus, which are related to the category. We test the entailment relation in a window of three consecutive phrases. The results obtained are encouraging.

4 Opinion Annotation Scheme

In order to train and test a system performing feature-based opinion mining, an annotated corpus is needed. However, in this field, of feature-based opinion mining, there are just two corpora available, annotated in a simple manner (for each sentence, the feature mentioned and the associated positive/negative score), both containing a small number of products and all pertaining to the same category [26, 27]. Another corpus developed for the more general opinion field is the Multi-Perspective Questioning Answering (MPQA) one [6], separating among the subjective and objective aspects of the annotated texts; however, this corpus only contains newspaper articles and does not take into consideration the aspects of product characteristic annotation that we are mostly interested in within the feature-based opinion mining task. Starting with this observation, we decided to develop and apply an annotation scheme that would be able to capture all the aspects involved in the opinion mining process. Further on, we could use this annotated information for three experiments – fact versus opinion sentence classification, polarity classification for opinionated sentences and automatic characteristic detection.

As in the case of MPQA, our annotation scheme is designed for the integration within the GATE (General Architecture for Text Engineering)[1] framework. The annotation elements and their attributes are described using XML Schema files.

We describe four types of annotations: fact, opinion, feature expression and modifier. The first element – *fact* - was created for the labeling of tokens and phrases containing factual information. The attributes we defined for this element are *source* (writer, a quote etc.), *target* (name of product capability it refers to), *id* (an identifier given by the annotator for future reference), *feature* (name of the feature that the fact describes), *type* (direct, indirect or implicit), *POS* (part of speech, for factual information expressed in individual tokens, which can be *adjective, noun, verb, adverb* or *preposition*), *phrase* (if the factual information is expressed in more than one token, a sentence or a group of sentences). The second element – *opinion* – is presented in Figure 1. Apart from the attributes that are common to the fact element, this label contains the attributes *polarity* (positive or negative), *intensity* (degree of the opinion expressed, which can be *low, medium* or *high*) and *affect* (as results from the formulation of the opinion statement). Another defined element – *feature expression,* with the same elements as the opinion tag, aims at identifying phrases that

---

[1] http://gate.ac.uk

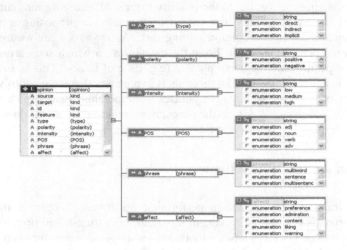

Fig. 1. Annotation scheme for the *"opinion"* element

indirectly express a feature of the product under review. Finally, the ___modifier___ element, whose structure is the same as for the opinion element, is included to spot the words whose use lead to a change in the polarity of the expressed opinion., but cannot be used alone to express an opinion (e.g. *"It's absurd"*). An example of annotation is given in Figure 2.

5 Review Corpus Annotation

With the created scheme, we annotated a corpus containing 100 reviews, each containing approximately 1000 words (they range from 700 to 2000 words per review), on a high variety of "objects", pertaining to the categories of travel (resorts, hotels, restaurants and touristic attractions), home & garden (furniture, washing machines), electronics (laptops, PDAs, mobile phones, digital cameras), computers and software, music, movies, books and cars. All these reviews were taken from the *epinions.com* site. The advantage in using this source is that reviewers tend to profoundly analyze each of the aspects of the products and the reviews are long and complicated. The idea behind the high variety of topics was, on the one hand, to detect similar manners to express opinions and, on the other hand, to test our opinion mining approaches at the coarse grained level, against the categories on which the users can punctuate the object using "stars" and at the fine-grained level, using our annotations. Moreover, choosing to label reviews on such a diversity of "objects" allowed us to test the applicability of the schema at the level of the feature-based opinion mining paradigm (i.e. identify, for any type of object, be it location, event, product etc., its features and the opinion words that are used to describe it). The annotation was done by two non-native speakers of English, one with a degree in Computer Science and the second a Linguistics student. Two examples of annotation are presented in Figure 3.

---

1. \<modifier gate:gateId="46" source ="w" target="restaurant" id="83" feature=""
type="direct" polarity="positive" intensity="medium" POS="" phrase="multiword"
affect="admiration">I was pleasantly surprised\</modifier> to learn that the food is
\<modifier gate:gateId="47" source ="w" target="excellent" id="84" feature=""
type="direct" polarity="positive" intensity="medium" POS="adverb" phrase="word"
affect="admiration">still\</modifier> \<opinion gate:gateId="48" source ="w"
target="food" id="85" feature="length" type="direct" polarity="positive"
intensity="high" POS="" phrase="word" affect="admiration">excellent\</opinion>,
and the staff very \<opinion gate:gateId="49" source ="w" target="staff" id="86"
feature="length" type="direct" polarity="positive" intensity="high" POS="adjective"
phrase="word"   affect="admiration">professional\</opinion>   and   \<opinion
gate:gateId="50" source ="w" target="staff" id="87" feature="length" type="direct"
polarity="positive"   intensity="high"   POS="adjective"   phrase="word"
affect="admiration"> gracious \</opinion>.

2. Closed, the laptop is \<fact gate:gateId="13" source ="w" target="laptop" id="17"
feature="length" type="direct" POS="" phrase="multiword">12.5" wide\</fact>, \<fact
gate:gateId="14" source ="w" target="laptop" id="18" feature="width" type="direct"
POS="" phrase="multiword">9.5" deep\</fact>, with the \<fact gate:gateId="15" source
="w"   target="laptop"   id="19"   feature="thickness"   type="direct"   POS=""
phrase="multiword">thickness ranging from 1" along the front edge, to 1.5" in
back\</fact>.

Fig. 2. Sample from the annotated corpus

As a result of the annotation process, we found some interesting phenomena that are worth mentioning and that justify our annotation schema:

1. References along the argumentation line must be resolved (e.g. *I went to another restaurant from this chain and they treated us horribly. These people were very nice.*). This is the reason for which our scheme contains the elements *"target"* and *"id"*. In this way, all factual or opinion expressions referring to an entity can be traced along the text.

2. An idea is many times expressed along a line of sentences, combining facts and opinions to make a point (e.g. *This camera belongs to the new generation. Quality has highly improved over the last generations.*). We did not have a problem with this phenomenon, since our annotation scheme was designed to allow the labeling of token or expression-level elements, as well as sentence and multi-sentence level.

3. Within a phrase, a reviewer presents facts in such a manner that the reader is able to extract the corresponding opinion (e.g. *They let us wait for half an hour in the lobby and finally, when they gave us the key, realized the room had not been cleaned yet.*). While the sentence is purely factual in nature, it contains the phrases *"let us wait for half an hour in the lobby"* , *"the room had not been cleaned yet"*, which we annotated as *feature expressions* of *implicit* type.

4. There is an extensive use of conditionals within reviews. (e.g. *If you dare, buy it! It was great for two weeks until it broke!; If you've got shaky hands, this is the camera for you and if you don't, there's no need to pay the extra $50...*) . We consider the sentence containing the conditional expression a *modifier*.

5. There are many rhetoric-related means of expressing opinion (e.g. *Are you for real? This is called a movie?*). We annotate these elements as implicit *feature expressions*.

At the time of performing our experiments, these phenomena are important and must be taken into consideration, since 27% of the annotated opinionated phrases in our corpus are composed of more than one sentence. More generally, these findings draw our attention upon the context in which opinion mining is done. Most of the work so far concentrated on sentence or text level, so our findings draw the attention upon the fact that more intermediate levels should also be considered.

6 Experiments and Evaluation

The first experiment we performed aimed at verifying the *quality and constancy of the annotation* as far as fact versus opinion phrases are concerned and, within the opinionated sentences, the performance obtained when classifying among positive and negative sentences. In the first phase, we lemmatize the annotated sentences using TreeTagger[2] and we represented each fact and opinion phrase as a vector of characteristics, measuring the n-gram similarity (with n ranging from 1 to 4) and overall similarity with each of the individual corpus annotated sentences, tokens and phrases. We perform a ten-fold cross validation using the Support Vector Machines Sequential Minimal Optimization (SVM SMO). The results for fact versus opinion and positive versus negative classifications are presented in Table 1.

Table 1. Evaluation of fact vs. opinion and positive vs. negative classification

	Precision	Recall	Kappa
Fact	0.72	0.6	0.53
Opinion	0.68	0.79	0.53
Positive	0.799	0.53	0.65
Negative	0.72	0.769	0.65

In the second phase, we consider for classification only the phrases containing two sentences. As in the first phase, we represent each fact and opinion phrase as a vector of characteristics, measuring the n-gram similarity (with n ranging from 1 to 4) and overall similarity with each of the individual corpus annotated sentences and with all other phrases containing two sentences, then perform a ten-fold cross validation using SVM SMO. We summarize the results obtained in Table 2.

---

[2] http://www.ims.uni-stuttgart.de/projekte/corplex/TreeTagger/

Table 2. Evaluation of fact versus opinion and positive versus negative 2-sentences phrase classification

	Precision	Recall	Kappa
Fact	0.88	0.76	0.43
Opinion	0.74	0.89	0.43
Positive	0.84	0.71	0.68
Negative	0.89	0.92	0.68

Table 3. Evaluation of fact versus opinion and positive versus negative 3-sentences phrase classification

	Precision	Recall	Kappa
Fact	0.76	0.6	0.80
Opinion	0.78	0.94	0.80
Positive	0.85	0.76	0.68
Negative	0.92	0.96	0.68

In the third phase, we consider for classification only the phrases containing three sentences. The fact and opinion phrases are represented as in the first experiments and a ten-fold cross validation using SVM SMO is done The results are shown in Table 3.

From the results obtained, we can notice that using longer phrases, we obtain an improved classification performance in both fact versus opinion classification, as well as positive versus negative classification. The only drop in classification performance is in the case of longer factual phrases. We explain this by the fact that in many of the cases, these types of phrases contain descriptions of opinionated sentences or represent combinations of factual and opinion sentences (e.g. *"They said it would be great. They gave their word that it would be the best investment ever made. It seems they were wrong"*). The results show that our annotation is constant and that labeled elements present similarity among them; this fact can be used to automate the annotation, as well as use the labeled corpus for the training of an opinion mining system or for its evaluation. Evaluation proved that the annotation schema and approach are general enough to be employed for labeling of reviews on any product.

7 Applying Textual Entailment for Opinion Classification

In this section, we study the manner in which feature-based opinion mining can be done at a coarse level, i.e., not discovering all the features of a product and the corresponding opinion words that are used to describe them, but the opinion on the main aspects of the "object" on which opinion is mined, which are also evaluated by the users with "stars" (from 1 to 5 stars, 1 being the lowest and 5 being the highest). Comparing against the number of stars given to these main product functionalities can also be a useful method to evaluate opinion mining systems. The approach taken in this section is motivated by the fact that many products on review sites have a "star" assigning system associated. In this manner, people are given the opportunity to value the product in question, besides using opinionated phrases, by employing, for a number of default defined features, from one to five stars. We consider this to be useful information at the time of review mining, both for the fact that we can thus overcome the problem of automatically discovering the distinct product features, as well as the problem of evaluating our approach without having a previously annotated corpus on the specific product.

Textual entailment recognition [28] is the task of deciding, given two text snippets, one entitled Text (T) and the other called Hypothesis (H), if H can be inferred from T.

For this task, we will use the information given by the stars category. We consider the given criteria and generate simple sentences (hypotheses) for each of the positive and negative categories. The idea behind this proposed method is to test whether a textual entailment system would be able to capture and better resolve the semantic variability given by opinionated text.

Our textual entailment system [29] is based on the tree edit distance algorithm. Each of the sentences is dependency parsed using Minipar[3] and passed through Lingpipe[4] in order to detect and classify the Named Entities it contains. Subsequently, syntactic, lexical and semantic similarities are computed, starting from the root of the dependency tree, between all the tokens and their corresponding dependency links in the hypothesis and the tokens in the text using the DIRT[5] collection, eXtended WordNet[6] and a set of rules for context modifiers.

From the annotated corpus, for each of the considered products, we selected the reviews containing sentences describing opinions on the criteria which users are also allowed to assess using the stars system. The categories which are punctuated with two or less stars are considered as negative and those punctuated with four or five stars are considered as having been viewed positively.

Table 4. Polarity classification accuracy against the number of stars per category

Name of product	Stars Category	Accuracy
Restaurant	Price	62%
	Service	58%
	Food	63%
	View	53%
	Atmosphere	58%
Digital camera	Ease of use	55%
	Durability	60%
	Battery life	80%
	Photo quality	65%
	Shutter lag	53%
Washing machine	Ease of Use	60%
	Durability	72%
	Ease of Cleaning	67%
	Style	65%

---

[3] http://www.cs.ualberta.ca/~lindek/minipar.htm
[4] http://alias-i.com/lingpipe/
[5] http://demo.patrickpantel.com/Content/LexSem/paraphrase.htm
[6] http://xwn.hlt.utdallas.edu/

We generated hypotheses under the form *"Category is good"* and *"Category is nice."*, *"Category is not good"* and *"Category is not nice."*,e.g. *"The price was good."*,*"The price was not good."* , *"The food was good."*, *"The food was not good."* *"The view was nice."*, *"The view was not nice."*. In case no entailment was found with such built sentences, we computed entailment with annotated sentences in the review corpus. The results obtained are shown in Table 4.

As we can see from the obtained results, textual entailment can be useful at the time of performing category based opinion mining. However, much remains to be done at the level of computing semantic similarity between opinionated texts. Such work may include the discovery of opinion paraphrases or opinion equivalence classes.

8 Conclusions and Future Work

In this paper we have described our present contribution to the task of feature-based opinion mining in reviews on different products.

We have defined an opinion annotation scheme that is *general* (because we could use it in the same form to annotate a large variety of object types, from restaurants to books, travel resorts and laptops), it is *proper for this type of texts* (because it is able to spot all the linguistic phenomena present in opinion expression, be it direct, indirect or implicit), and that *takes into consideration all the characteristics* that must be observed by systems automating the task of opinion mining. We first made a distinction between the fact and opinion categories and subsequently between positive and negative opinions. We showed how the elements of the defined scheme can be employed to annotate a corpus of reviews on a whole variety of products and described the problematic phenomena encountered in the labelling process, which, however, we could overcome with our annotation scheme. We suggested corresponding changes to the opinion mining approach and evaluated our findings over the entire corpus. Last, but not least, we proposed a method to employ textual entailment recognition for the task of opinion mining, considering the review phrases (single sentences, 2 and 3 consecutive sentences) as texts and sentences generated by using the categories on which stars are given and corpus extracted sentences as hypotheses. We obtained promising results. Future work includes the use of the annotated corpus for automatic feature discovery and the testing of alternative machine learning algorithms and classification methods.

Acknowledgements. This research has been partially funded by the Spanish Government under the project TEXT-MESS (TIN 2006-15265-C06-01), by the European project QALL-ME (FP6 IST 033860) and by the University of Alicante, through its doctoral scholarship.

References

1. Strapparava, C., Valitutti, A.: WordNet-Affect: an affective extension of WordNet. In: Proceedings of the 4th International Conference on Language (2004)
2. Esuli, A., Sebastiani, F.: SentiWordNet: A Publicly Available Resource for Opinion Mining. In: Proceedings of the 6th International Conference on Language Resources and Evaluation, LREC 2006, Genoa, Italy (2006)

3. Cerini, S., Compagnoni, V., Demontis, A., Formentelli, M., Gandini, G.: Micro-WNOp: A gold standard for the evaluation of automatically compiled lexical resources for opinion mining. In: Angeli, F. (ed.) Language resources and linguistic theory: Typology, second language acquisition, English linguistics, Milano, Italy (2007)

4. Balahur, A., Montoyo, A.: Applying a Culture Dependent Emotion Triggers Database for Text Valence and Emotion Classification. In: Proceedings of the Symposium on Affective Language in Human and Machine, Aberdeen, Scotland (2008)

5. Wiebe, J.M.: Tracking point of view in narrative. Computational Linguistics 20, 233–287 (1994)

6. Wiebe, J., Wilson, T., Cardie, C.: Annotating expressions of opinions and emotions in language. Language Resources and Evaluation 39(2-3), 165–210 (2005)

7. Riloff, E., Wiebe, J., Phillips, W.: Exploiting Subjectivity Classification to Improve Information Extraction. In: Proceedings of the 20th National Conference on Artificial Intelligence, AAAI 2005 (2005)

8. Stoyanov, V., Cardie, C.: Toward Opinion Summarization: Linking the Sources. In: COLING-ACL 2006 Workshop on Sentiment and Subjectivity in Text (2006)

9. Turney, P.: Thumbs Up or Thumbs Down? Semantic Orientation Applied to Unsupervised Classification of Reviews. In: Proceedings of ACL 2002, pp. 417–424 (2002)

10. Pang, B., Lee, L., Vaithyanathan, S.: Thumbs up? Sentiment classification using machine learning techniques. In: Proceedings of EMNLP 2002, the Conference on Empirical Methods in Natural Language Processing (2002)

11. Dave, K., Lawrence, S., Pennock, D.: Mining the Peanut Gallery: Opinion Extraction and Semantic Classification of Product Reviews. In: Proceedings of WWW 2003 (2003)

12. Mullen, T., Collier, N.: Sentiment Analysis Using Support Vector Machines with Diverse Information Sources. In: Proceedings of EMNLP (2004)

13. Chaovalit, P., Zhou, L.: Movie Review Mining: a Comparison between Supervised and Unsupervised Classification Approaches. In: Proceedings of HICSS 2005, the 38th Hawaii International Conference on System Sciences (2005)

14. Goldberg, A.B., Zhu, J.: Seeing stars when there aren't many stars: Graph-based semi-supervised learning for sentiment categorization. In: HLT-NAACL Workshop on Textgraphs: Graph-based Algorithms for Natural Language Processing (2006)

15. Ng, V., Dasgupta, S., Arifin, S.M.: Examining the Role of Linguistic Knowledge Sources in the Identification and Classification of Reviews. In: Proceedings of the COLING/ACL 2006 Main Conference Poster Sessions (2006)

16. Gamon, M., Aue, S., Corston-Oliver, S., Ringger, E.: Mining Customer Opinions from Free Text. LNCS. Springer, Heidelberg (2005)

17. Cui, H., Mittal, V., Datar, M.: Comparative Experiments on Sentiment Classification for Online Product Reviews. In: Proceedings of the 21st National Conference on Artificial Intelligence AAAI (2006)

18. Riloff, E., Wiebe, J.: Learning Extraction Patterns for Subjective Expressions. In: Proceedings of the 2003 Conference on Empirical Methods in Natural Language Processing (2003)

19. Hatzivassiloglou, V., Wiebe, J.: Effects of adjective orientation and gradability on sentence subjectivity. In: Proceedings of COLING (2000)

20. Wiebe, J., Riloff, E.: Creating Subjective and Objective Sentence Classifiers from Unannotated Texts. In: Gelbukh, A. (ed.) CICLing 2005. LNCS, vol. 3406, pp. 486–497. Springer, Heidelberg (2005)

21. Wilson, T., Wiebe, J., Hwa, R.: Just how mad are you? Finding strong and weak opinion clauses. In: Proceedings of AAAI (2004)

22. Kim, S.M., Hovy, E.: Determining the Sentiment of Opinions. In: Proceedings of COLING (2004)
23. Lin, W.H., Wilson, T., Wiebe, J., Hauptman, A.: Which Side are You On? Identifying Perspectives at the Document and Sentence Levels. In: Proceedings of the Tenth Conference on Natural Language Learning CoNLL (2006)
24. Turney, P., Littman, M.: Measuring praise and criticism: Inference of semantic orientation from association. ACM Transactions on Information Systems 21 (2003)
25. Stoyanov, V., Cardie, C., Litman, D., Wiebe, J.: Evaluating an Opinion Annotation Scheme Using a New Multi-Perspective Question and Answer Corpus. In: AAAI Spring Symposium on Exploring Attitude and Affect in Text: Theories and Applications
26. Hu, M., Liu, B.: Mining Opinion Features in Customer Reviews. In: Proceedings of Nineteenth National Conference on Artificial Intelligence AAAI (2004)
27. Ding, X., Liu, B., Yu, P.: A Holistic Lexicon-Based Approach to Opinion Mining. In: Proceedings of First ACM International Conference on Web Search and Data Mining (WSDM 2008), Stanford, California, USA (2008)
28. Dagan, I., Glickman, O., Magnini, B.: The PASCAL Recognising Textual Entailment Challenge. In: Quiñonero-Candela, J., Dagan, I., Magnini, B., d'Alché-Buc, F. (eds.) MLCW 2005. LNCS (LNAI), vol. 3944, pp. 177–190. Springer, Heidelberg (2006)
29. Iftene, A., Balahur-Dobrescu, A.: Hypothesis transformation and semantic variability rules for recognizing textual entailment. In: Proceedings of the ACL Workshop on Entailment and Paraphrasis, ACL (2007)

Analyzing Document Collections via Context-Aware Term Extraction

Daniel A. Keim, Daniela Oelke, and Christian Rohrdantz

University of Konstanz, Germany
firstname.lastname@uni-konstanz.de

Abstract. In large collections of documents that are divided into predefined classes, the differences and similarities of those classes are of special interest. This paper presents an approach that is able to automatically extract terms from such document collections which describe what topics discriminate a single class from the others (discriminating terms) and which topics discriminate a subset of the classes against the remaining ones (overlap terms). The importance for real world applications and the effectiveness of our approach are demonstrated by two out of practice examples. In a first application our predefined classes correspond to different scientific conferences. By extracting terms from collections of papers published on these conferences, we determine automatically the topical differences and similarities of the conferences. In our second application task we extract terms out of a collection of product reviews which show what features reviewers commented on. We get these terms by discriminating the product review class against a suitable counter-balance class. Finally, our method is evaluated comparing it to alternative approaches.

1 Introduction

With the growing amount of textual data available in digital form, also methods and techniques for exploring these resources are increasingly attracting attention. In many cases classes (or clusters) of documents can be distinguished and topical differences and similarities among those classes are of interest. Depending on the concrete task it can also be worthwhile to explore stylistic or linguistic differences.

In this paper we present an approach that helps in analyzing a set of classes of documents with respect to the question what one class of documents discriminates from the rest - by extracting discriminating terms. The technique also determines so-called overlap terms that discriminate a subset of the classes from the remaining ones. The classes of documents e.g. could correspond to different scientific conferences with their published papers as documents. The extracted discriminating terms then show the topics that are unique for the specific conference. See figure 1 for an example that was generated by our new approach. In the venn diagram each circle represents one conference. All three conferences deal with graphical representations and visualization. Yet, each conference has its own specific orientation in the field. In an outer section of the diagram that is unique for one of the conferences the terms that discriminate this conference from all the others are displayed. In the case of the Vis and the comparison to Siggraph and InfoVis those terms are {flow field, scalar field, volume data, volume

H. Horacek et al. (Eds.): NLDB 2009, LNCS 5723, pp. 154–168, 2010.

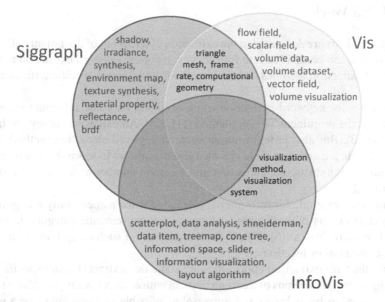

Fig. 1. Discriminating and overlapping terms for the three conferences Siggraph, Vis, and Info-Vis (generated with about 100 papers of each conference). Terms in the overlapping areas are shared by two classes and discriminate them against the third one while the rest of the terms discriminates one specific class against the others.

dataset, vector field, volume visualization}. Furthermore, you can see the terms that are shared by two conferences and discriminate them against the third conference in the overlap regions of the diagram. Apparently, there is no overlap of the Siggraph and the InfoVis conference. While this might not be surprising for an expert in the area of these conferences (as the Vis conference is topically somewhere in between Siggraph and In-foVis), it provides quite useful information to non-experts without requiring substantial reading efforts. The overlap area of all three conferences remains empty, because our approach only extracts discriminating terms and in this case there is nothing to discrim-inate against. Please note that this is a very small introductory example. In section 4.1 we present the result of an analysis with more conferences and more terms.

But the application area of our method is much wider than that. The technique can be applied to any application task in which discriminating and / or overlapping terms between different classes are of interest. It turns out to be a powerful tool in any scenario in which terms that cover a certain topic or aspect have to be separated not only from general stopwords but also from terms covering other aspects that are currently not of interest. An example for such a scenario is given in section 4.2 where we use our technique to extract product attributes from a set of printer reviews. The challenge here is to extract the terms that hold the information about what the customers were satisfied or dissatisfied with, but filter out the review-typical words that they use to convey their message.

2   Related Work

Numerous methods are dedicated to the extraction of terms out of document collections. These methods can be divided into four main categories: keyword extraction methods, information extraction methods, labeling methods and domain specific term extraction methods.

Approaches for keyword extraction often originate from the information retrieval field like e.g. the prominent TFIDF method ([1], [2]). An extensive survey on that can be found in [3]. But also in text mining research keyword extraction methods play a role ([4], [5]). In a usual case there is a measure that allows to score terms with respect to a document or a document collection and a certain number of top scored terms are then extracted.

An important example for information extraction is the named entity recognition. It is aimed at extracting proper names, that have a certain semantic category, in order to construct semantic lexica ([6], [7]). Typical examples for such categories are names of persons, companies or locations.

Among the term extraction approaches are some that extract domain specific terms comparing an analysis corpus of a certain domain with a reference corpus. The reference corpus is aimed to be as broad and universal as possible and can either be a general language corpus ([8], [9]) or composed of several other domain corpora [10]. Another approach takes a large collection of heterogeneous newspaper articles as a reference corpus [11]. Those approaches are useful for example to support terminology extraction or ontology construction.

Methods for labeling are mainly used for visualization tasks. Usually they extract very few terms that describe (the documents that constitute) a certain area of a visualization. In the ThemeScape^{TM} visualization [12] a common TFIDF approach is used for the labeling of the distinct document clusters. A similar labeling approach is done in the WEBSOM visualization [13] where the relative frequencies of terms in the different nodes of the self-organizing map are compared [14] [15].

Our method is similar to the approaches that do domain specific term extraction because we also compare the scores of a certain term for different domains/classes. Yet, in contrast to those methods we compare several class corpora with and to each other instead of using a general reference corpus for comparison. Furthermore, we use a novel measure called TFICF which is an adaption of the popular TFIDF measure to assess the importance of a term within a class. This allows us to determine discriminating terms for single classes or sets of classes in the concrete context of other interesting classes. By doing so, we are able to figure out the topical coherences and distinctions among a whole set of particular classes and thus satisfy a very specific information need.

In [16] a term frequency inverse cluster frequency value is calculated to get feature vectors of previously attained clusters of document paragraphs. In contrast to our approach the cluster simply can be seen as a concatenation of all of its documents so that actually there is no difference to the common TFIDF formula.

Our approach is also situated in the context of contrastive summarization [17] and comparative text mining [18] which is a subtask of contextual text mining [19]. Contrastive summarization has a rather narrow application field, it only regards the binary case (two classes) and is focused on opinion mining. Having reviews of two products, the aim is to automatically generate summaries for each product, that highlight the difference in opinion between the two products.

While the fundamental idea of comparative text mining is closely related to our work, the outcome of the cross-collection mixture model proposed in [18] is rather orthogonal to our approach. The process is subdivided in two steps "(1) discovering the common themes along all collections; (2) for each discovered theme, characterize what is in common among all the collections and what is unique to each collection". Whereas this kind of analysis is based on the themes common to *all* classes, our method does explicitly not account for those themes, but for themes that discriminate one or several classes from the remaining ones.

The rest of this paper is organized as follows. First, in section 3 we motivate and introduce our new technique. Then in section 4 we present two application domains with concrete examples. In section 5 we provide an evaluation of our methods and compare it with other techniques. Finally, in section 6 we give a conclusion.

3 Technique

In order to determine discriminating or overlap terms, first of all we need to be able to quantify how important a certain term is for a certain class. It would be straight forward to use the standard TFIDF method. But unfortunately that approach is not suitable in this case. The TFIDF value determines an importance value for a certain term with respect to a document within a document collection. But what we need is an importance value for a certain term with respect to a whole document collection within the context of other document collections. Subsection 3.1 introduces the concept of TFICF (term frequency inverse class frequency) an extension of TFIDF that fulfills our criteria. In subsection 3.2, we explain how to use this new measure to extract discriminating and overlap terms. Finally, some notes on parameter tuning and preprocessing are given in the subsections 3.3 and 3.4.

3.1 Term Frequency Inverse Class Frequency (TFICF)

TFICF (term frequency inverse class frequency) is an extension of the classical TFIDF measure. The formula for weighted TFICF is composed of three factors (see equation 1).

$\forall$ terms $t_i \wedge \forall$ classes C_j with $i \in \{1 \cdots \#terms\}$ and $j \in \{1 \cdots \#classes\}$: [1]

$$weighted - tficf(t_i, C_j) = distr\_weight(t_i, C_j) \cdot tf(t_i, C_j) \cdot icf(t_i) \tag{1}$$

The tf value reflects the normalized overall frequency of a term within a collection. The icf (inverse class frequency) value takes into account in how many classes the term is present. The distr_weight value depends on the distribution of a term over the documents of a collection.

---

[1] "#" stands for "number of all..."

The tf value is calculated dividing the overall frequency of a term among the documents of a collection by the overall number of tokens in the collection (see equation 2).

$$tf(t_i, C_j) = \frac{\sum_{k=1}^{\#docs_j} freq(t_i, doc_{jk})}{\sum_{k=1}^{\#docs_j} \#tokens(doc_{jk})} \qquad (2)$$

The rationale is that longer documents exert a stronger influence on the tf value than shorter documents, which is appropriate in most application scenarios. However, the influences of all documents may be adjusted to be similar by considering only the relative frequency of terms in documents.

In contrast to the standard idf formula our icf formula has to operate on classes of documents instead of single documents. A straight forward application of the idf formula would be to say that a term t is an element of a class C, if it occurs in at least one of the corresponding documents. However, that means that outlier documents get a high influence on the result. Therefore, we propose to define that t is only considered element of a class C if at least X percent of the documents d contain the term - where X is a user-defined parameter (see equation 3).

$$icf(t) = log\left(\frac{\#classes}{\left|\{C \in classes : \frac{|\{d \in C : t \in d\}|}{|\{d \in C\}|} > X\}\right|}\right) \qquad (3)$$

The icf value plays an important role in filtering stopwords (in the broadest sense, see section 4.2). This is due to the fact that it becomes 0 if all classes are considered as containing the term and in this case the term cannot be considered as being discriminating for any class.

The distribution of a term over the documents of a class also can reveal something about its importance for the class. There are a number of possible distribution weights that could be included into the multiplication - even several at once. We made good experiences using the standard deviation of a term's frequency as such a distribution weight but the χ^2 significance value may be used as well (both are suggested in [4] as term weights).

Another valuable choice can be the integration of a term relevance weight which was defined by Salton & Buckley [20] as "the proportion of relevant documents in which a term occurs divided by the proportion of nonrelevant items in which the term occurs". In contrast to a typical information retrieval task where the division into relevant and nonrelevant documents is not given apriori, the term relevance weight can easily be evaluated here. To calculate the term relevance weight for a term t and a class C, we simply consider all documents out of C as relevant and all documents contained by the other classes as nonrelevant. The higher the percentage of documents in C containing t and the lower the corresponding percentage for the other classes, the higher is our weight (see equation 4 and 5).

$$term\_relevance\_weight(t_i, C_j) = \frac{support(t_i, C_j)}{\sum_{k \neq j} support(t_i, C_k)} \qquad (4)$$

with

$$support(t_x, C_y) = \frac{|\{D_z \in C_y : t_x \in D_z\}|}{|\{D \in C_y\}|} \qquad (5)$$

3.2 Determining Discriminating and Overlap Terms

The weighted tficf measure provides a term score that is comparable among several classes. So the next logical step is to use it for comparison. For any term we get as many scores as there are classes: For each individual class, there is a particular score. We now define that a term is discriminating for one of these classes if its score is much higher for this class than its scores for the other classes. To determine the discriminating terms for a class, we use a threshold called discrimination factor by which a score for one class must outnumber the scores for all other classes (see definition 1).

Definition 1. *Discriminating terms*

> *A term t is discriminating for a single class C_k if:*
>
> $\forall i \in \{1 \cdots n\} \setminus k:$
> *weighted_tficf(t, C_k) > discrimination-factor · weighted_tficf(t, C_i).*

The same approach can be applied to determine if a term is discriminating for the overlap of several classes. This is precisely the case if the lowest term score for one of the overlap classes outnumbers the highest term score of the remaining classes at least by the threshold factor (see definition 2).

Definition 2. *Overlap terms*

> *For the overlap area of several classes $\{C_k, C_l, \cdots, C_m\}$ a term t is discriminating if:*
>
> $\forall i \in \{1 \cdots n\} \setminus \{k, l, \cdots, m\}:$
> *min(weighted_tficf(t, C_k), weighted_tficf(t, C_l), $\cdots$, weighted_tficf(t, C_m))*
> *> discrimination-factor · weighted_tficf(t, C_i).*

In practice both discriminating terms and overlap terms can be determined in a single scan through the database.

3.3 Parameter Tuning

Our algorithm for determining the discriminating and overlapping terms has two parameters: a minimum percentage and the discrimination factor. The minimum percentage is used to specify the minimum number of documents of a class that must contain the term to allow it to be chosen as discriminative. Without that parameter all terms that only occur in one class would most certainly be considered as being discriminative even if they only occur once in that class (because $X > 0$ * factor would always be true, no matter how small the value of X is).

While the minimum percentage can easily be set by the user (e.g. 0.2 if at least 20% of the documents shall contain a term), the discrimination factor threshold cannot easily be fixed without prior experience. In our experiments reasonable thresholds showed to lie typically in the interval between 1.5 and 3.0. In our implementation the exact threshold is set by using a dynamic slider, which allows the user to get the desired amount of discriminating terms.

3.4 Preprocessing

Like in many text mining applications careful preprocessing is valuable. In our case we applied a base form reduction algorithm [21] to all words in order to get singular forms for nouns and infinitive forms for verbs. In addition we used a POS-tagger ([22], [23], [24]) and a NP-chunker ([25], [26]) to identify nouns respectively noun phrases. This allows us to focus only on nouns and noun phrases if this is desired. Numbers and short strings with less than 3 characters were deleted in the preprocessing step, since they often correspond to punctuation marks or special characters that do not need to be considered.

One interesting advantage of our method is that we do NOT use any stopword lists. High-frequent stopwords like "the" or "and" are ignored with very high probability because their icf values become 0. Stopwords with a lower frequency in a regular case should not appear considerably more often in one class than in the others and thus are filtered out.

4   Application Examples

As mentioned in section 1 of this paper, our method can be used to explore the characteristics of predefined classes. The extracted discriminating and overlap terms enable users to gain insight into the hidden underlying topical structure of sets of document classes.

4.1   Characteristic Terms for Conferences

One concrete example for the application of our method could be motivated by the questions: If we take different conferences in the computer science area, can we detect automatically by processing all of the papers published in these conferences: (a) How they differ from each other? (b) What single conferences focus on or what makes them special? (c) What several conferences have in common, respectively what distinguishes them from the other conferences?

We tried to answer these questions for a set of 9 different conferences by regarding about 100 recently published papers for each of these conferences. Besides the NLDB conference we decided to focus on other conferences that we know well, dealing with:

 – Information Retrieval (SIGIR),
 – Database and Data Storage (VLDB and SIGMOD),
 – Knowledge Discovery and Data Mining (KDD),
 – Visual Analytics (VAST),
 – Information Visualization (InfoVis),
 – Visualization (VIS), and
 – Computer Graphics (SIGGRAPH).

The results of our approach can be found in figure 2:

As can be seen the discriminating terms of the NLDB relate very much to natural language. Database-related vocabulary does not appear in the list as it is also covered

Siggraph	Vis	InfoVis	VAST	KDD	SIGMOD	VLDB	NLDB	SIGIR	Terms
									diffuse, input image, scene, irradiance, environment map, brdf, radiance, silhouette, parameterization, light source, material property, reflectance, lighting condition, shadow, illumination, eye, scatter, texture synthesis
									opacity, streamline, voxel, volume data, terrain, transfer function, vector field, scalar field, volume dataset, flow field, volume visualization, isosurface, scalar value, time step
									information space, shneiderman, draw, treemap, cone tree, information visualization, layout algorithm, layout
									workspace, card, story, traffic, pacific northwest national laboratory pnnl u.s. department, time range, decision make, analysis method, network traffic, intelligence analysis, analytic, analysis technique, national visualization, pacific northwest national laboratory pnnl, intelligence analyst, analytics application, network data, u.s. department, science laboratory, workflow, energy office, analytics center, analysis algorithm, analytics system, thought, nvac, analytics tool, homeland security program
									support vector machine, a. mccallum, kdd, uci repository, machine learn, decision tree
									database application, database management, sql statement, skew, keyword search, database engine
									memory usage, path query, input stream
									dictionary, semantic web, noun phrase, method, wordnet, noun, auto, parse, ontology, verb, adjective, english, document
									trec topic, retrieval performance, retrieval result, relevance judgment, retrieval effectiveness, retrieval information search, information need, retrieval model, pseudo-relevance feedback, information storage, pool, trec, relevance feedback, average precision, retrieval, information search, retrieval system, test collection
									computer graphic, discontinuity, plane, camera, realism, particle, computer graphics computational geometry, curvature, velocity, triangulation, frame rate, texture map, convolution, image plane, vertex, mesh, coefficient, graphics hardware, sample point, render, texture, scalar, triangle mesh, ray, hole, deformation, coherence
									scatterplot, slider, information visualization, metaphor, layout
									knowledge discovery
									knowledge base
									query term, query expansion, corpora
									method
									query process, query optimization, xpath, query workload, vldb, xml data, insert, query processor, query execution, optimizer, query plan, response time, tuple, selectivity, xml, xquery, query optimizer, data warehouse, database system, xml document, vldb page, dbm, cost model
									scene, frame, distortion
									tool
									animation, screen
									effort

Fig. 2. On the left side the set of conferences is listed for which a set of terms is discriminating. A conference is contained in this set if its corresponding matrix entry is marked in a blue color tone. The more conferences a set contains, the darker is the blue. The corresponding terms can be found on the right side. The combinations of conferences that do not appear, simply do not jointly discriminate against the others in a certain topic.

by other conferences and thus not discriminating for NLDB in this context. NLDB has a discriminating overlap with SIGIR conference, because only those two deal with query terms and corpora. In contrast, everything related to information or document retrieval apparently is significantly more covered by the papers of the SIGIR conference. NLDB has also small discriminating overlaps with VLDB and VAST but there is no overlap with the conferences that focus on visualization and computer graphics.

Also the extracted terms for overlaps between two or more conferences fit nicely and are reasonable: E.g. SIGGRAPH and VIS share a lot of computer graphics vocabulary, and InfoVis and VAST the topic of visualizing information. SIGGRAPH, VIS and Info-Vis still share some vocabulary related to graphical representations, while VIS, InfoVis and VAST all deal with visualizations. Finally, while SIGMOD and VLDB are both database conferences that share many database-related topics our method reveals that there are also differences in topic coverage. The term "database management", for example, only occurs in the SIGMOD term list, while VLDB papers seem to focus more on topics such as "memory usage".

One nice particularity of our method is that if a term is important for every class then it is not extracted: Although NLDB surely shares topic terms such as e.g. "algorithm" or "data" with the visualization conferences, they are not extracted, as all the other considered conferences also contain these topics. Within the context of these specific other conferences such terms are not of interest as they do not provide any discrimination power. Another interesting issue is that some proper names appear in result sets. This is an indication that certain persons and institutions seem to have strong influences on specific conferences.

4.2   Characteristic Terms in Customer Reviews (Amazon)

In a different project we worked on a data set of printer reviews from amazon.com. We were interested in the attributes that the customers frequently commented on (such as the paper tray of the printer, the cartridges etc.). However, in those reviews not only the terms describing printer attributes occur frequently, but also the review-related vocabulary. Widely used stopword lists contain only very general terms like conjunctions, determiners, pronouns etc. and thus were not suitable to separate the printer terms from the rest. We had to apply a special term filtering that extracted the printer terms while it did not consider the review terms. For this purpose we applied our discrimination-based term extracting method: We used a counter-balance class containing book reviews and discriminated the printer review class against it. As both classes shared the review specific terms, only printer related terms were discriminating the printer class and hence got extracted. Figure 3 compares a simple approach that just extracts the 40 most frequent terms after filtering stopwords out (top) with the result of our technique using the book reviews as a counter-balance class (bottom). It is easy to see that the quality of the second list is much higher since lots of review-related terms such as "good", "like" or "need" that are uninteresting in our case are not contained in the list.

As you can see, besides getting deeper insight into the commonalities and differences of document collections our approach also allows us to do domain-specific term filtering without the usage of an ontology or a specialized knowledge base. To apply the technique a set of documents has to be provided that contains the words that we would like to be

40 terms with highest frequencies (stopwords have been removed):

printer, print, use, good, work, scan, buy, problem, install, software, great, time, easy, like, need, try, machine, ink cartridge, fax, ink, set, purchase, make, hp printer, copy, paper, run, product, come, price, look, say, want, photo, new, quality, real, page, wireless, think

40 discriminating terms:

network, product, ink cartridge, fax, jam, paper, scan, print quality, print, download, printer, cartridge, software, mac, unit, function, month, all-in-one, installation, machine, scanner, install, box, model, use, hp, feature, replace, easy, black, document, fix, support, driver, ink, color, wireless, photo, expensive, hp printer

Fig. 3. 40 most frequent terms (top) compared to the Top-40 discriminating terms. It can easily be seen that the list of discriminating terms is more dense with respect to the question what the customers frequently comment on while the list of the most frequent terms also contains many terms that are typically used in reviews but do not convey the desired information (e.g. need, like, good, etc).

filtered out but does not contain (or does less often contain) the type of words that we are interested in. Our method is then used to extract the terms that discriminate the class of documents that we are interested in from this counter-balance class. As only terms are selected as discriminating terms that are significantly more important for one class than for the other, the aspects that the documents of both classes share are automatically filtered out. Sometimes it can be helpful to use more than one class as a counter-balance class. This is the case when there are several undesired aspects to be filtered out and there does not exist a single counter-balance class that contains all of those aspects.

5   Experimental Evaluation

To evaluate how well the extracted terms are able to discriminate one class of documents from the others we used the extracted terms in a classification task. This was done as follows: Given three different classes of documents we used 4 different methods to extract (in average) 15 terms per class (the different methods are described in detail below). As classes we used the three conferences InfoVis, Siggraph, and Vis and each class was made up of 100 papers of the conference. The extracted terms were then used to classify a set of 60 test documents (20 of each class) that were different from the training set. Each of the 60 documents was assigned to the class that it shared most discriminating terms with. If there was more than one winning class the document was assigned to the class whose absolute frequency was largest (counting all the occurrences of discriminating terms instead of just every term once). If the document still could not be assigned unambiguously it was assigned to the class of ambiguous documents. In that classification task a method performs best if it extracts terms that discriminate a class from the others but yet also chooses terms that are characteristic for the class they have been extracted for (i.e. that they are shared by many documents of the specific class instead of being only significant for a small subset of documents of the class).

ground truth →

predicted ↓

TFIDF avg	InfoVis	Siggraph	Vis
InfoVis	19	1	4
Siggraph	1	18	9
Vis	0	1	3
ambiguous	0	0	4

TFIDF max	InfoVis	Siggraph	Vis
InfoVis	3	1	2
Siggraph	0	3	0
Vis	0	0	4
ambiguous	17	16	14

diff. analysis	InfoVis	Siggraph	Vis
InfoVis	20	1	10
Siggraph	0	18	1
Vis	0	1	8
ambiguous	0	0	1

Our approach	InfoVis	Siggraph	Vis
InfoVis	16	0	1
Siggraph	1	18	1
Vis	0	2	16
ambiguous	3	0	2

Fig. 4. Confusion matrices for the four different methods classifying 60 documents

We used the following four methods for term extraction:

- TFIDF average: Given the training corpus of 300 documents for each document and each term in the corpus a TFIDF value was calculated. Afterwards the documents were sorted into classes and for each class the average TFIDF of each term was calculated. Next, the terms were sorted according to their average value. Finally, for each class the 15 top terms were chosen.
- TFIDF max: The second method is very similar to the first one. The only difference is that instead of calculating the average TFIDF value the maximum TFIDF value of the class is chosen for each term. Then, again the terms are sorted according to their TFIDF values and the 15 top terms for each class were chosen. We included this method, too, since it has been proposed in other publications ([4], [27]).
- Differential Analysis: This is a general technique that extracts technical terms from a corpus by comparing the probability of occurrence in the given corpus to a general reference corpus [9]. We used the authors online tool to extract the terms for our paper [28]. There are two main differences to our method: First, instead of comparing the different classes against each other a general reference corpus is used. Secondly, a different term weighting approach is used. As before, for each class we extracted the top 15 terms.
- Our approach: To extract terms with the approach that is proposed in this paper we set the parameter values as follows: The minimum percentage was set to 0.11 (that means that more than 10% of the documents have to contain the term) and the discrimination factor to 2.0. Since our method does not extract a given number of terms but automatically determines the number of terms that well discriminate one class from the others we do not have exactly 15 terms per class but 14 terms for InfoVis, 15 for Vis and 16 for Siggraph.

The evaluation result: The following accuracy values were calculated for the four methods (accuracy = number of correctly classified documents divided by the total number of documents)[1]: TFIDF avg: 0.71, TFIDF max: 0.77, Diff. analysis: 0.78, Our approach: 0.91.

Figure 4 shows the result in more detail. The large number of documents in the class "ambiguous" shows that the relatively good result for TFIDF max is misleading. Almost

---

[1] Ambiguous documents were ignored in the accuracy calculation.

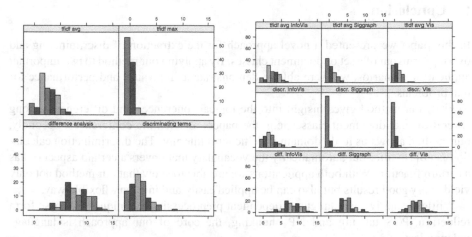

Fig. 5. Analysis of the distribution of the terms, comparing the three methods TFIDF avg, TFIDF max and Differential Analysis to our method (Discriminating Terms). Left: Distribution across the documents of the class that the terms were extracted for (InfoVis). The height of each bar in the graphic represents the number of documents in the training corpus that contain k extracted terms (with k being mapped to the x-axis). Right: Distribution across the documents of the other two classes that the terms were *not* extracted for.

80% of the documents could not be classified unambiguously. The results for the other 3 techniques are more meaningful. It can easily be seen in the confusion matrix that all the methods performed well on the classes InfoVis and Siggraph but that TFIDF avg and the Differential Analysis had problems with the class Vis. An explanation for that might be that the Vis conference is thematically somehow in between the two other conferences. The closer the classes are related to each other the more important it is that the applied method is able to find terms that are really discriminating and not only characteristic for the class as our method does.

In order to get some deeper insight we conducted a more extensive evaluation where we also analyzed the distribution of the extracted terms visually. The left graphic of figure 5 shows the distribution across the documents of the class that the terms were extracted for (we used the terms and documents of class InfoVis). The height of each bar in the graphic represents the number of documents in the training corpus that contain k extracted terms. Obviously, the distribution for TFIDF max falls apart. More than 90% of the documents contain only 1 or even 0 of the extracted terms! That means that the method extracts many terms that can only be found in very few documents of the class (which means that they cannot be considered as characteristic for the class). The three other methods show distributions that are similar to each other. The right graphic of figure 5 reveals the difference between those three methods. This time not only the distribution of the terms across the class that the terms were extracted for has been analyzed but also the distribution across the two other classes. As can clearly be seen our approach is the only one that favors terms that often occur in the corresponding class but rarely in other classes.

6 Conclusion

In this paper we presented a novel approach for the extraction of discriminating and overlap terms out of a set of document classes. By applying our method to two important application scenarios, we were able to demonstrate its relevance and performance for real problems.

First, our method gives insight into the topical coherences and differences among several distinct document classes, e.g. the papers of scientific conferences. Secondly, our method allows us to do domain-specific term filtering. The discrimination calculation is able to filter out automatically the vocabulary that covers a certain aspect or has a certain function. With both applications we are able to show that our method not only yields very good results but also can be applied easily and in a very flexible way.

While we apply some language dependent preprocessing techniques like base form reduction, POS tagging and NP chunking, the core of our approach is language independent.

Finally, we evaluate our method using the extracted terms in a classification task, where it yields better results than a number of other methods for term extraction. We assure that our method extracts discriminating terms that at the same time have a high relevance for a class.

A wider range of promising application scenarios is easily imaginable - wherever classes of documents differ in topical, stylistic or linguistic features, and those differences on their part are of interest.

Acknowledgment. This work has partly been funded by the Research Initiative "Computational Analysis of Linguistic Development" at the University of Konstanz and by the German Research Society (DFG) under the grant GK-1042, Explorative Analysis and Visualization of Large Information Spaces, Konstanz.

We thank the anonymous reviewers of the NLDB 2009 for their valuable comments.

References

1. Spärck Jones, K.: A statistical interpretation of term specificity and its application in retrieval. Journal of Documentation 28(1), 11–21 (1972)
2. Salton, G., Wong, A., Yang, C.S.: A Vector Space Model for Automatic Indexing. Communications of the ACM 18(11), 613–620 (1975)
3. Kageura, K., Umino, B.: Methods of automatic term recognition: A review. Terminology 3(2), 259 (1996)
4. Feldman, R., Fresko, M., Kinar, Y., Lindell, Y., Liphstat, O., Rajman, M., Schler, Y., Zamir, O.: Text mining at the term level. In: Proceedings of the Second European Symposium on Principles of Data Mining and Knowledge Discovery, pp. 65–73 (1998)
5. Matsuo, Y., Ishizuka, M.: Keyword Extraction from a Single Document using Word Co-occurrence Statistical Information. In: Proceedings of the 16th International Florida AI Research Society, pp. 392–396 (2003)
6. Riloff, E., Jones, R.: Learning dictionaries for information extraction by multi-level bootstrapping. In: Proceedings of the Sixteenth National Conference on Artificial Intelligence, pp. 474–479 (1999)

7. Collins, M., Singer, Y.: Unsupervised models for named entity classification. In: Proceedings of the Joint SIGDAT Conference on Empirical Methods in Natural Language Processing and Very Large Corpora (1999)
8. Brunzel, M., Spiliopoulou, M.: Domain Relevance on Term Weighting. In: 12th International Conference on Applications of Natural Language to Information Systems, pp. 427–432 (2007)
9. Witschel, H.F.: Terminologie-Extraktion: Möglichkeiten der Kombination statistischer und musterbasierter Verfahren. In: Content and Communication: Terminology, Language Resources and Semantic Interoperability. Ergon Verlag, Würzburg (2004)
10. Velardi, P., Missikoff, M., Basili, R.: Identification of relevant terms to support the construction of domain ontologies. In: Proceedings of the workshop on Human Language Technology and Knowledge Management, pp. 1–8 (2001)
11. Drouin, P.: Detection of Domain Specifc Terminology Using Corpora Comparison. In: Proceedings of the International Language Resources Conference, pp. 79–82 (2004)
12. Wise, J.A.: The ecological approach to text visualization. Journal of the American Society for Information Science, 1224–1233 (1999)
13. Kaski, S., Honkela, T., Lagus, K., Kohonen, T.: WEBSOM - Selforganizing maps of document collections. Neurocomputing 21, 101–117 (1998)
14. Lagus, K., Kaski, S.: Keyword selection method for characterizing text document maps. In: Proceedings of ICANN 1999, Ninth International Conference on Artificial Neural Networks, pp. 371–376 (1999)
15. Azcarraga, A.P., Yap, T.N., Tan, J., Chua, T.S.: Evaluating Keyword Selection Methods for WEBSOM Text Archives. IEEE Transactions on Knowledge and Data Engineering 16(3), 380–383 (2004)
16. Seki, Y., Eguchi, K., Kando, N.: Multi-Document Viewpoint Summarization Focused on Facts, Opinion and Knowledge. In: Computing Attitude and Affect in Text: Theory and Applications. The Information Retrieval Series, pp. 317–336. Springer, Heidelberg (2005)
17. Lerman, K., McDonald, R.: Contrastive Summarization: An Experiment with Consumer Reviews. In: Proceedings of the North American Association for Computational Linguistics, NAACL (2009)
18. Zhai, C., Velivelli, A., Yu, B.: A Cross-Collection Mixture Model for Comparative Text Mining. In: Proceedings of the ACM SIGKDD international conference on Knowledge discovery and data mining (KDD), pp. 743–748 (2004)
19. Mei, Q., Zhai, C.: A mixture model for contextual text mining. In: Proceedings of the 12th ACM SIGKDD international conference on Knowledge discovery and data mining (KDD), pp. 649–655 (2006)
20. Salton, G., Buckley, C.: Term weighting approaches in automatic text retrieval. Information Processing and Management: an International Journal 24(5), 513–523 (1988)
21. Kuhlen, R.: Experimentelle Morphologie in der Informationswissenschaft. Verlag Dokumentation (1977)
22. Toutanova, K., Manning, C.: Enriching the knowledge sources used in a maximum entropy part-of-speech tagger. In: Proceedings of the Joint SIGDAT Conference on Empirical Methods in Natural Language Processing and Very Large Corpora (EMNLP/VLC), pp. 63–70 (2000)
23. Toutanova, K., Klein, D., Manning, C., Singer, Y.: Feature-rich part-of-speech tagging with a cyclic dependency network. In: Proceedings of the Conference of the North American Chapter of the Association for Computational Linguistics on Human Language Technology (NAACL), pp. 173–180 (2003)

24. Stanford Log-linear Part-Of-Speech Tagger,
    http://nlp.stanford.edu/software/tagger.shtml
25. Ramshaw, L., Marcus, M.: Text Chunking Using Transformation-Based Learning. In: Proceedings of the Third ACL Workshop on Very Large Corpora (1995)
26. Greenwood, M.: Noun Phrase Chunker Version 1.1,
    http://www.dcs.shef.ac.uk/~mark/phd/software/chunker.html
27. Thiel, K., Dill, F., Kötter, T., Berthold, M.R.: Towards Visual Exploration of Topic Shifts. In: IEEE International Conference on Systems, Man and Cybernetics, pp. 522–527 (2007)
28. Online tool for terminology extraction,
    http://wortschatz.uni-leipzig.de/~fwitschel/terminology.html

AIR: A Semi-Automatic System for Archiving Institutional Repositories

Natalia Ponomareva[1], Jose M. Gomez[2], and Viktor Pekar[3]

[1] University of Wolverhampton, UK
nata.ponomareva@wlv.ac.uk
[2] University of Alicante, Spain
jmgomez@ua.es
[3] Oxford University Press
viktor.pekar@oup.com

Abstract. Manual population of institutional repositories with citation data is an extremely time- and resource-consuming process. These costs act as a bottleneck on the fast growth and update of large repositories. This paper aims to describe the AIR system developed at the university of Wolverhampton to address this problem. The system implements a semi-automatic approach for archiving institutional repositories: firstly, it automatically discovers and extracts bibliographical data from the university web site, and, secondly, it interacts with users, authors or librarians, who verify and correct extracted data. The system is integrated with the Wolverhampton Intellectual Repository and E-theses (WIRE), which was designed on the basis of standard software adopted by many UK universities. In this paper we demonstrate that the system can considerably increase the intake of new publication data into an institutional repository without any compromise to its quality.

1 Introduction

The continued development of OAI-PMH (Open Archive Initiative Protocol for Metadata Harvesting) compliant institutional repositories is of vital importance to education and research institutions and to the economy in general[1]. Many institutions that have implemented repositories have encountered cultural barriers to embedding repositories and challenges surrounding deposit. The university of Wolverhampton has successfully established WIRE - an open access institutional repository for research output[2]. Open Repository is a service from BioMed Central to build, launch, host and maintain institutional repositories for organisations. Built upon the latest DSpace repository software[3] the service has been designed to be flexible and cost-effective.

The WIRE repository was launched two years ago and it is already populated with over 1500 articles and theses. However, we are acutely aware both from our

---

[1] http://www.jisc.ac.uk/news/stories/2007/06/news_repos.aspx
[2] http://wlv.openrepository.com/wlv/
[3] http://www.dspace.org/

H. Horacek et al. (Eds.): NLDB 2009, LNCS 5723, pp. 169–181, 2010.

own experience and through talking to WIRE librarians that we are still far away from comprehensive archiving (or as close as we can get to this while remaining within the constraints of copyright, licensing and embargo restrictions). Common difficulties facing many repositories include the identification and capture of their institution's research output, time spent providing mediated deposition services or resistance to, and/or lack of engagement with, self-archiving and copyright/licence clearing of material for open access.

The AIR (Automatic Archiving for an Institutional Repository) system was developed to address the problems mentioned above and was implemented based on the essential principals of software development: platform independence, reliability, portability, easy customisation, scalability, etc. It employs an automatic approach to identify publications and extract bibliographical metadata from them in order to assure completeness of the research data produced by the university staff. The quality of the bibliographical metadata uploaded to WIRE is ensured by authors and WIRE librarians who can access, verify and, if necessary, correct the extracted information with the help of a user friendly interface.

The paper is organised as follows. In Section 2 an overview of related works is provided. General system architecture and its main characteristics are reported in Section 3. Section 4 is devoted to the detailed description of the AIR core module, whose evaluation is given in Section 5. Finally, in Section 6 we draw conclusions and discuss the future work.

2   Related Works

In response to time- and resource constraints of manual population of institutional repositories, a number of research projects have been recently carried out. There can be distinguished two main approaches to create and populate digital repositories. One of them exploits the principal of collaborative sharing bibliographical metadata. For example, CiteULike[4] is based on social bookmarking: researchers create their own databases and share them with other users. This system has a facility of importing articles from different online resources with automatic detection of bibliographical metadata. A similar principal is used by Bibster[5] - an ontology based Peer-to-Peer system for exchanging bibliographical data among researchers. It uses two different ontologies: one contains some generic aspects of bibliographical metadata and the other describes specific categories for the Computer Science domain. The use of these ontologies enables automatic extraction of metadata from bibliographical entries and provides an effective search for required publications ranking the results according to their semantic similarity with a query [1].

Another approach that was adopted by many existing systems exploits an automatic search of bibliographical metadata on the Web. For example, Symplectic Publications Management System[6] (PMS) automatically harvests

---

[4] http://www.citeulike.org/
[5] http://bibster.semanticweb.org/
[6] http://www.symplectic.co.uk/products/publications.html

journal citation data from some online sources such as ISI Proceedings, PubMed and Web of Science. Nonetheless, most of these systems can only make use of data that has already been encoded in the format with which they have been adapted to deal. As a result, the coverage of such systems is greatly limited. Symplectic PMS, for example, processes only journal papers of specified publishers, missing out such publications as conference proceedings, books and book chapters.

To our knowledge, the only systems that actually extract bibliographical data from the unrestricted Web are CiteSeer[7] and Google Scholar[8].

3   System Architecture

General architecture of the AIR system is presented in Fig. 1. The process of extracting bibliographical metadata can be described as follows: first, a crawler populates the database with all web pages belonging to the university domain. The extracted web pages go to the AIR core module, which classifies pages into relevant (containing bibliographical entries) and irrelevant ones, then finds references and maps their parts with Dublin Core Metadata tags. We propose a semi-automatic approach that ensures the reliability of information transferring to WIRE: automatically extracted data must be verified by users in order to be uploaded to WIRE. The verification process is carried out through user-friendly web interfaces. The final step of the system consists in transferring the fully revised data to WIRE, which was implemented using SWORD[9] protocol.

3.1   Principal Modules

In order to crawl web pages belonging to the university domain we used Nutch software[10]: an open source Java implementation of a search engine. It provides all necessary tools to run a search engine satisfying some user requirements. Nutch is built on top of Lucene, which is an API for text indexing and searching, and it is naturally divided into two pieces: the crawler and the searcher. The crawler fetches pages and turns them into an inverted index, which the searcher uses to answer users' search queries. The crawler can fetch billions of web pages quickly and efficiently, using a distributed filesystem via Hadoop[11]. We employed Nutch mostly because it claims to be a transparent and stable tool with an open source implementation and it is widely used by many researchers.

The AIR core module (Fig. 2) is the principal module of the AIR system as it accomplishes an automatic search for publications in the university web site. Due to the complex structure of the AIR core module it will be described in detail in a separate section (see Section 4).

---

[7] http://citeseer.ist.psu.edu/

[8] http://scholar.google.co.uk/

[9] http://www.ukoln.ac.uk/repositories/digirep/index/SWORD_guide

[10] http://lucene.apache.org/nutch/

[11] http://hadoop.apache.org/core/

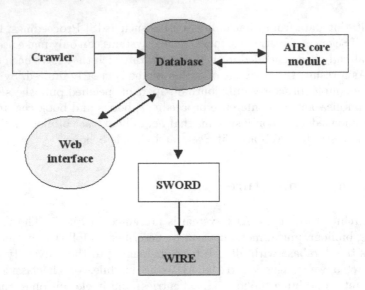

Fig. 1. General system architecture

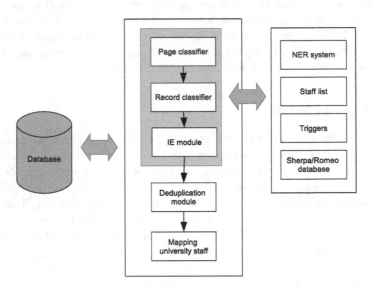

Fig. 2. The AIR core module design

The Web interfaces were developed in order to provide user interaction with the AIR system. The author of each publication needs to verify the automatically extracted data, post-edit it if necessary, and approve the deposit of the data into the institutional repository. Once authors log into the system (after being notified by email that their new publications are ready to be deposited into the repository), bibliographical data from the database is displayed in the web interface form that authors would normally fill in manually. For authors' convenience, they can see the actual document from which the data has been extracted. Authors can change the text in the form and then submit it for further approval by WIRE librarians. The underlying idea is that instead of typing in whole entries in the web form, authors only have to edit the automatically extracted data that is already inserted into the appropriate fields of the form, in this way speeding up the process.

The web interface via which WIRE librarians verify the data submitted by authors has additional features that can facilitate the process of verifying and uploading information to WIRE. Firstly, it provides a facility that automatically queries the SHERPA/Romeo database[12] to determine the copyright status of publications. Secondly, it carries out an automatic web search of publication's identifiers (ISSN, ISBN, etc.) using Yahoo APIs. The search is based on four elements: identifier format, publication's title, closeness of identifiers (that satisfied the format) to the title and frequency of identifier appearance in retrieved pages.

The possibility of transferring validated data to WIRE is delivered by SWORD facility which requires the bibliographical data to be encoded according to Dublin Core Metadata format and written to an XML-file of a special structure dictated by SWORD. An XML-file is created separately for every publication, then it is compressed to a ZIP file and sent to WIRE.

3.2 System Characteristics

The AIR system was developed and implemented to satisfy the following requirements:

1.Platform independence. The system was implemented in such a manner that it can be easily installed and run under any platform with minimum effort from the user.

2.Reliability. The reliability of the AIR system was ensured through careful definition of its requirements, separating the processes of software design and implementation and careful testing.

3.Easy customisation. The AIR system was designed in such a way that all its parameters are set in external XML-files that can be easily changed without any code modification.

4.Strong object orientation and code clearness. This is an important feature if the code has to be modified and adjusted for some requirements. Object orientation allows the code to be flexible and simplifies modifications and additions.

5.Portability. Portability is a very important issue as it refers to an easy adaptation of the AIR system to other institutional repositories. This characteristic has to be taken into account due to the same standards used for implementation of institutional repositories in the UK. In order to archive portability the data extraction components of the system were not fine-tuned to web page layouts and formats established at the university of Wolverhampton. Moreover, machine

---

[12] http://www.sherpa.ac.uk/

learning algorithms, we employed in order to train the extraction components, also assure easy system adjustment to other formats or even domains.

6.Scalability. Experiments with the system showed that it is scalable and can easily cope with large amounts of data (i.e. web pages and publications). Currently the only restrictions identified are due to the Nutch engine used for crawling web pages and MySQL database where all publications are stored. Both programs are mature pieces of software that can be employed to process large quantities of data.

7.Security issues. The system must be secure and prevent the transfer of invalid or junk information to the WIRE. Therefore, librarians, as the only users who have the right to upload data to WIRE, are provided with logins and passwords. The passwords are coded and securely stored on the server. For the sake of security librarians are given different access rights to the system. Only librarian-administrators can create new users and allow them to enter the system.

8.Extracting information types corresponding to Dublin Core Metadata. The system aims to extract the following types of information: Author name, Title, Year of Issue, Identifier, Publisher, Citation (which includes Conference/Journal/Book Name, Volume, Pages, Venue, etc.). It should be highlighted that the system can be easily adapted to other types of information due to its machine-learning based implementation.

4   AIR Core Module

The AIR core module consists of 5 submodules:

1. Page classifier that extracts web pages containing bibliography.
2. Record classifier that selects bibliographical entries from all document records.
3. Information Extraction (IE) module which aims to map bibliographical data with Dublin Core Metadata fields, i.e. title, publisher, identifier, etc.
4. Deduplication module which filters out repeated references.
5. Mapping authors of extracted bibliographical references with members of the university staff. As we are interested only in scientific contribution by university researchers we decided to limit the scope of bibliographical entries to those belonging only to staff members.

We should point out that the first three modules address different types of classification problems and, henceforth, they will be called classification components.

4.1   General Implementation of Classification Components

During the system development, we explored two approaches: rule-based and machine learning-based. Rule-based approaches usually give high precision but rather low recall. In addition, their development is time-consuming as rules need to be manually produced, making them domain-dependent and difficult to adapt to other domains. In contrast, machine learning approaches usually ensure high recall and easy portability to other domains, with the downside that a large amount of annotated data is required for training in order to achieve sufficient accuracy. As we are mostly interested in high recall rather than in very precise results, we decided to adopt a machine learning approach and to increase accuracy by increment of training data and experimenting with different sets of features and machine learning techniques.

The development of classification components requires the following steps:

1. Establishing a set of suitable machine-learning techniques.

Machine-learning methods must be chosen according to the task they are solving. Selection of web pages containing publications is a regular classification task, while the extraction of bibliographical fields can be seen as a sequence labeling procedure. The latter should be treated differently applying methods, which take into account not only the features of the element but also its neighborhood and location in a sequence.

2. Establishing a set of relevant features.

Due to the same subject of the three tasks we use a similar set of features while constructing corresponding classifiers:

- **Named Entities,** such as PERSONS, LOCATIONS, ORGANIZATIONS and DATES that were tagged using Annie[13] parser, a part of GATE[14] package.

- **Staff list**. A list of all university members of research staff was collected and stored in the database. The list is regularly renewed in order to capture any alterations in staff complement.

- **Sherpa/Romeo publishers** - a list of publishers stored in Sherpa/Romeo database.

- **Sherpa/Romeo journals** - a list of journals stored in Sherpa/Romeo database.

- **Presence of year**. This feature indicates whether a corresponding element contains a year.

- **Publication triggers**. We built different lists of triggers: for example, header triggers (the most frequent words that occur in <h> tags of files containing publications), publication triggers (words appearing in a bibliographical entry itself), citation triggers (words appeared in citation of an entry).

- **Orthographic features**, which contain capitalization, digits, punctuation marks, etc. This feature was only used for implementation of the IE module.

- **Parts of speech (POS)**. Preliminary experiments revealed many cases when automatically annotated bibliographical fields ended with articles, conjunctions or prepositions. In order to correct this situation we incorporated these parts of speech into the IE module.

3. Experimenting with feature adjusting and classifier selection.

It is a well-known fact that not all the features contribute positively into the final result of the classification and that it is a very important to select

---

[13] http://gate.ac.uk/ie/annie.html
[14] http://gate.ac.uk/

a set of features that gives the highest accuracy. Moreover, a great amount of noisy features increases the time costs without any improvement in the system performance. In Section 5 the experiments conducted with different numbers of features are described and discussed.

4.2   Page Classifier

Web pages with publications can have different formats: HTML, PDF or just plain text. It is obvious that not only the content but the structure of a document must be used in order to locate bibliographical entries. Concerning HTML format we distinguish 3 types of structured text that can be useful for revealing publications:

- Metadata - text appeared in meta tags like <keywords>, <title> and <description>. We also consider anchor text as it usually contains an explicit description of document content.
- Headers - text contained in header tags: <h1>, <h2>, etc. The text between these tags should be processed separately because if words like "publications" or "books" appear in headers the possibility that the file contain bibliographical entries is much higher than if those words occur in another part of a document.
- Content - the rest of a document.

As far as PDF and plain text formats are concerned, we only consider the content of documents due to absence of structural elements for these formats. .

The choice of machine learning methods was mostly defined by a set of methods contained in WEKA[15] - a well-known open source software implementing a number of machine learning algorithms for data mining. We employ 5 different methods: JRIP (rule learner algorithm described in [2]), Naive Bayes, Decision table (a simple decision table majority classifier [3]), J48(C4.5 decision tree classifier) and SMO (Support vector machine classifier as described in [4]).

4.3   Record Classifier and IE Module

When classifying document records it is very important to incorporate information about a document structure together with features of separate records. One such structural element that can reveal enumerations or a list of equivalent records is an HTML-tag of a record. As an order of records can also help to correctly determine the start and end point of a publication list, we decided to apply a finite state transducer - Conditional Random Fields (CRF) [5] that are proved to be one of the most effective methods for sequence labelling tasks. In order to implement CRF-based IE module an open source Mallet package [6] was used.

The IE module needs to identify 5 different metadata types: author, title, citation, date and publisher (extraction of identifiers was not carried out due to the lack of training data). As bibliographical reference represents a logical consequence of metadata tags, the use of CRF is indisputable.

---

[15] http://www.cs.waikato.ac.nz/ml/weka/

One of the important parameters of CRF classifier is a Markov order, characterising a number of previous elements which can influence the state of a pending one. In Section 5 we experiment with different values of Markov orders and show the effect of this on the evaluation results.

4.4 Deduplication Module

Once bibliographical entries are extracted and stored in the database, the deduplication analysis has to be carried out. References can be repeated in different web pages of the university, moreover, some spelling mistakes can occur, which makes the deduplication process more complicated. In our implementation the deduplication module is based only on the first author's name and a title, which means that if they are equal or similar the corresponding references are considered to be equal. For measuring similarity we use Levenshtein edition distance: it was applied at word level for titles and at letter level for author's names. Different thresholds for both titles and names were established in order to allow spelling mistakes to be introduced into the records.

The process of comparing database entries is a very time-consuming, moreover, in the case of similar references the problem of choosing one of them as the target reference arises. In order to make the process more efficient we employed Lucene[16] search engine that indexes first author's name and publication title for each database entry. Each new entry is compared with the stored ones and the number of times it appears is counted. At the end if there are several similar references we choose the one with the larger number of occurrences. In case of an equal number of occurrences the entry with the longer title is selected. This process ensures that we always obtain more complete titles without abbreviations and more complete names instead of only initials.

4.5 Mapping Authors with Members of the University Staff

Once all repeated references are removed researchers belonging to the university staff should be associated with their publications. Obviously one publication can be connected with different members of staff if it has multiple authors. Our method of mapping authors with members of staff consists in comparing author's last and first names and if only initials are available the first names are shortened to compare only first letters. One of the problems that can occur during this procedure is an ambiguity problem caused when two or more members of staff have the same name. In this case we associate a bibliographical entry with all of them, giving the authors responsibility to clarify the ambiguity.

5   Evaluation of the AIR Core Module

Evaluation of the AIR core module was carried out separately for each of its classification components. In all experiments we used cross validation with

---

[16] http://lucene.apache.org

number of folds equal to 5 in order to obtain a reliable estimation of the system performance.

5.1 Evaluation of the Page Classifier

The data consists of about 1850 manually annotated web pages, 150 of them containing publications. Table 1 gives the results obtained by all employed methods. The best performance was achieved by the JRIP method which outperformed other methods both in recall and F-score reaching 90% of recall. Although 90% is quite a good result it is not sufficient for our purposes because we lose 10% of relevant web pages, which means that 10% of authors are missed by the system. In order to avoid missing relevant data we decided to skip this stage in the final implementation of the AIR system, assuming that all web pages contain publications and passing them directly to the record classifier. Due to rather small size of the university web site (approximately 80 000 web pages) the classification of records of all university web pages is not a time- or resource-intensive process.

Table 1. Performance of the page classifier

N	Method	Precision	Recall	F-score
1	JRIP	0.864	**0.901**	**0.882**
2	Decision table	0.863	0.801	0.831
3	Naive Bayes	0.43	0.894	0.583
4	SMO	0.853	0.823	0.838
5	J48	**0.870**	0.809	0.838

5.2 Evaluation of the Record Classifier

The data consists of 55 web pages, total number of records is equal to 4200, 520 of which are bibliographical entries. We conducted our experiments for different Markov orders: 1, 2, 3. The results of the experiments are shown in Table 2.

Table 2. Performance of the record classifier

Markov order	Precision	Recall	F-score
1	0.929	**0.914**	**0.919**
2	0.929	0.898	0.910
3	0.929	0.859	0.888

The model based on bigrams obtained the best performance which means that knowledge about more distant neighbours of a bibliographical entry is not necessary and only brings noise into input data.

The overall performance of this component can be estimated as good although we have the same problem as for the previous module: in spite of quite high recall (91.4%) the risk of skipping bibliographical references still exists. In any case, as the classification is conducted at record level, we can miss only separate entries, not the whole pages.

5.3   Evaluation of IE Module

The data represents 520 publication references manually annotated with bibliographical metadata. In Tables 3-4 results of evaluation for Markov orders equal to 1 and 2 respectively are presented. Our baseline corresponds to the model where a set of standard features used for previous procedures was used: words, staff list, citation triggers, presence of year, etc. We compare our baseline with richer models adding also POS and orthographic features. In order to estimate statistical significance of differences in performance shown by different models the results are given with their confidence intervals calculated for $\alpha = 0.05$ (confidence level equal to 0.95). Comparison of the results for different Markov orders shows a slightly better performance of the more complex model, which is statistically significant for the fields "title", "citation" and "publisher".

Table 3. Performance (F-score) of IE module for Markov order 1

Model	Author	Date	Title	Citation	Publisher
baseline	0.968 ± 0.007	0.872 ± 0.033	0.929 ± 0.008	0.878 ± 0.009	0.438 ± 0.047
bas+POS	0.968 ± 0.007	0.859 ± 0.034	0.918 ± 0.009	0.860 ± 0.010	0.379 ± 0.046
bas+ortho	0.984 ± 0.005	0.865 ± 0.034	0.940 ± 0.008	0.895 ± 0.009	0.538 ± 0.047
bas+POS+ortho	0.984 ± 0.005	0.870 ± 0.033	0.942 ± 0.008	0.899 ± 0.009	0.556 ± 0.047

Table 4. Performance (F-score) of IE module for Markov order 2

Model	Author	Date	Title	Citation	Publisher
baseline	0.974 ± 0.006	0.860 ± 0.034	0.928 ± 0.009	0.877 ± 0.009	0.379 ± 0.046
bas+POS	0.971 ± 0.007	0.857 ± 0.035	0.922 ± 0.009	0.870 ± 0.010	0.352 ± 0.046
bas+ortho	0.975 ± 0.006	0.869 ± 0.033	0.944 ± 0.008	0.899 ± 0.009	0.562 ± 0.047
bas+POS+ortho	0.982 ± 0.005	0.878 ± 0.033	0.956 ± 0.007	0.918 ± 0.008	0.616 ± 0.046

As far as feature contribution is concerned it is obvious that orthographic features improve the model performance especially for the fields "author", "citation" and "publisher" as these are the fields with the greatest number of punctuation marks. The positive contribution of POS tags is not that evident, they even worsen the results for the majority of fields if they are taken without orthographic features. The best result obtained for Markov order equal to 2 and the richest set of features demonstrates slight contribution of POS tags, mostly for extraction of citations and publishers.

In general, the evaluation of the IE module shows its good performance with very high F-score obtained for extraction of authors and titles. Low F-score for the field "publisher" can be explained by the lack of training data as many authors do not introduce this information while posting their publications. Another difficulty of publisher's extraction is caused by the presence of geographical entities (or, more precisely, cities) in their names. Therefore, they can be easily confused with citations.

6   Conclusions and Future Work

The presented work introduced a semi-automatic system for archiving institutional repositories that was developed at the university of Wolverhampton to foster the dissemination of research output produced by the university staff. Manual population of institutional repositories is an extremely time- and resource-intensive process and, therefore, their growth is restricted by limits dictated by human resources. The AIR system aims to automate the data archiving process, so that librarians will be involved only at the final stage of the system for data verification and correction.

The main challenge the system copes with is caused by the need to deal with unstructured Web data that could have rather diverse formats. The evaluation results demonstrate high system performance and encourages us to improve the system further in order to make it more convenient for users.

As future work, we plan to extend the volume of the training data set reusing already verified references obtained at the end of the system cycle. We expect to improve the system performance not only because of the growth of training data but also due to the fact that the accumulation of references' formats used at the university adjusts the system to the university standards and requirements. Another direction of our future research will be related to conducting more experiments with other Machine Learning algorithms and other feature sets in order to raise the system accuracy.

Better access to the bibliographical metadata is currently being explored using the QALL-ME framework[17], a domain independent architecture that can be used to easily implement question answering systems. By using the framework we envisage to be able to give users access to the contents of the database in a natural way, by answering questions about publications such as "What authors published papers in the Journal of Natural Language Engineering in 2002?".

Acknowledgements

This work has been partially supported by the Joint Information Systems Committee (JISC)[18] through the AIR project. Some system modules have been developed in the framework of the project QALL-ME[19], which is a 6th Framework Research Programme of the European Union (EU), contract number: FP6-IST-033860 and the TEXT-MESS project. The authors also want to acknowledge Javier Espinosa de los Monteros for his large contribution to the development of the Web interfaces.

---

[17] http://qallme.sourceforge.net/
[18] http://www.jisc.ac.uk/
[19] http://qallme.itc.it/

References

1. Haase, P., Broekstra, J., Ehrig, M., Menken, M., Plechawski, M., Pyszlak, P., Schnizler, B., Siebes, R., Staab, S., Tempich, C.: Bibster - a semantics-based bibliographic peer-to-peer system. In: Proceedings of the Third International Semantic Web Conference, pp. 122–136 (2004)
2. Cohen, W.W.: Fast effective rule induction. In: Proceedings of the Twelfth International Conference on Machine Learning, pp. 115–123. Morgan Kaufmann, San Francisco (1995)
3. Kohavi, R.: The power of decision tables. In: Proceedings of the European Conference on Machine Learning, pp. 174–189. Springer, Heidelberg (1995)
4. Platt, J.C.: Fast training of support vector machines using sequential minimal optimization. In: Advances in kernel methods: support vector learning, pp. 185–208. MIT Press, Cambridge (1999)
5. Lafferty, J., McCallum, A., Pereira, F.: Conditional random fields: Probabilistic models for segmenting and labeling sequence data. In: ICML 2001: Proceedings of the Eighteenth International Conference on Machine Learning, pp. 282–289. Morgan Kaufmann Publishers Inc., San Francisco (2001)
6. McCallum, A.K.: Mallet: A machine learning for language toolkit (2002), http://mallet.cs.umass.edu

Protein-Protein Interactions Classification from Text via Local Learning with Class Priors

Yulan He and Chenghua Lin

School of Engineering, Computing and Mathematics
University of Exeter, North Park Road, Exeter EX4 4QF
{y.he,cl322}@exeter.ac.uk

Abstract. Text classification is essential for narrowing down the number of documents relevant to a particular topic for further pursual, especially when searching through large biomedical databases. Protein-protein interactions are an example of such a topic with databases being devoted specifically to them. This paper proposed a semi-supervised learning algorithm via local learning with class priors (LL-CP) for biomedical text classification where unlabeled data points are classified in a vector space based on their proximity to labeled nodes. The algorithm has been evaluated on a corpus of biomedical documents to identify abstracts containing information about protein-protein interactions with promising results. Experimental results show that LL-CP outperforms the traditional semi-supervised learning algorithms such as SVM and it also performs better than local learning without incorporating class priors.

Keywords: Text classification, Protein-protein interactions, Semi-supervised learning, Local learning.

1 Introduction

Text classification is the process of categorizing documents into different classes using predefined category labels. It is a difficult task because of the complexity and ambiguity of natural language, where a word may have different meanings or multiple phrases can be used to express the same idea. The task of classifying biomedical literature is made more complex than even standard text classification by the fact that such papers use a varied and specialized vocabulary. The corpus also tends to be very large because of the vast number of papers available, with online repositories such as PubMed[1] containing over 16 million citations alone. This makes it an uphill task to get the data that one needs.

One area where biomedical text classification would be useful is in the study of protein-protein interactions (PPI). Analyzing these interactions is invaluable in learning more about cellular function, which in turn paves the way for breakthroughs in medicine and biochemistry. As such, the cataloguing of interactions between proteins is an essential facet of data mining in biomedical literature, which has led to the creation of online databases specifically devoted to this task [1,2]. In order to narrow down the search for human and automated curators alike, text classification could be performed to separate the documents that describe protein-protein interactions from those that do not. It

---

[1] http://www.ncbi.nlm.nih.gov/entrez/query.fcgi?DB=pubmed

H. Horacek et al. (Eds.): NLDB 2009, LNCS 5723, pp. 182–191, 2010.

has been shown to be possible to identify Medline papers containing such interactions by examining the word frequencies in their abstracts instead of the entire documents [3]. This would speed up the rate at which classification is performed, making it more cost-effective to use before searching for interactions within the documents.

So far, there has not been much work done on text classification of biomedical literature, or at least that focus on protein interactions. One popular method of classification in use appears to be support vector machines. It has been used in general classification of biomedical literature along with clustering [4], and has also been applied specifically to classifying PPI-related documents [5]. Another approach uses a Bayesian method to calculate the relevance of documents to PPI [3]. In this method, documents containing PPI were used as a training set and their frequencies compared with a dictionary of the most common words in the corpus. From this, a list of discriminating words were obtained that might differentiate other relevant documents as well. Each abstract was then ranked using Bayesian probabilities that were converted into log likelihood scores.

In the area of identifying significant terms before classification, substring matching has been proposed [6]. This method involves indexing all substrings of the words in the corpus and ranking them based on relevance to classification. This results in a greatly enlarged vocabulary, but is able to identify word parts like *acety* and *peptide* that are meaningful in biology, that traditional stemming algorithms are unable to find.

Traditionally, classification problems have been handled by supervised learning, in which the entire training set consists of labeled data. However, such data is difficult and tedious to obtain, making it impractical for real-life situations. This has resulted in a growing interest in semi-supervised learning, where the training set only has a small proportion of labeled data in comparison with the large amount of unlabeled data in the set. The fact that training a classifier using both labeled and easily obtained unlabeled data makes semi-supervised learning much more flexible.

This paper proposed a semi-supervised learning algorithm based on local learning with class priors (LL-CP) for biomedical text classification. The LL-CP algorithm represents labeled and unlabeled examples as vertices in a connected graph. The label information from the labeled vertices is then propagated to the whole dataset using the linear neighborhoods with sufficient smoothness. The class prior has been incorporated to force the class distribution of the unlabeled set to be similar to that of the labeled set. Experiments have been extensively studied to identify text documents containing protein-protein interactions with only a limited number of label documents.

The rest of the paper is organized as follows. Section 2 presents the semi-supervised learning algorithm based on local learning with class priors for protein-protein interactions classification. Experimental setup and results are discussed in Section 3 and Section 4 respectively. Finally, Section 5 concludes the paper and outlines the possible future work.

2   Local Learning with Class Priors (LL-CP) for PPI Classification

In the local learning framework, data objects are represented as vertices in a fully connected graph with weighted edges. Each vertex has soft labels (i.e. the value of the label can be continuous) associated with it, which stand for the distribution over the various classes for that vertex. The larger the weight is on an edge, the closer the vertices connected by that edge are to each other, and the easier it is for labels to propagate through that edge. Most graph-based semi-supervised learning methods [7,8,9] adopted a Gaussian function to calculate the edge weights of the graph and as a result, they are sensitive to the setting of the variance σ of the Gaussian function. A small variation of σ could affect the classification accuracy dramatically. More recently, several algorithms [10,11] have been proposed to overcome this problem where the predicted label at an unlabeled point x_i is the weighted average of its neighbors' solutions.

We propose a semi-supervised learning algorithm based on local learning with class priors (LL-CP). Assume that a class prior conditional probability is given in the form of $\tilde{P}(y|x_i)$ where $y \in \{-1, 1\}$ is the binary class label and x_i is a input instance. This prior knowledge essentially expresses our belief about the conditional distribution of the labels given the input features. It could be obtained in various ways. In the simplest case, it could be obtained from human prior knowledge. If such knowledge is not available, it could either be the maximum entropy prior $\forall x, y : \tilde{P}(y|x) = 0.5$ or the class prior estimated from the labeled data only. In this paper, we are particularly interested in obtaining the class prior in the later case. This essentially enforce that all unlabeled data are not put in the same class [12].

Thus, our goal is to find a model which minimizes the prediction error for each document as much as possible while at the same time its probabilistic predictions over the unlabeled data resembles the given class prior. Let $D = \{d_1, d_2, ..., d_{|D|}\}$ be a set of $|D|$ cosine-normalized document vectors with N dimensions each, with the first l documents being labeled and the remaining u documents left unlabeled. We have two classes here, either positive or negative. The document space is represented as a fully-connected graph where each node represents a document, and the edge between any two nodes represent a relationship between them.

Assume each document can be optimally reconstructed using a linear combination of its neighbors. Thus, for each document d_i, the objective is to minimize the least square error

$$\varepsilon_i = \frac{1}{n_i} \sum_{d_j \in \mathcal{N}_i} \|\mathbf{w}_j^T d_j - f_j\|^2 + \lambda_i \|\mathbf{w}_i\|^2 \tag{1}$$

$$s.t. \sum_j w_{ij} = 1, w_{ij} \geqslant 0 \tag{2}$$

where $\mathcal{N}_i$ represents the neighborhood of d_i, $n_i = |\mathcal{N}_i|$ is the cardinality of $\mathcal{N}_i$, f_j indicates whether d_j belongs to a positive or negative class, and w_{ij} is the contribution of d_j to d_i with larger w_{ij} indicating closeness of the documents, and λ_i is a regularization parameter.

It has been shown in [13] that the optimal solution is

$$\mathbf{w}_i^* = \mathbf{D}_i(\mathbf{D}_i^T \mathbf{D}_i + \lambda_i n_i \mathbf{I}_i)^{-1}\mathbf{f}_i \tag{3}$$

where $\mathbf{D}_i = [d_i^1, d_i^2, ..., d_i^{n_i}]$ in which d_i^k denotes the k-th nearest neighbor of d_i, $\mathbf{f}_i \in \mathcal{R}^{n_i}$ is the vector $[f_j]^T$ for $d_j \in \mathcal{N}_i$, and $\mathbf{I}$ is an $n_i \times n_i$ identity matrix.

An iterative procedure is then performed to propagate labels of the labeled data to the remaining unlabeled data using the graph constructed in the above step. In each iteration, the label information of a document object is updated by the label information from its neighborhood. At time $t + 1$, the label of d_i becomes

$$f_i^{t+1} = \alpha \sum_{j:d_j \in \mathcal{N}_i} w_{ij} f_j^t + (1 - \alpha) y_i \tag{4}$$

where $0 < \alpha < 1$ determines the amount of the label information that d_i receives from its neighbors. y_i is the label of d_i at the initial state. That is, if d_i is initially labeled, then y_i is its original label; if d_i is initially unlabeled, then $y_i = 0$. f_i^t is the predicted label at iteration t.

The label of each document object is updated iteratively until the predicted labels of the data do not change in several successive iterations.

There are several ways to incorporate the class prior knowledge into the local learning process. First, the class prior information can be added as an additional constraint into the objective function. Let P_l be the multinomial distribution of class proportion in the labeled set, and $\tilde{P}_W$ be the class proportion produced by the current model parameterized by $\mathbf{W}$, P_l and $\tilde{P}_W$ are defined as:

$$P_l = \frac{1}{l} \sum_{j=1}^{l} f_j \tag{5}$$

$$\tilde{P}_W = \frac{1}{u} \sum_{i=l+1}^{l+u} f_i \tag{6}$$

where l and u are the number of documents in the labeled set and unlabeled set respectively. An additional constraint $P_l = \tilde{P}_W$ could be added.

It is also possible to add the class prior itself as a regularizer to the objective function by minimizing the KL-divergence of P_l and P_W [14]. We leave this as future work for further exploration.

We follow a simple procedure called class mass normalization (CMN) proposed in [15] to adjust the class distributions to match the priors. Let P^+ and P^- denote the class prior probability estimated from the labeled set for the positive and negative class respectively. The estimated class label f_i for an unlabeled document d_i is readjusted by incorporating the class prior probabilities and is classified as positive class iff

$$P^+ \frac{f_i}{\sum_{i=l+1}^{l+u} f_i} > P^- \frac{1 - f_i}{\sum_{i=l+1}^{l+u} (1 - f_i)} \tag{7}$$

3   Experimental Setup

For all experiments, the LL-CP algorithm was evaluated using data provided by the second BioCreAtIvE (Critical Assessment for Information Extraction in Biology) challenge[2]. The BioCreAtIvE challenge evaluation was set up in order to apply approaches in information retrieval and text mining to biomedical literature, and to evaluate them

---

[2] http://biocreative.sourceforge.net/

against a standard set of data for comparison. One of the tracks in the second BioCre-AtIvE challenge in 2006 was the extraction of PPIs from text, which includes the retrieval of documents containing information about protein interactions (Protein Interaction Article Sub-task 1).

The documents in the training set provided for the Protein Interaction Article Sub-Task 1 was used as the corpus. This set consists of biomedical publications from the PubMed database, and the documents are split into two categories: those that contain information about protein interactions and those that do not. In total, there were 3536 true positive examples and 1959 true negative examples available. There were also 18930 positive but noisy examples which were not used in the experiments.

Training sets of labeled examples were obtained from the corpus using different sizes, with each containing 10, 25, 50, 75 or 100 documents. In addition, different proportions of positive/negative examples were also used, with 25%, 50% or 75% of the documents in the training sets being positive. Altogether, there were 15 different kinds of training sets used. There were also 5 test sets of unlabeled documents created, each unique set consisting of 250 relevant and 250 irrelevant documents.

3.1 Preprocessing

The documents to be classified were first read in by an XML parser. Only the CURATION_RELEVANCE (indicates whether the specified document is in the relevant set), TITLE and ABSTRACT child elements were saved into memory, while the other elements were ignored. During the parsing of the XML documents, stemming was performed using Porter's algorithm [16]. Stop-words were also removed by comparing words to a list of common words. Punctuation, numbers and other non-alphabet characters were ignored. After parsing, the tf-idf of the document vectors was computed, which in turn was passed into a matrix for performing singular value decomposition (SVD) along with the integer k, which is the reduced number of dimensions required. In order to perform the SVD needed for latent semantic indexing (LSI), JAMA (a Java matrix package)[3] was used. After decomposition, the resultant right-singular matrix V' was then saved as the set of column vectors of reduced dimensionality, and cosine-normalized.

3.2 Evaluation Metrics

In the area of information retrieval, the set of relevant documents matched (or retrieved) by the classifier is normally not exactly the same as the set of relevant documents in the corpus. Correct matches of relevant and irrelevant documents are known as true positives and true negatives respectively, while incorrectly classified documents are known as false negatives or positives. The most commonly used measures to evaluate the effectiveness of an algorithm are precision and recall [17]. Precision is the proportion of documents retrieved by the classifier that are relevant, while recall is the proportion of relevant documents in the entire corpus that were retrieved. The F-measure (or F-score) is a combination of both precision and recall, F-measure= $(2\times$Precision$\times$Recall$)/($Precision$+$Recall$)$. The ranges of all 3 values fall in the range $[0, 1]$, with a higher value indicating a better classification result.

---

[3] http://math.nist.gov/javanumerics/jama/

4   Results

This section presents the experimental results on the Protein Interaction Article Sub-task 1 of the second BioCreAtIvE challenge.

4.1   Number of Vector Dimensions

Experiments were conducted to find out if LL-CP performs better with fewer vector dimensions, or whether there is an optimal number of dimensions across all data sets.

The LSI of each pair of training and test sets was first computed, then the LL-CP algorithm was run on the resultant document matrix V_k, with k starting from the original length of the document vectors in V and decreasing in intervals of 10 until k was equal to either 10 or 5. For each pair of sets, the number of dimensions that resulted in the highest F-measure was recorded.

As can be seen from Figure 1, the majority of optimal results occurred when the number of vector dimensions was below 20, with the most frequently occurring number of dimensions being 10. This indicates that the top few dimensions returned from LSI are most important in correctly classifying the documents. In addition, all F-measures of more than 0.7 occurred only when the number of dimensions used was less than 400. As a comparison, the average F-measure of all trials run using vector dimensions of 5 or 10 was 0.693, while none of the trials with all dimensions included had a score above 0.6.

The average performance of the training sets was evaluated with the results displayed in Figure 2. The size of the training set varies between 10 documents to 100 documents and the axes in dash lines show the F-measure values 0.66, 0.7, and 0.74. It was found that the best overall size for a training set was at least 50, with better results for larger training sets. From this experiment, the most consistent and best overall split of positive and negative examples was 50%-50%, while the training sets with only 25% positive examples fared the worst. This implies that LL-CP either gives better performance with equal numbers of labeled examples from all classes, or with labeled data that mimics the proportions of class labels in the test set. Also, the size of the training set did not appear to make much difference in the classification quality, except where the proportion of positive labeled examples was 75%.

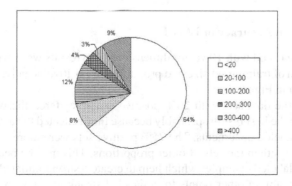

Fig. 1. A breakdown of optimal vector dimensions

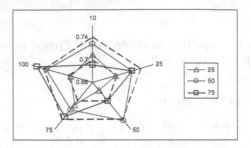

Fig. 2. Relationship between F-measure, size of training set and number of positive examples

The scores of an arbitrarily selected trial against the number of vector dimensions are charted in Figure 3, and are typical of most trials in this experiment. As can be seen from the graph, the scores usually stayed around 0.5 for the most part, indicating that some of the dimensions used were probably too noisy to aid in classification. The precision scores tended towards more gradual changes than the recall scores, which tended to be either 0 or at least 0.5, while rarely being in between the 2 values. Precision was hardly ever higher than recall.

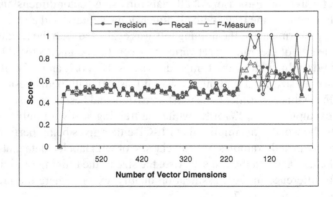

Fig. 3. The effect of vector dimensions on precision, recall and F-measure

4.2   Classification Accuracy of LL-CP

All data sets were tested with 10 vector dimensions. The results were averaged out over each combination of training set size and proportion of positive/negative examples. The results are shown in Figure 4.

As expected, the data sets with 25% positive examples fared the worst, The 50% positive sets gave the best results, probably because they reflected the actual class distribution of the unlabeled documents. The 75% positive sets were more consistent across all the data set sizes than the sets of other proportions. This may be because of the additional positive labeled examples which help to create a strong initial cluster of known positive points, enabling other points to be labeled strongly as positive. Negative examples are less likely to cluster together because irrelevant documents are not likely to

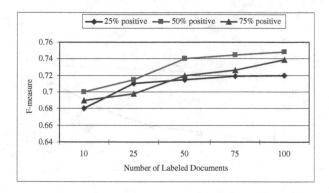

Fig. 4. The F-measures of each group of data with 25%, 50% and 75% positive examples versus different sizes of labeled documents

be similar to other irrelevant documents, whereas most relevant documents will have some features in common. In addition, the focus of this classifier is on labeling *relevant* documents correctly, and the F-measure reflects this.

This implies that if the class distribution of the unlabeled documents are known, then the documents in the labeled set should be chosen to reflect this. However, if it is unknown, then the labeled set should contain more positive than negative examples in order to strongly label the relevant documents.

4.3   Comparison with SVM and Local Learning without Class Priors

This set of experiments compares the performance of three different models, our proposed LL-CP, local learning without incorporating class priors (LL), and the Support Vector Machines (SVMs). Based on previous experiments carried out in this area [5], a radial-basis kernel function was used for SVM with σ set to 0.01 and C set to 2. Since these parameters were obtained from the optimization from PubMed abstracts like those in the BioCreAtIvE corpus, it was assumed that these parameters could be reused for the purposes of our experiments here. The SVM was run only on the 50% positive training sets so that differences in the class distribution between the labeled and unlabeled data would not affect its accuracy.

It can be observed from Figure 5 that the classification performance using SVM was mediocre, with the F-measure never rising above 0.7. Both LL and LL-CP outperforms SVM. The performance difference between LL and LL-CP is negligible when there are only 25 or less labeled documents in the training set, suggesting that incorporating class priors does not help with a limited number of labeled documents. However, LL-CP outperforms LL when the number of labeled documents increases.

The results for the SVM also contrast sharply with the purported accuracy of the SVM used in text classification [18,19,20], which had a recall of 90%, precision of 92% and an overall F-measure of 92%. However, it should be noted that the SVM in that case was trained in several rounds, using articles classified by an expert and by an already trained SVM. Their method of classification also included user feedback and training. The fact that LL-CP could outperform SVM in the absence of such close

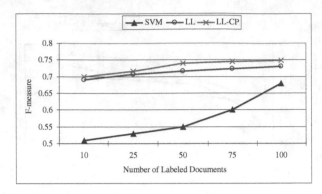

Fig. 5. Average performance of LL-CP and SVM performance over all data set sizes

supervision suggests that it is robust enough to hold its own against more established algorithms such as SVM.

5   Conclusions and Future Work

This paper has investigated a semi-supervised learning algorithm based on local learning with class priors for protein-protein interactions classification from text. Experiments have been carried out on the algorithm to determine the effect of incorporating the class prior knowledge. It was discovered that LL-CP performed better than the local learning method without incorporating class priors.

The algorithm has been applied successfully to the problem of document classification of biomedical literature with promising results, and a brief comparison of LL-CP with SVM has shown that LL-CP performs better in this area. However, biomedical literature is quite different from general text because of its specialized, complex vocabulary, so the findings from this study may not apply to document classification in general. Thus, it is suggested that the use of LL-CP in document classification should be further investigated, and perhaps put to practical use if proven to be superior to other available algorithms.

One disadvantage of using LSI is that it requires a lot of computation time and memory due to the calculation of matrix inverses. In addition, information about individual terms is lost in the calculation of the reduced document matrix, so it is impossible to determine which words were most significant in the classification process. An alternative is to use some other form of feature reduction. For example, substring matching [6] could be explored further in place of performing stemming and LSI, and is ideal for biomedical literature because it would be able to identify and group together biomedical affixes that a normal stemming algorithm would miss.

References

1. Hermjakob, H., Montecchi-Palazzi, L., Lewington, C., Mudali, S., Kerrien, S., Orchard, S., Vingron, M., Roechert, B., Roepstorff, P., Valencia, A., Margalit, H., Armstrong, J., Bairoch, A., Cesareni, G., Sherman, D., Apweiler, R.: IntAct: an open source molecular interaction database. Nucleic Acids Research 32(1) (2004)

2. Xenarios, I., Rice, D., Salwinski, L., Baron, M., Marcotte, E., Eisenberg, D.: DIP: the database of interacting proteins. Nucleic Acids Research 28(1), 289–291 (2000)
3. Marcotte, E., Xenarios, I., Eisenberg, D.: Mining literature for protein-protein interactions. Bioinformatics 17(4), 359–363 (2001)
4. Chen, D., Muller, H.M., Sternberg, P.W.: Automatic document classification of biological literature. BMC Bioinformatics 7 (2006)
5. Donaldson, I., Martin, J., de Bruijn, B., Wolting, C., Lay, V., Tuekam, B., Zhang, S., Baskin, B., Bader, G., Michalickova, K., et al.: PreBIND and Textomy – mining the biomedical literature for protein protein interactions using a support vector machine. BMC Bioinformatics 11(4) (2003)
6. Han, B., Obradovic, Z., Hu, Z., Wu, C., Vucetic, S.: Substring selection for biomedical document classification. Bioinformatics 22(17), 2136–2142 (2006)
7. Szummer, M., Jaakkola, T.: Partially labeled classification with markov random walks. In: Advances in Neural Information Processing Systems, vol. 14 (2002)
8. Zhou, D., Bousquet, O., Lal, T., Weston, J., Schölkopf, B.: Learning with local and global consistency. In: 18th Annual Conf. on Neural Information Processing Systems (2003)
9. Zhu, X., Ghahramani, Z., Lafferty, J.: Semi-supervised learning using gaussian fields and harmonic functions. In: Proceedings of the 20th International Conference on Machine Learning (2003)
10. Wang, F., Zhang, C.: Label propagation through linear neighborhoods. In: ICML 2006: Proceedings of the 23rd international conference on Machine learning, pp. 985–992 (2006)
11. Wu, M., Scholkopf, B.: Transductive classification via local learning regularization. In: Proceedings of the 11th International Conference on Artificial Intelligence and Statistics (AISTATS 2007), pp. 628–635 (2007)
12. Chapelle, O., Zien, A.: Semi-supervised classification by low density separation. In: Proceedings of the 9th International Conference on Artificial Intelligence and Statistics, AISTATS 2005 (2005)
13. Wang, F., Zhang, C., Li, T.: Regularized clustering for documents. In: SIGIR 2007: Proceedings of the 30th annual international ACM SIGIR conference on Research and development in information retrieval, pp. 95–102. ACM, New York (2007)
14. Mann, G.S., McCallum, A.: Simple, robust, scalable semi-supervised learning via expectation regularization. In: Proceedings of the 24th international conference on Machine learning, pp. 593–600. ACM, New York (2007)
15. Zhu, X., Ghahramani, Z., Lafferty, J.: Semi-supervised learning using gaussian fields and harmonic functions. In: The 20th International Conference on Machine Learning, pp. 912–919 (2003)
16. Porter, M.: An algorithm for suffix stripping. Program 14(3), 130–137 (1980)
17. Manning, C.D., Raghavan, P., Schütze, H.: Introduction to Information Retrieval. Cambridge University Press, Cambridge (2007)
18. Yu, H., Han, J., Chang, K.C.C.: PEBL: Positive Example-Based Learning for Web Page Classification Using SVM. In: ACM SIGKDD International Conference in Knowledge Discovery in Databases (KDD 2002). ACM Press, New York (2002)
19. Liu, B., Dai, Y., Li, X., Lee, W.S., Yu, P.S.: Building text classifiers using positive and unlabeled examples. In: Third IEEE International Conference on Data Mining, pp. 179–188 (2003)
20. Li, X., Liu, B.: Learning to classify texts using positive and unlabeled data. In: Eighteenth International Joint Conference on Artificial Intelligence, pp. 587–594 (2003)

Natural Language Interfaces: What Is the Problem? – A Data-Driven Quantitative Analysis

Philipp Cimiano[1] and Michael Minock[2]

[1] WIS, TU Delft
[2] University of Umea

Abstract. While qualitative analyses of the problems involved in building natural language interfaces (NLIs) have been available, a quantitative grounding in empirical data has been missing. We fill this gap by providing a quantitative analysis on the basis of the Geobase dataset. We hope that this analysis can guide further research in NLIs.

1 Introduction

So far, there has been an impressive amount of research on natural language interfaces (NLIs), i.e. on interfaces allowing users to interact with a certain information system in natural language. While NLIs are not inherently restricted only to the task of answering questions on the basis of a given database or knowledge base, most of the NLIs developed so far have been designed for this purpose. Along these lines, as in most other research on natural language interfaces, we limit ourselves to this restricted view of natural language interfaces essentially as systems providing answers to natural language questions in this paper. Research on NLIs dates back to the 70s and 80s (see [1], [6]) and has yielded increased attention in recent years with a plethora of systems emerging: PRECISE [13], STEP [11], ORAKEL [3], Aqualog [10], GINSENG [2], just to name a few of the very recent systems. What seems missing so far is a description of the problem, in particular a quantitative analysis of the problems inherent in the task of building natural language interfaces. While there have been qualitative analyses of the problems involved in constructing NLIs ([1], [6]), to our knowledge there has been no quantitative analysis grounding the qualitative characteristics of the problem in real data. This is crucial in our view as it can and should guide the development of NLIs in the future, focusing them on the challenging problems. It would also help system developers to focus on a specific phenomenon encountered in NLIs (e.g. resolution of ambiguities) and foster progress in the field by clearly designing and evaluating the solution to a specific phenomenon which would ideally not be specific to one particular approach but reusable across systems. In our view, no real progress can be expected in NLI research only from charts hiding the interesting details and solutions to characteristic problems involved in the task behind top performing precision and recall measures.

H. Horacek et al. (Eds.): NLDB 2009, LNCS 5723, pp. 192–206, 2010.

The structure of this paper is as follows: in the next Section 2 we describe the dataset we have used to provide a quantitative analysis and describe our methodology. Then, in Section 3 we describe our interesting findings and derive conclusions in terms of requirements on NLIs. In addition, we include some comments on portability (see Section 4) and on the issue whether deep syntactic and semantic processing is needed for an NLI (see Section 5). We conclude in Section 6 with a summary of our findings and implications for the development on NLIs.

2 Datasets and Methodology

To provide a quantitative analysis of the problem of constructing an NLI we proceeded as follows: we downloaded a dataset which has been frequently used for the evaluation of natural language interfaces, i.e. the Geobase dataset collected by Mooney and his students[1]. The Geobase dataset describes states, cities, mountains, lakes, rivers and roads in the U.S., together with attributes such as area (state, lake), population (state, city), length (river), height (mountain, location) etc.

The datasets consists of a set of 880 test questions (actually 883 questions) and was collected through a web interface hosted at the University of Austin in Texas[2]. We used the 883 test questions for our analysis. After downloading the dataset (in Prolog), we converted the whole dataset into the ontology languages F-Logic [9] and OWL[3]. The datasets are available from http://www.cimiano.de → Projects → Datasets and other Material → ORAKEL.

When converting the dataset into OWL and F-Logic, we used 7 concepts with a total of 17 different relations. We give below the concepts used together with their relations:

Concepts	Relations
state	name, abbreviation, capital, density, population, area, code hasCity, border, highest_point,lowest_point
city	name, area, inState
river	name,length, flowsThrough
mountain	name, inState, height
road	number, passesThrough
lake	name, area, inState
location	name, inState, height

The design above slightly deviates from the original schema in Mooney's dataset, consisting of 8 relations (state, city, river, border, highlow, mountain, road and lake). We have essentially merged some of the information into one class (the class state thus containing the border as well as highest and lowest point information), removed some redundancies (e.g. the name of the state appearing in various relations) and added the location class which includes a height attribute for the location in question.

The original dataset of Mooney et al. consists of the following 7 relations:

---

[1] This dataset is available from: http://www.cs.utexas.edu/users/ml/nldata.html
[2] There is also a dataset consisting of 250 questions available from the University of Texas, but this is merely a subset of the larger dataset.
[3] http://www.w3.org/TR/owl-features/

Relation	Attributes
state	name, abbreviation, capital, population, area, state_number, city1, city2, city3, city4
city	state, state_abbreviation, name, population
river	name, length, [states through which it flows]
border	state, state_abbreviation, [states that border it]
highlow	state, state_abbreviation, highest_point, highest_elevation, lowest_point, lowest_elevation
mountain	state, state_abbreviation, name, height
road	number, [states it passes through]
lake	name, area, [states it is in]

The main differences between our and the original dataset are the following:

- Some redundancies have been removed, keeping the name and abbreviation of a state only in the state class.
- The border and highlow relations have been removed and the information added to the state class.
- The height of the highest and lowest points has been modeled in the class location.
- The 4 cities in the state relation have not been modeled explicitly. We suppose these were the "major" cities in each state. In that case we can recover this information from the area/population information, thus defining "major" through one of these attributes rather than by its extension as in Mooney's dataset.

For the 883 questions, we manually created F-Logic queries yielding the appropriate answers as result when evaluated with the OntoBroker [7] system, but also queries in generic logical form. All these settings do not restrict the general case in any way. Any other database or knowledge representation and query language could have been used in principle. The benefit of the F-Logic language as implemented by the OntoBroker system is that it provides built-in functionality for numerical comparisons as well as aggregation operators for calculating minima, maxima, sums etc., which, as we will see below, are crucial for the Geobase dateset.

We proceeded by manually annotating each of the questions together with characteristics that we regarded as relevant to our quantitative analysis. All annotations have been performed by one of the authors and suffer unavoidably from some subjective bias. The interesting findings obtained through these "annotations" are mentioned below. All cases are illustrated with example questions from Mooney's dataset and the numbers directly refer to the questions as they are ordered in the dataset. The annotated dateset can be downloaded at http://www.cimiano.de → Projects → Datasets and other Material → ORAKEL.

3   Quantitative Analysis

3.1   Question Types

The questions were annotated with 4 question types:

- **wh-questions** (77.29%): the standard type of question starting with a *wh*-pronoun such as *What are major rivers in texas? (201)* or *What is the lowest point in texas?* (472)
- **how (adj/many) -questions** (17.38%), such as *How big is massachusetts?* (19) or *How many big cities are in pennsylvania? (50)*.
- **requests** (3.59%): direct requests such as *Give me the cities in virginia. (1)* or *Can you tell me the capital of texas? (5)*.
- **topicalized questions** (1.16%) where a certain entity is topicalized for the purpose of emphasis, as in *Iowa borders how many states? (173)* or a prepositional phrase is topicalized, as in *Of the states washed by the mississippi river which has the lowest point? (186)*
- **nominal** (0.58%): these are "questions" consisting of a single noun phrase, such as *people in boulder? (187)* or *rivers in new york (189)*

A system which therefore only supports standard wh-questions can only reach a recall of 77.56% on the Geobase dataset. This might partially explain the recall of significantly below 80% of the PRECISE system on this dataset (see [13]) as it requires that some *'wh-value'* maps to a database element.

3.2 Language "Light"

Our findings suggest that the language used in the questions is rather simple, containing a lot of 'light syntactic constructions' such as:

- copula verb *'be'* (appearing in 59.43% of the sentences): e.g. *What are the high points of states surrounding mississippi? (1)*, or *What is the major cities in montana?* [sic.] *(482)*
- light preposition *'of'* (appearing in 21.52% of the sentences): *What are the highest points of all the states? (210)* or *What are the major cities of texas ? (235)*
- light verb *'have'*: (appearing in 12.46% of the sentences), e.g. *How many capitals does rhode island have ? (51)*, or *How many rivers are in the state that has the most rivers? (118)*
- light preposition *'with'* (appearing in 7.36% of the sentences): *How many people live in the state with the largest population density? (104)*, or *What are the cities of the state with the highest point? (209)*

Thus, in many cases the relevant relations are not expressed directly in the text, but are hidden implicitly behind light constructions involving the verb *"to have"* as well as light prepositions such as *'with'* and *'of'*. This is probably the reason why shallow approaches which ignore the linguistic details (for example ignoring non-content words in the input as in the PRECISE system), essentially relying on the structure of the knowledge base or data base to perform interpretation (such as PRECISE [13] or Aqualog [10]) are so successful on the type of questions the Geobase dataset consists of. Nevertheless, any NLI should implement techniques to deal with such a kind of lightweight (or semantically weak) constructions which require to infer the appropriate relation implicit in the surface realization.

3.3   Lexical Ambiguities

While the problem of lexical ambiguities is mentioned in many overviews on natural language interfaces ([1], [6]), our findings suggest that classical lexical ambiguities are typically not the problem. For example, we did not find any single case of 'classical' noun homonymy or polysemy where the same word can convey completely different meanings in the dataset. Most of the 'lexical' ambiguities are actually introduced artificially by the knowledge base. For example, the adjectives *'big'* and *'small'* or the superlatives *'smallest'* and *'largest'* are ambiguous when modifying a state, as 'size' can be measured either in terms of area or inhabitants in the Geobase knowledge base. We found around 60 cases (6.80% of the questions) of such artificial ambiguities, e.g. *What is the largest state?* (422).

The fact that these cases are indeed ambiguous is corroborated by the fact that in some cases the users actually tried to disambiguate by specifying the property with respect to which size should be measured, as in (4 sentences, 0.45% of the questions): *What is the smallest state by area?* (579).

Further, there are further examples of reference ambiguities in the dataset e.g. *"How many people are there in new york"?* (77). In the latter case, it is not clear whether *New York* refers here to the city or to the state. There are many other and in some cases people do indeed try to disambiguate by adding the state for instance (in case of multiple cities with the same name in different states):

- What is the population of springfield missouri? (536)
- What is the population of springfield south dakota? (537)

In case no explicit disambiguation is provided, any NLI should request the user to disambiguate the input, in case of question 77 for example by asking: *"Do you mean New York city or the state of New York?"* or in cases like 422: *"Do you mean largest in terms of area or population?"*. The Aqualog system, for example, recognizes such reference ambiguities and asks a user for reference disambiguation.

3.4   Syntactic Ambiguities

Frequently, literature on NLI research also mentions syntactic ambiguities, especially attachment ambiguities, as problematic. To explore this phenomenon, we have annotated modifying prepositional phrases (PPs), relative clauses and modifying gerund constructions making explicit whether they i) attach to the only possible antecedent (noun phrase or verb phrase), ii) to the last one or iii) to a non-preceding constituent. In the case of prepositional phrases, we also distinguish the case of a PP providing essentially the predicate in a copula construct. We give examples for each of these cases below:

- PP attachment (only attachment point) (48.07%), i.e. *How many states in the us does the shortest river run through?* (166) or *Where is the highest point in hawaii?* (4)
- PP attachment (last attachment point) (40.62%), i.e. *How many people reside in utah?* (109) or *What is the capital of the state with the highest elevation?* (346)

- PP attachment (copula) (10.34%): *Tell me what cities are* <u>*in texas*</u>*?* (196), or *What mountains are* <u>*in alaska*</u>*?* (602)
- PP attachment (non-preceding attachment point) (0.48%): *What is the city in texas* <u>*with the largest population*</u>*?* (355) or *What is the state with the largest density* <u>*in usa*</u>*?* (588)
- relative clause (last attachment point) (72.92%): *What river traverses the state* <u>*which borders the most states*</u>*?* (612), or *What states border states* <u>*that border colorado?*</u>
- relative clause (only attachment point) (27.08%): *Give me the cities* <u>*which are in texas*</u>*?* (12) or *What is the smallest state* <u>*that the mississippi river runs through?*</u> (583)

So this means that by simply attaching PPs or relative clauses to the last constituent, we will take a correct decision in 99.27% of the cases for PP attachment (including last and only attachment point cases as well as the copula case where the PP functions as predicate) and in 100% of the cases for relative clauses (last and only attachment point). While we have not listed the gerund data explicitly, we get similar results, with last and only attachment points representing 100% of the cases. In other words, we found no gerund attaching to a non-preceding constituent.

As a consequence, a very simple baseline strategy which attaches every PP or relative clause to the last constituent will be difficult to beat. Thus, at least what the Geobase dataset is concerned, no substantial effort in syntactic disambiguation is needed. While this can not be claimed in the general case, we hypothesize that people try to produce less ambiguities when interacting with a natural language interface.

3.5 Scope Ambiguities

Many natural language interfaces ignore scope ambiguities and even determiners (representing quantifiers) completely (e.g. PRECISE [13], Aqualog [10] etc.). In what follows we give examples for determiners/quantifiers appearing in the dataset:

- **question operator**: <u>*What*</u> *are major rivers in texas?* (201)
- **definites**: *What are* <u>*the capitals of states that border missouri*</u>*?* (250)
- **most**: *What is the capital of the state that borders* <u>*the most states*</u>*?* (344)
- **a**: *What is the largest city in* <u>*a*</u> *state that borders texas?* (407)
- **negation**: *How many rivers do* <u>*not*</u> *traverse the state with the capital albany?* (124)
- **all**: *Show me* <u>*all*</u> *the major lakes in the us?* (193)
- **the least**: *What city has* <u>*the least population*</u>*?* (272)
- **at least**: *How many states border* <u>*at least one other state*</u>*?* (139)
- **fewest**: *Which rivers run through states with* <u>*fewest cities*</u>*?* (807)
- **each**: *What are the population densities of* <u>*each us state*</u>*?* (243)

Some statistics are given below:

Scope taking elements	#Occurrences (sentences)	Rel. Percentage
question op. (all other than requests)	832 (94.22%)	51.81%
definites (incl. superlatives)	697 (78.94%)	43.40%
most	32 (3.62%)	1.99%
a	13 (1.47%)	0.81%
negation	12 (1.36%)	0.75%
all	11 (1.25%)	0.68%
least	7 (0.79%)	0.44%
fewest	1 (0.11%)	0.06%
each	1 (0.11%)	0.06%
Total	1606	100%

Clearly, there is a high occurrence of question operators, realized at the surface by wh-pronouns (not surprising), but there is also a very high number of definite noun phrases. However, it is important to mention that we have subsumed entity descriptions such as *'the state of oregon'* under the category definites in the above table, but we have not taken them into account when calculating the number of scope-bearing elements in each question, which is almost 2 (Avg. 1.97, Std. Dev. 0.81), including plural NPs and excluding the uses of *'the'* which are part of a named entity expression as mentioned above. The minimum number of scope bearing elements per sentence is 1 and the maximum 5. This shows that in principle there is one scoping decision to take per sentence. While in many cases it is true that this can be solved via some heuristics, i.e. *'each'* outscopes every other quantifier in the question, definites are accommodated as high as possible, the question quantifier outscopes the rest of the quantifiers, etc. (compare the heuristics used in the TEAM system [8]), we are making a principled point here: strategies for disambiguation are needed, no matter how adhoc they are (as long as they are effective).

Table 1. Some statistics about spatial prepositions (left) and aggregation operators (right)

Preposition	Occurrences (questions)	Rel. Percentage
in	280 (32.71%)	73.11%
through	100 (11.33%)	36.11%
next to	3 (0.34%)	0.78%
Total	383	100%

Operator	Occurrences (sent.)	Rel. Percentage
max	278 (31.48%)	55.71%
count	107 (12.12%)	21.44%
min	91 (10.31%)	18.24%
negation	12 (1.36%)	2.40%
sum	7 (0.79%)	1.40%
average	4 (0.45%)	0.89%
Total	499	100%

3.6 Spatial Prepositions

While there is a significant number of light or vague prepositions in the dataset (see results in Section 3.2), spatial prepositions also tend to appear frequently, in particular the prepositions *in*, *next* and *through* (see Table 1 - left).

Examples are the following:

- **in:** *Name the rivers in arkansas.* (3), or *what is the biggest american city in a state with a river?* (301)

- **through:** *How many rivers run <u>through</u> texas?* (129), *what is the largest city in smallest state <u>through</u> which the mississippi runs?* (414)
- **next to:** *what states are <u>next to</u> arizona?* (694) *how many states are <u>next to</u> major rivers ?* (134)

The presence of spatial prepositions can be explained by the fact that the Geobase dataset is modeling locations as well as their spatial relationships. While some systems have shown that one can perform very well by essentially ignoring prepositions, we would like to make the point that the more principled solution would be to capture the domain-independent meaning of such spatial prepositions, allowing to reuse their meaning across domains. For example, *'in'* has definitely a meaning in terms of spatial inclusion which is compatible with many domains if modeled appropriately, e.g. at the level of a foundational ontology as suggested in [4]. Taking into account the specific semantic contribution of spatial (and also temporal) prepositions gets important in many domains, especially in those including temporal knowledge (see Section 5).

3.7 Adjective Modifiers and Superlatives

There are at least 105 adjectives in the Geobase dataset (appearing in 11.89% of the questions) as well as 316 superlatives (appearing in 35.79% of the questions). The adjectives are distributed among the following cases: i) modifiers (6.9% of the questions) as in *'How many major cities are in arizona?'* (67), ii) most/least+adj (0.91% of the sentences), as in *'What is the most populous city?'* (487) and as iii) attribute selectors in how+adj questions (3.9% of the questions), as in *How big is alaska?* (18). Clearly, any NLI needs to handle adjective modification as well as superlatives. The challenge here is certainly that the interpretation of adjectives (and in consequence also of superlatives) is domain-specific and needs to be specified for each domain. For each adjective, an NLI needs information about the predicate in the knowledge base it represents (e.g. area/population in the case of small) as well as the polarity of the adjective, which is crucial to handle superlatives. It is also important to specify the conditions under which some object will fulfill the property represented by the adjective, e.g. specifying which is the minimum population for a city to be counted as *'big'*. An appropriate mechanism to provide such crucial and basic data about the meaning of adjectives is important to allow portability of NLIs. Many NLIs have 'cheated' in this respect, hardcoding the meaning of adjectives (e.g. *'major'*) and superlatives (e.g. *'shortest'*) in the backend system (see the evaluation engine in Prolog by Mooney where the appropriate 'meaning' has been hard-coded[4]). With respect to the PRECISE system, it is unclear at all how it can handle superlatives or modifying adjectives as they are clearly not semantically tractable, i.e. there is no column in the database corresponding to the superlative or to the adjective as modifier (requiring a specific value), so that additional mechanisms are needed beyond the algorithm described in [13] to handle adjectives and superlatives.

---

[4] ftp://ftp.cs.utexas.edu/pub/mooney/nl-ilp-data/geosystem/geoquery

3.8   Aggregation, Comparison and Negation Operators

We define aggregation operators as those calculating a minimum, maximum or a sum over a given set of values as well as those allowing us to count the number of individuals fullfilling a certain property. Comparison operators are those that allow to compare numbers w.r.t. to a given order (e.g. the one between integers).

Such aggregation operators are crucial to evaluate certain queries on the Geobase dataset. Before giving a few statistics, we will first give a few examples of questions requiring aggregation, comparison and negation operators. As we will see, such operators are "hidden" behind certain quantifiers and only appear in the translation into logical form:

- **counting:** e.g. *How many major cities are in arizona?* (67), requiring to count all those x which are major cities and are located in Arizona
- **maximum** (no counting involved): e.g. What cities in texas have the highest number of citizens? (269), returning those cities with a maximum number of inhabitants, where the number of inhabitants is given explicitly in the form of the population and does not need to be counted.
- **counting + maximum:** e.g. *What is the length of the river that runs through the most states?* (441), requiring to count, for each river x, the number of states that it flows through, taking the maximum over these and returning the length of the x flowing through the maximum number of states.
- **comparison:** e.g. *What states high point are higher than that of colorado?* (754), comparing the height of the high points of all states to the height of the highest point of colorado and returning those states with a higher point.
- **negation:** e.g. *Name the states which have no surrounding states?*
- **negation with counting:** *How many rivers do not traverse the state with the capital albany?* (124)
- **sum:** *What is the area of all the states combined?* (280)
- **average:** *Which state has the smallest average urban population?* (840)

It is clear that the occurrence of these 'operators' is highly correlated with the appearance of a certain quantifier in the surface form (see Section 3.5 and the corresponding analysis of quantifier frequencies). However, the challenge here is to predict how a certain quantifier will be realized as, i.e. as which set of logical operators. 'How many', for example can involve a counting operation or only a look-up, depending whether the information is modeled as a datatype property or an object property thus requiring to count objects standing in the relation in question:

- *How many capitals does rhode island have?* (51) (counting)
- *How many inhabitants does montgomery have?* (66) (no counting, only lookup)

The same holds for the quantifier most, also appearing in two different forms, one requiring a maximization only, the other one requiring a counting and a maximization operator:

- *What is the capital of the state with <u>the most inhabitants</u>?* (351) (max.)
- *What is the length of the river that runs through <u>the most states</u>?* (441) (counting + max.)

This means that the correct interpretation of *'how many'* and *'the most'* depends on the way the relevant information has been modelled in the knowledge base (e.g. as datatype or object property). In other cases, the aggregator is in many cases only very loosely selected by the surface form of the question. In the sentence: *'What is the area of all the states combined?'* (289), "all *x* combined" actually maps to a *sum* operator in the logical form but there is no reference to a sum in the surface form.). Some statistics about the occurrence of such "aggregators" are given in Table 1 (right).

3.9 Non-compositionality

In the Geobase dataset there is a high number of questions which with respect to their mapping into logical form are non-compositional, i.e. the logical form of the question is not exactly the composition of the meaning of the parts of the question. In particular, in many cases there are "parts" of the question which do not correspond to any element of the logical form. This is what we refer to as "non-compositionality". According to our analysis, at least 12% (11.89%) of the questions in the Geobase dataset are non-compositional. Most of these questions include a reference to the USA which of course is not explicitly mentioned in the dataset as it only models information about the USA but never mentions this explicitly, as it is clear from the overall scope of the knowledge base.

A few examples of non-compositional elements in questions are given below:

- <u>Give me</u> the cities in virginia. (1)
- What are the major cities <u>in the usa</u> ? (232)
- What states <u>in the united states</u> have a city of springfield ? (755)
- What is the biggest <u>american</u> city in a state with a river? (301)

Of course, besides having elements of the query which do not appear in the logical form, we encounter also this situation the other way round, i.e. in many cases elements which appear in the logical form are not mentioned explicitly in the question. This is partly a byproduct of the light language used in the questions and essentially amounts to the cases we have discussed above, so that we do not analyze this any further here. So there is a need for NLI systems to ignore part of the input. This is for example accomplished by the 'fudging' operator in the STEP system [11].

3.10 Variability

One of the greatest challenges for any natural language interface is to handle the large variability in the way that a certain fact or relation can be expressed. It is certainly difficult to quantify variability and we will not even try to do so. The important observation is that the Geobase dataset is certainly no exception to the above. As an example, let us consider various forms in which one can ask for the population of a certain state:

- *How many inhabitants does montgomery have?* (66)
- *How big is the city of new york?* (23)
- *How many citizens in alabama?* (62)
- *How many residents live in texas?* (111)
- *How many citizens live in california?* (64)
- *How many people reside in utah?* (109)
- *What is the population of alaska?* (505)
- *population of boulder?* (188)

Any NLI should clearly somehow handle this variability, either by allowing people adapting the system to a new domain to encode it explicitly (as in ORAKEL [3] or STEP [11]) or implicitly in the way the question is mapped to a logical query by using the schema of the database or knowledge base.

3.11   Out of Scope

In the Geobase dataset we find at least 17 (1.93%) questions which are definitely out of scope of the knowledge base. A few examples are:

- *How many states border the mississippi river?* (147): There is no information about which states border a river, but only about which rivers flow through a state.
- *What is the biggest state in continental us?* (317): There is no information about which states are on the continent and which not (e.g. Hawaii)
- *What is the length of the colorado river in texas?* (435): There is no information about the length of a river in a particular state (only the absolute length)
- *What is the maximum elevation of san francisco?*: No information about the highest points of cities is available (only for states).

A NLI should definitely inform the user in some way about the fact that the question is out of scope and not simply return no answer.

4   Portability: No Free Lunch!

While we have no data we could analyse with respect to portability issues, we think it is important to mention this issue as one of the most challening problems in NLI development. Seldom have the efforts and resources needed to port a system been made explicit or compared accross systems (an exception being [12]). This would enhance our understanding of the problems and potential solutions to the portability problem. The most important issue here is to have some mechanism to specify how content words (verbs, nouns, adjectives etc.) map to predicates in the knowledge base. For sure, portability does not constitute a free lunch. ORAKEL and STEP rely on manual mappings created by a lexicon engineer on the basis of the data schema. Systems which seem to require no effort to be ported at first sight (e.g. [13]) require at least a lexicon which needs to be handcrafted or derived using general lexical resources such as WordNet in combination with the natural language labels available in the ontology or database schema. But WordNet clearly has limitations in scope, especially in technical domains, so that porting to technical domains also comes at the cost of manually enhancing the lexica. Other systems which learn from training data require pairs of questions and queries in logical form [14], the provision of which constitutes a huge effort. The effort that is needed to customize an NLI to a new domain has been rarely quantified and compared across paradigms.

5   To Deepen or Not to Deepen? That Is the Question

As mentioned already, there have been many shallow approaches to NLIs with very impressive results on the Geobase dataset we have examined here (as well as other datasets collected by Mooney and colleagues). When looking at NLIs just from a narrow perspective, i.e. looking at their input (a question) and output (a formal query or the result of its evaluation), a shallow approach with almost 100% precision might look like the right solution. However, when expanding the capabilities of the NLI, we will encounter a number of obstacles when using a shallow system. We mention some of these capabilities below:

Paraphrase Generation: As argued recently by Minock [11] as well as others before [5], paraphrase generation is crucial to make sure that the system has captured the intended meaning of the user's question. Paraphrase generation capabilities require some level of deeper representation (syntactic and semantic) to generate a non-ambiguous form of the question that a user can confirm or reject. Generating a non-ambiguous representation of the input sentence presupposes that the system is able to detect, represent, reason with and resolve ambiguities, which requires some level of deeper processing.

Discourse Processing: While most NLIs process each question in isolation (and they have been evaluated also in this mode), in real systems we expect that users will use pronouns to refer to previous entities and that they will provide fragmentary input (i.e. in the form of ellipsis). Any NLI should thus be designed in such a principled way that allows to extend it with some discourse processing capabilities. While the resolution of pronouns could be arguably carried out by pure statistical approaches neglecting discourse structure, the resolution of ellipsis would clearly benefit from some structural representation which allows to compute which gap the new content can fill. In any case, it seems that more sophisticated NLIs going beyond analyzing each sentence in isolation will require some deeper analysis and make explicit the discourse structure. This is also important when introducing new modalities such as gestures, prosody etc. While the integration of various modalities can also be approached in a "shallow" way (meaning statistical or adhoc), we claim that the integration could profit from an explicit and deep syntactic and semantic representation of the questions and their context.

Temporal Aspects: While many successful systems ignore non-content words such as determiners or prepositions, in some domains prepositions might indeed matter. In the context of an ontology about american presidents, it makes a difference whether we ask for *'Who was president after WWII'*, vs. *'Who was president during WWII'*. In general, systems would thus profit from capturing the meaning of prepositions explicitly, possibly even assigning them a domain-independent meaning (as suggested in [4]).

Guidance: Guidance to a user in the task of formulating a query which is in line with the capabilities of the system is a very important feature of NLIs. It requires, however, at least the existence of a grammar (in whatever form) which can be used to generate to suggest possible completions to the user. Ideally, the system would have one and only one grammar to be used in analysis and generation.

In summary, while it is tempting to explore shallow approaches which neglect deep linguistic and semantic structure, we think that this is the wrong way to go as it makes the extension of the system with respect to aspects mentioned above very hard, possibly requiring adhoc and principled extensions. For example, a system which ignores quantifiers in the semantic representation will have to integrate procedural attachments to simulate the 'semantics' of the quantifiers in an adhoc way. A system which ignores non-content words such as prepositions might have to incorporate prepositions in an adhoc manner when confronted with a domain in which they indeed matter (e.g. domains involving temporal aspects). Paraphrase generation and discourse processing are hard to achieve building on adhoc and shallow representations rather than on principled semantic representations. Building on sound theories of semantic representation and discourse would for example allow the import of all those insights on discourse processing from linguistics into our systems, while this "import" is not as straightforward when systems lack principled semantic representations. Overall, we think that building systems with deep syntactic and semantic processing capabilities will pay off in the long term, at least for the above mentioned reasons.

6   Conclusion and Future Work

We have started this research by wondering why there was no systematic quantitative analysis of the problem of providing NLIs to databases or knowledge bases. While there have been many qualitative descriptions of the problems, a data-driven analysis has been missing so far. We have aimed at closing this gap. We can conclude that tackling the following issues is crucial for any natural language interface:

- handling other questions than wh-questions (otherwise we will miss about a quarter of the questions in the Geobaset dateset for instance)
- dealing with light syntactic/semantic constructions and vague prepositions (more than 60% of the questions contain such "light" constructions)
- including some approach for handling scope, in particular for representation and disambiguation (one disambiguation decision needed per question),

- generally, an adequate treatment of quantifiers, in particular those translated into aggregation, comparison and negation operators, which are very frequent in the dataset (over 40% of the questions involve some aggregation operator)
- accounting for the domain-specific meaning of adjectives (in various constructions) as well as superlatives
- treating prepositions (in particular spatial and temporal) in an appropriate way, ideally defining their domain-independent semantics
- handling non-compositional input
- handling and reacting appropriately to out-of-scope questions

Of course, to have a more substantial analysis, we would need to consider further datasets, e.g. Mooney's restaurant and jobs datasets[5]. A first look at these datasets confirmed the prevalence of light language. Nevertheless, it also seems that these datasets do not involve as many aggregation operators, but contain mainly questions which can be answered directly from the database without any additional operations (by mere "look-up" so to speak). Nevertheless, considering more datasets could substantiate or not the conclusions derived from the analysis of the Geobase dataset. Nevertheless, we regard the Geobase as the most interesting one and it is also the one used and cited most frequently, so that we have started our analysis on the Geobase dataset. The fact that the jobs and restaurant domains are inherently "easier" is demonstrated by the higher performance in terms of recall on these datasets by PRECISE (see [13]).

As a next step, it would be interesting to consider in how far the different recent systems do indeed fulfill the above requirements. This would allow a meaningful qualitative comparison of the different systems and help to work out their strengths and weaknesses abstracting from the concrete results they yield on the dataset. Overall, we hope that with this analysis we will also be able to guide future research in NLIs and help to focus on the challenging problems as well as to focus researchers on specific aspects of an NLI rather than trying to solve all problems with one single "black box" approach from which it is hard to tell how and why it solves the challenging issues mentioned in this paper and thus do not contribute much to our understanding of the task. With our analysis we also open the possibility for researchers to concentrate on a particular subset of the Geobase data to study a specific phenomenon (e.g. by selecting from our data the subset of questions with superlatives, the direct requests etc.) This allows exactly the type of research on a specific phenomenon that in our view is needed to push the field further. Past research has shown that we can get very good results with shallow approaches which neglect many of the difficulties we have mentioned here. However, we believe that such solutions will have i) problems to scale to mode demanding domains (e.g. involving temporal knowledge) as well as ii) handle the small percentage of remaining questions which possibly require deeper processing (e.g. the PRECISE system, while being maximally precise, fails on more than 20% of the questions in the Geobase dataset).

On a more general note, while it is tempting to explore shallow approaches which neglect deep linguistic and semantic structure, we think that this is the

---

[5] see http://www.cs.utexas.edu/users/ml/nldata.html

wrong way to go as it makes the extension of the system with "advanced capabilities" harder, possibly requiring adhoc and non-principled extensions. Overall, we think that building systems with deep syntactic and semantic processing capabilities will pay off in the long term.

Acknowledgements. This research has been partly supported by the MULTIPLA project, funded by the Deutsche Forschungsgemeinschaft (DFG) under grant number 38457858.

References

1. Androutsopoulos, I., Ritchie, G.D., Thanisch, P.: Natural language interfaces to databases–an introduction. Journal of Language Engineering 1(1), 29–81 (1995)
2. Bernstein, A., Kaufmann, E., Kaiser, C., Kiefer, C.: Ginseng: A guided input natural language search engine for querying ontologies. In: Proceedings of the JENA User Coference (2006)
3. Cimiano, P., Haase, P., Heizmann, J., Mantel, M., Studer, R.: Towards portable natural language interfaces to knowledge bases: The case of the ORAKEL system. Data and Knowledge Engineering (DKE) 62(2), 325–354 (2007)
4. Cimiano, P., Reyle, U.: Towards foundational semantics - ontological semantics revisited. In: Proceedings of the International Conference on Formal Ontology in Information Systems (FOIS), vol. 150, pp. 51–62. IOS Press, Amsterdam (2006)
5. Codd, E.F.: Seven steps to rendezvous with the casual user. In: Proceedings of the IFIP Working Conference on Data Base Management, pp. 179–200 (1974)
6. Copestake, A., Sparck Jones, K.: Natural language interfaces to databases. Knowledge Engineering Review 5(4), 225–249 (1989); Special Issue on the Applications of Natural Language Processing Techniques
7. Decker, S., Erdmann, M., Fensel, D., Studer, R.: Ontobroker: Ontology Based Access to Distributed and Semi-Structured Information. In: Database Semantics: Semantic Issues in Multimedia Systems, pp. 351–369. Kluwer, Dordrecht (1999)
8. Grosz, B.J., Appelt, D.E., Martin, P.A., Pereira, F.C.N.: Team: An experiment in the design of transportable natural language interfaces. Artificial Intelligence 32, 173–243 (1987)
9. Kifer, M., Lausen, G., Wu, J.: Logical foundations of object-oriented and frame-based languages. Journal of the ACM 42, 741–843 (1995)
10. Lopez, V., Motta, E.: Ontology-driven question answering in aqualog. In: Meziane, F., Métais, E. (eds.) NLDB 2004. LNCS, vol. 3136, pp. 89–102. Springer, Heidelberg (2004)
11. Minock, M.: A phrasal approach to natural language interfaces over databases. In: Montoyo, A., Muñoz, R., Métais, E. (eds.) NLDB 2005. LNCS, vol. 3513, pp. 333–336. Springer, Heidelberg (2005)
12. Minock, M., Olofsson, P., Naslund, A.: Towards building robust natural language interfaces to databases. In: Kapetanios, E., Sugumaran, V., Spiliopoulou, M. (eds.) NLDB 2008. LNCS, vol. 5039, pp. 187–198. Springer, Heidelberg (2008)
13. Popescu, A., Etzioni, O., Kautz, H.: Towards a theory of natural language interfaces to databases. In: Proceedings IUI 2003, pp. 149–157 (2003)
14. Thompson, C., Mooney, R., Tang, L.: Learning to parse natural language database queries into logical form. In: Proceedings of the Workshop on Automata Induction, Grammatical Inference and Language Acquisition (1997)

Semantic Mapping between Natural Language Questions and SQL Queries via Syntactic Pairing

Alessandra Giordani and Alessandro Moschitti

Department of Computer Science and Engineering
University of Trento
Via Sommarive 14, 38100 POVO (TN) - Italy
{agiordani,moschitti}@disi.unitn.it

Abstract. Automatically mapping natural language semantics into programming languages has always been a major and interesting challenge in Computer Science. In this paper, we approach such problem by carrying out mapping at syntactic level and then applying machine learning algorithms to derive an automatic translator of natural language questions into their associated SQL queries. To build the required training and test sets, we designed an algorithm, which, given an initial corpus of questions and their answers, semi-automatically generates the set of possible incorrect and correct pairs.

We encode such relational pairs in Support Vector Machines by means of kernel functions applied to the syntactic trees of questions and queries. The accurate results on automatic classification of the above pairs above, suggest that our approach captures the shared semantics between the two languages.

Keywords: Natural Language Processing; Kernel Methods; Support Vector Machines.

1 Introduction

The design of models for automatically mapping natural language semantics into programming languages has been always a major and interesting challenge in Computer Science since it would have a direct impact on industrial and social worlds. For example, accessing a database requires machine-readable instructions that common users are not supposed to know. Ideally, they should only pose a question in natural language without knowing either the underlying database schema or any complex machine language. The development of natural language interfaces to databases (NLIDBs) that translate the human intent into SQL instructions is indeed a classic problem, which is becoming of greater importance in today's world.

This could be addressed by finding a mapping between natural language and the database programming language. If we knew how to convert natural language questions into their associated SQL queries, it would be straightforward to obtain the answers by just executing a query. Unfortunately, previous work in

H. Horacek et al. (Eds.): NLDB 2009, LNCS 5723, pp. 207–221, 2010.

Natural Language Understanding has shown the inadequacy of logic and rule-based approaches to this problem; in contrast shallow and statistical methods appear to be promising.

In this paper, we exploit mapping at syntactic level between the two languages and apply machine learning models to derive the shared shallow semantics. Such approach requires the design of a dataset of relational pairs containing syntactic trees of questions and queries. For this purpose, we used syntactic parsers to obtain natural language and SQL trees and we designed an effective algorithm, which, given an initial corpus of correct question and query pairs, semi-automatically generates the labeled set of possible correct and incorrect instances.

We used the above dataset to train classifiers based on Support Vector Machines and kernel functions over pairs. Such functions are combinations of tree kernels applied to syntactic trees and linear kernels applied to bag-of-words. The cross-validation experiments on the task of selecting correct queries given a target set of questions show that our best kernel improves the baseline model of about 32%. The latter is the typical approach based on a linear kernel applied to the union of the bag-of-words from question and query texts. The most interesting finding is that the product between the two kernels representing questions and queries provides feature pairs, which can express the relational features between the syntactic/semantic representation of the two languages.

In the remainder, Section 2 introduces the problem of mapping questions into queries and illustrates the idea of our solution whereas Section 3 describes the technology to implement it, i.e. kernel methods. Section 4 shows our proposed algorithm to generate a training set of question and query pairs, Section 5 discusses our results and finally, Section 6 draws conclusions.

2   Automatic Mapping of Questions into SQL Queries

Studying the automatic mapping of questions into SQL queries is important for two main reasons: (a) it allows to design interesting applications based on databases and (b) it offers the possibility to understand the role of syntax in deriving a shared semantics between a natural language and an artificial language.

Given the complexity of theoretically modeling such relationship we use a statistical and shallow model. We consider a dataset of natural language questions $\mathcal{N}$ and SQL queries $\mathcal{S}$ related to a specific domain/database[1] and we automatically learn such mapping from the set of pairs $\mathcal{P} = \mathcal{N} \times \mathcal{S}$. More in detail, (a) we assume that pairs are annotated as correct when the SQL query answers to the question and incorrect otherwise and (b) we train a classifier on the pairs above for selecting the correct queries for a question. Then, to map new questions in the set of the available queries, (c) we rank the latter by means of the question classifier score and select the top one. In the following we provide the formal definition of our learning approach.

---

[1] We assume that for any database there is a core set $\mathcal{S}$ of queries, which are frequently asked. $\mathcal{S}$ should at least represent the syntactic structures of the most part of the frequent queries.

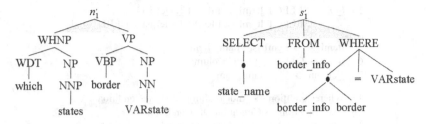

Fig. 1. Question/Query Syntactic trees

2.1 Pair Ranking

The problem of assigning a query (with its result) to a question, can be formally described as the following ranking problem: (i) given a question $n \in \mathcal{N}$ and a set of possible useful queries $\mathcal{S}$, we generate the set of possible pairs $P(n) = \{\langle n, s \rangle : s \in \mathcal{S}\}$; (ii) we classify them with an automatic categorizer; (iii) we use the score/probability output by such model to rank $P(n)$; (vi) we select the top ranked pair.

For example, let us consider question n_1:"*Which states border Texas?*" and the queries s_1: `SELECT state_name FROM border_info WHERE border='texas'` and s_2: `SELECT COUNT(state_name) FROM border_info WHERE border='texas'`. Since s_1 is a correct and s_2 is an incorrect interpretation of the question, the classifier should assign a higher score to the former, thus our ranker will output the $\langle n_1, s_1 \rangle$ pair. Note that both s_1 and s_2 share three terms, *state, border and texas*, with n_1 but $\langle n_1, s_2 \rangle$ is not correct. This suggests that we cannot only rely on the common terms but we should also take into account the syntax of both languages.

2.2 Pair Representation

The aim of our research is to derive the shared shallow semantics within pairs by means of syntax. Thus we represent questions and queries using their syntactic trees[2], as shown in Figure 1: for the question (a) we use the output of the Charniak's syntactic parser [2], whereas for the query (b) we use a modification of the SQL derivation tree.

To build the SQL tree we implemented an ad-hoc parser that follows the syntactic derivation of a query according to our grammar. Since our database system embeds a MySQL server, we use the production rules of MySQL, shown at the top of Figure 2, slightly modified to manage punctuation, i.e. rules 5*, 6* and 20* related to comma and dot, as shown at the bottom.

More in detail, we change the non-terminals *Item* and *SelectItem* with the symbol • to have an uniform representation of the relationship between a table and its column in both the `SELECT` and `WHERE` clauses. This allows for matching between the subtrees containing table, column or both also when they appear in different clause types of two queries.

---

[2] Early work on the use of syntax for text categorization were based on part-of-speech tags, e.g. [1]. The efficiency of modern syntactic parsers allows us to use the complete parse tree.

1   SQL →SELECT ItemList FROM TableList |
       SELECT ItemList FROM TableList WHERE Cond

...

5   ItemList → ItemList , Item | Item
6   Item → Table . Column | Column
7   Column → * | ColumnName

...

15   WhereCondition → AndCondition ANDAndCondition
16   AndCondition → Condition OR Condition
17   Condition → NOT Condition | Operand | ...
18   Operand → Factor | Operand + Factor | ...
19   Factor → Term | Factor * Term | ...
20   Term → Value | SelectItem | SQL
21   SelectItem → Table . ColumnName | ColumnName

5*   ItemList → ItemList • | •
6*   • → Table | Column | Table Column
20* Term → Value | • | SQL

Fig. 2. Modified MySQL Grammar

It is worth noting that rule 20* still allows to parse nested queries and that the overall grammar, in general, is very expressive and powerful enough to express complex SQL queries involving nesting, aggregation, conjunctions and disjunctions in the WHERE clause.

Note that, although we eliminated comma and dot from the original SQL grammar, it is still possible to obtain the original SQL query by just performing a preorder traversal of the tree.

To represent the above structures in a learning algorithm we use tree kernels described in the following section.

3   Tree Kernels

Kernel Methods refer to a large class of learning algorithms based on inner product vector spaces, among which Support Vector Machines (SVMs) are one of the most well-known approaches. The main idea is that the parameter model vector w generated by SVMs (or by other kernel-based machines) can be rewritten as

$$\sum_{i=1..l} y_i \alpha_i x_i, \tag{1}$$

where y_i is equal to 1 for positive and -1 for negative examples, $\alpha_i \in \Re$ with $\alpha_i \geq 0$, $\forall i \in \{1, .., l\}$ x_i are the training instances.

Therefore we can express the classification function as

$$Sgn(\sum_{i=1..l} y_i \alpha_i x_i \cdot x + b) = Sgn(\sum_{i=1..l} y_i \alpha_i \phi(o_i) \cdot \phi(o) + b), \tag{2}$$

where x is a classifying object, b is a threshold and the product $K(o_i, o) = \langle \phi(o_i) \cdot \phi(o) \rangle$ is the kernel function associated with the mapping ϕ.

Note that it is not necessary to apply the mapping ϕ, we can use $K(o_i, o)$ directly. This allows, under the Mercer's conditions [3], for defining abstract functions, which generate implicit feature spaces. The latter allow for an easier feature extraction and the use of huge feature spaces (possibly infinite), where the scalar product (i.e. $K(\cdot, \cdot)$) is implicitly evaluated.

In the remainder of this section, we illustrate some kernels for structured data: the Syntactic Tree Kernel (STK) [4], which computes the number of syntactic tree fragments and the Extended Syntactic Tree Kernel (STK$_e$) [5], which includes leaves in STK. In the last subsection, we show how to engineer new kernels from them.

3.1 Syntactic Tree Kernel (STK) and Its Extension (STK$_e$)

The main underlying idea of tree kernels is to compute the number of common substructures between two trees T_1 and T_2 without explicitly considering the whole fragment space. Let $\mathcal{F} = \{f_1, f_2, \ldots, f_{|\mathcal{F}|}\}$ be the set of tree fragments and $\chi_i(n)$ an indicator function equal to 1 if the target f_i is rooted at node n and equal to 0 otherwise. A tree kernel function over T_1 and T_2 is defined as

$$TK(T_1, T_2) = \sum_{n_1 \in N_{T_1}} \sum_{n_2 \in N_{T_2}} \Delta(n_1, n_2), \tag{3}$$

where N_{T_1} and N_{T_2} are the sets of nodes in T_1 and T_2, respectively, and $\Delta(n_1, n_2) = \sum_{i=1}^{|\mathcal{F}|} \chi_i(n_1)\chi_i(n_2)$.

The Δ function is equal to the number of common fragments rooted in nodes n_1 and n_2, and thus, depends on the fragment type. We report its algorithm for the evaluation of the number of syntactic tree fragments (STFs) [4].

A syntactic tree fragment (STF) is a set of nodes and edges from the original tree, which is still a tree and with the constraint that any node must have all or none of its children. This is equivalent to state that the production rules contained in the STF cannot be partial.

To compute the number of common STFs rooted in n_1 and n_2, the STK uses the following Δ function [4]:

1. if the productions at n_1 and n_2 are different then $\Delta(n_1, n_2) = 0$;
2. if the productions at n_1 and n_2 are the same, and n_1 and n_2 have only leaf children (i.e. they are pre-terminal symbols) then $\Delta(n_1, n_2) = \lambda$;
3. if the productions at n_1 and n_2 are the same, and n_1 and n_2 are not pre-terminals then $\Delta(n_1, n_2) = \lambda \prod_{j=1}^{l(n_1)}(1 + \Delta(c_{n_1}(j), c_{n_2}(j)))$, where $l(n_1)$ is the number of children of n_1, $c_n(j)$ is the j-th child of node n and λ is a decay factor penalizing larger structures.

Figure 3.a shows some STFs of the left tree in Figure 1. STFs satisfy the constraint that grammatical rules cannot be broken.

STK does not include individual nodes as features. As shown in [5] we can include at least the leaves, (which in constituency trees correspond to words) by simply inserting the following step 0 in the algorithm above:

0. if n_1 and n_2 are leaf nodes and their labels are identical then $\Delta(n_1, n_2) = \lambda$;

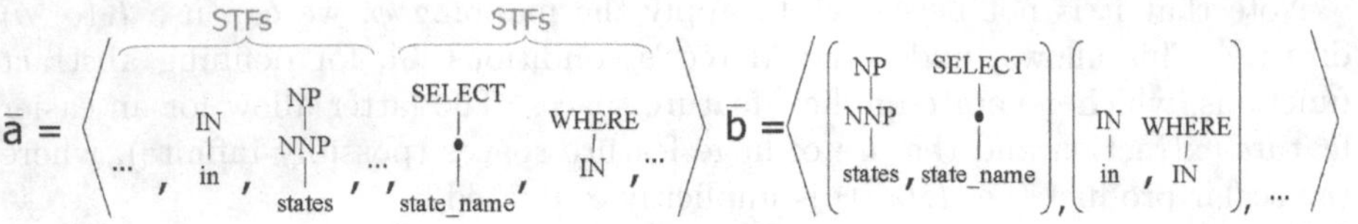

$$\bar{a} = \left\langle \ldots, \overbrace{\underset{in}{\overset{NP}{\underset{\displaystyle \underset{states}{\mid}}{\overset{\mid}{IN}}}} \, , \, \overset{STFs}{\overbrace{\underset{state\_name}{\overset{SELECT}{\mid}}}} \ldots, \, \overset{WHERE}{\underset{IN}{\mid}} , \ldots \right\rangle \quad \bar{b} = \left\langle \ldots \left(\underset{states \, , \, state\_name}{\overset{NP \quad SELECT}{\underset{\displaystyle \mid \quad \mid}{NNP \quad \bullet}}} \right) \left(\underset{in \, , \, IN}{\overset{IN \quad WHERE}{\underset{\mid}{IN}}} \right) , \ldots \right\rangle$$

Fig. 3. Feature spaces for the tree pair in Figure 1 a) joint space STK+STK b) Cartesian product STK×STK

3.2 Kernel Engineering

Kernel engineering [6,7,8] can be carried out by combining basic kernels with additive or multiplicative operators or by designing specific data objects, e.g. the tree representation for the SQL syntax, to which standard kernels are applied. Since our data is a set of pairs, we need to represent the members of a pair and their interdependencies. For this purpose, given two kernel functions, $k_1(.,.)$ and $k_2(.,.)$, and two pairs, $p_1 = \langle n_1, s_1 \rangle$ and $p_2 = \langle n_2, s_2 \rangle$, a first approximation is given by summing the kernels applied to the components: $K(p_1, p_2) = k_1(n_1, n_2) + k_2(s_1, s_2)$. This kernel will produce the union of the feature spaces of questions and queries (e.g. see [9,10]). For example, the explicit vector representation of the STK + STK space of the pair in Figure 1 is shown in Figure 3.a. The Syntactic Tree Fragments of the question will be in the same space of the Syntactic Tree Fragments of the query.

In theory a more effective kernel is the product $k(n_1, n_2) \times k(s_1, s_2)$ since it generates pairs of fragments as features, where the overall space is the Cartesian product of the used kernel spaces. For example Figure 3.b shows pairs of STF fragments, which are essential to capture the relational semantics between the syntactic tree subparts of the two languages (see [11]). In particular, the first fragment pair of the figure may suggest that a noun phrase composed by *state* expresses similar semantics of the syntactic construct SELECT state_name.

As additional feature and kernel engineering, we also exploit the ability of the polynomial kernel to add feature conjunctions. By simply applying the function $(1 + K(p_1, p_2))^d$, we can generate conjunction up to d features. Thus, we can obtain tree fragment conjunctions and conjunctions of pairs of tree fragments.

4   Dataset Generation

In the previous sections we have defined models to automatically learn the mapping between questions and queries. To apply such models we need training data, i.e. correct and incorrect pairs of questions and queries. Since acquiring such kind of data is the most costly aspect of our design, we should generate the learning set in a smart way. In this perspective, we assume that, in real world domains, we may find examples of questions and the associated queries answering to their information need. Such pairs may have been collected when users and operators of the database worked together for the accomplishment of some tasks. In contrast, we cannot assume to have available pairs of incorrect examples, since (a) the operator tends to just provide the correct query and (b) both users and operators do not really understand the use of negative examples and the need to have unbiased distribution of them.

Therefore, we need techniques to generate negative examples from an initial set of correct pairs. Unfortunately, this is not a trivial task since when mixing a question and a query belonging to different pairs we cannot assume to only generate incorrect pairs, e.g. when swapping x with y in the two pairs $\langle Which\ states\ border\ Texas?,\ x \rangle$ and $\langle What\ are\ the\ states\ bordering\ Texas?,\ y \rangle$, we obtain other two correct pairs.

To generate a gold standard dataset we would need to manually check this aspect thus we design an algorithm to limit the human supervision. It consists of the following steps:

- Generalizing question and query instances: substitute the involved concepts in questions and their related field values in the SQL queries by means of variables (expressing the category of such values).
- Clustering the generalized pairs: intuitively each cluster represents the information need about a target semantic concept, e.g. "bordering state", common to questions and queries. This requires a limited manual intervention.
- Pairing questions and queries of distinct clusters, i.e. the Cartesian product between the set of questions and the set of queries belonging to the pairs of a target cluster. This allows to find new positive examples that were not present in the initial corpus.
- Final dataset annotation: consider all possible pairs, i.e. Cartesian product between all the questions and queries of the dataset, and annotate them as negatives if they have not been annotated as positives in the previous step.

We use the GEOQUERIES250[3] corpus translated by Popescu et al. [12] as our initial dataset. It consists of 250 pairs of NL questions and SQL queries over a small database about United States geography. In the following we describe in detail all the steps through which the final dataset is generated.

4.1 Pair Generalization

Our approach to automatically annotate pairs relies on automatically detecting if swapping the members of different pairs produces correct or incorrect examples. For this purpose, we detect similar syntactic structures of questions and queries by generalizing concept instances with variables.

In a database concepts are represented as tables' fields. These are used in SQL to select data satisfying some conditions, i.e. concept constrained to a value.

Typically these values are natural language terms so we substitute them with variables if they appear in both questions and queries. For example, consider s_1 in Figure 1. The condition is WHERE state_name = 'Texas' and 'Texas' is

---

[3] http://www.cs.utexas.edu/users/ml/nldata.html

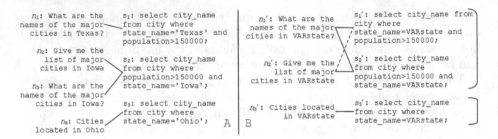

Fig. 4. Example of the initial corpus (*A, on the left*) and the generalized version (*B, on the right*). The latter is divided in two clusters (*identified by the two brackets*).

the value of the concept state_name. Since 'Texas' is also present in the related question we can substitute it with a variable *VARstate* (one variable for each different concept). Our assumption is that questions whose answer can be retrieved in a database tend to use the same terms stored in the database.

An example of the generalization phase is shown in Figure 4. On the left there is a set of four pairs containing four distinct questions and three related queries (connected by the lines) whereas on the right four generalized pairs are shown. We note that, after substituting instances with variables, both n_1 and n_3 are generalized into n_1', which is thus paired with two distinct SQL queries, i.e. s_1' and s_2'. This is not surprising since there can be multiple SQL queries that correctly retrieve an answer to a NL question. In this case we define them to be *semantically equivalent*, i.e. $s_1' \equiv s_2'$. At the same time it is possible to write many NL questions that map to the same query.

It is worth noting that with the generalization process, we introduce redundancy that we eliminate by removing duplicated questions and queries. Thus, the output dataset is usually smaller than the initial one. However the number of training examples will be larger, not only because of the introduction of negatives but also due to the automatic discovering of new positives.

4.2   Pair Clustering and Final Dataset Annotation

Once the pairs have been generalized, we cluster them according to their semantic equivalence so that we can automatically derive new positive examples by swapping their members. We define semantically equivalent pairs those correct pairs with (a) equivalent NL questions or (b) equivalent SQL queries. Given that two equivalent queries should retrieve the same result set, we can automatically test their equivalence by simply executing them. Unfortunately, this is just a necessary condition (e.g. 'Texas' can be the answer of two different queries) therefore we manually evaluate new pairings satisfying this condition.

Note that automatically detecting semantic equivalence of natural language questions with perfect accuracy is a hard task, so we consider as semantically equivalent either identical questions or those associated with semantic equivalent queries. We also apply transitivity closure to both members of pairs to extend the set of equivalent pairs.

```
function CreateFinalDataset(LIST OF PAIRS I) returns MATRIX M
M[|N|][|S|] = -1  // stores the cluster id for each pair; it is initialized to -1
k=0               // at the beginnig the number of cluster is zero
begin
∀ ⟨n,s⟩ ∈ I s.t. M[n][s] = -1 do    // for all initial pairs not clustered yet
    begin
    k:=k+1                          // add a new cluster
    M[n][s]:=k                      // the pair ⟨n,s⟩ belongs to the new cluster
    ∀ ⟨n',s'⟩ ∈ I do                // find other pairs that belong to it
    if (n = n') OR (s = s')         // if identical (indeed equivalent)
        then M[n'][s']:=k           // also ⟨n',s'⟩ belongs to the cluster
        else                        // check equivalence if similar result set
          if (fields(s) = fields(s')) AND (min(s) = min(s'))
            then if res(s) = res(s')  // if s and s' retrieve the same result set
              then if n ≡ n'          // if the associated sentences are equivalent
                then M[n'][s']:=k      // label also the pair ⟨n',s'⟩ with this cluster id
    end
∀ k do                              // find all pairings within a cluster
    ∀ n, n', s, s' s.t. M[n][s]=k AND M[n'][s']=k
    begin
    M[n][s']:=k                     // swap all queries
    M[n'][s]:=k                     // swap all questions
    end
    return M
end
```

Fig. 5. Clustering Algorithm

For example, in Figure 4.b s'_1 and s'_2 retrieve the same result set so we verify that they are semantically equivalent queries and we assign them to the same cluster (CL1), i.e. information need about the large cities of a state (with a population larger than 150,000 people). Alternatively, we can also consider that n'_1 and n'_2 are both paired with s'_2 to derive that they are equivalent, avoiding the human intervention. Concerning s'_3, it retrieves a result set different form the previous one so we can automatically assign it to a different cluster (CL2), i.e. involving questions about any city of a state. Note that, once n'_2 is shown to be semantically equivalent to n'_1 we can pair them with s'_1 to create the new pair highlighted with the dashed relation $\langle n'_2, s'_1 \rangle$. Thus the negative instance set is $\langle n'_3, s'_1 \rangle, \langle n'_3, s'_2 \rangle, \langle n'_1, s'_3 \rangle, \langle n'_2, s'_3 \rangle$.

The above steps are formally described by the algorithm in Figure 5. It takes as input the generalized dataset as a list of correct pairs $I \subset \{\langle n, s \rangle : n \in \mathcal{N}, s \in \mathcal{S}\}$ and returns a matrix M storing all positive and negative pairs $\mathcal{P} = \mathcal{N} \times \mathcal{S}$. M is obtained by (a) dividing I in k clusters of semantically related pairs and (b) applying the transitive closure to the semantic relationship between member pairs. More in detail, we first initialize its entries with a negative value, i.e. $M[n, s] = -1 \forall n \in \mathcal{N}, s \in \mathcal{S}$.

Second, we group together each $\langle n, s \rangle$ and $\langle n', s' \rangle \in I$, if at least two of their members are identical. If not we test if the two query members, s and s' retrieve the same result set. Since this may be time consuming we run this test only if the selected columns in their SELECT clause are the same and if the two result sets share the same minimum.

Third, since the condition above is only necessary for semantic equivalence, in case we find the same result sets, we manually check if the natural language question members are semantically equivalent. This is faster and easier than checking the SQL queries.

Finally, once the initial clusters have been created, we apply the transitive closure to the cluster c_k to include all possible pairing between questions and queries belonging to c_k, i.e. $c_k = \{\langle n, s \rangle : n, s \in c_k\}$. We store in M the id of the clusters in the related pairs, i.e. $M[n][s] = k$ for each c_k. As a side effect all entries of M still containing -1 will be negative examples.

5   The Experiments

In these experiments, we study the effectiveness of our algorithm for automatically mapping questions into queries by testing the accuracy of selecting for each question of the test set its correct query. For this purpose, we learn a classifier of correct and incorrect pairs and we use it as a ranker for the possible queries as described in Section 2.

5.1   Setup

The query ranker[4] consists in an SVM using advanced kernels for representing question and query pairs. We used SVM-Light-TK[5], which extends the SVM-Light optimizer [14] with tree kernels. i.e. Syntactic Tree Kernel (STK) and its extension (STK$_e$) as described in Section 3.1. We model many different combination described in the next section. We used the default parameters, i.e. the cost and trade-off parameters $= 1$ (for normalized kernels) and $\lambda = 0.4$ (see Section 3.1).

As test set, we use our dataset obtained from GEOQUERIES250 by applying our algorithm described in Section 4. After the generalization process the initial 250 pairs of questions/queries are reduced to 155 pairs containing 154 different NL question and 80 different SQL queries. We found 76 clusters, from which we generated 165 positive and 12,001 negative examples for a total of 154×80 pairs. Since the number of negatives is much larger than the positives, we eliminate some of them to make the learning more efficient. For this purpose we only keep the pairs whose members share at least two stems. Our assumption is that a positive pair should share at least one variable and one concept (e.g. *VARstate cities*). We applied such heuristics to both training and test data by considering the false negatives in the final computation of the system accuracy[6].

---

[4] We simply used the score of the SVM-based classifiers. More effective approaches have been proposed [4,13].

[5] http://disi.unitn.it/~moschitt/Tree-Kernel.htm

[6] However, our results show that the above pre-processing in our data does not lead to any errors.

Table 1. Kernel combination accuracy

Table 2. Kernel engineering results

K1	K2	K1×K2	K1+K2
LIN	LIN	70.7±12.0	57.3±10.4
POLY	POLY	71.9±11.5	55.1±8.4
STK	STK	70.3±9.3	54.9±10.1
STK_e	STK_e	70.1±10.9	56.7±12.0
LIN	STK_e	75.6±13.1	56.6±12.4

Advanced Kernels	Accuracy
$STK_e^2+POLY^2$	73.2±11.4
$(1+LIN^2)^2$	73.6±9.4
$(1+POLY^2)^2$	73.2±10.9
$(1+STK_e^2)^2$	70.0±12.2
$(1+LIN^2)^2+STK_e^2$	75.0±10.8

We excluded 10.685 negative examples from our initial set, reducing it to 1.316 elements.

We evaluated our automatic mapping by applying a standard 10-fold cross validation and measuring the average accuracy and the Std. Dev. in selecting the correct query for each question of the test sets.

5.2 Results

We tested several models for ranking based on different kernel combinations whose results are reported in tables Table 1 and Table 2. The first two columns of Table 1 show the kernels used for the question and the query, respectively. More specifically, our basic kernels are: (1) linear kernel (LIN) built on the bag-of-words (BOW) of the questions or of the query, e.g. SELECT is considered a feature for the query; (2) a polynomial kernel of degree 3 on the above BOW (POLY); (3) the Syntactic Tree Kernel (STK) on the parse tree of the question or the query and (4) STK extended with leaf features (STK_e).

Columns 3 and 4 show the average accuracy (over 10 folds) ± Std. Dev. of two main kernel combinations by means of product and sum. Note that we can also sum or multiply different kernels, e.g. LIN×STK_e.

An examination of the reported tables suggests that: first, the basic traditional model based on linear kernel and BOW, i.e. LIN + LIN, provides an accuracy of only 57.3%, which is greatly improved by LIN×LIN=LIN^2, i.e. by 13.5 points. The explanation is that the kernel sum cannot express the relational feature pairs coming from questions and queries, thus LIN does not capture the underlying shared semantics between them. It should be noted that only kernel methods allow for an efficient and easy design of LIN^2, since the traditional approach would have required to build the Cartesian product of the question BOW by query BOW. This can be very large, e.g. 10K features for both spaces leads to a pair space of 100M features.

Second, the K_1+K_2 column confirms that the feature pair space is essential since the accuracy of all kernels implementing individual spaces (e.g. kernels which are sums of kernels) is much lower than the baseline model for feature pairs, i.e. LIN^2.

Third, if we include conjunctions in the BOW representation by using POLY, we improve the LIN model, when we use the feature pair space, i.e. 71.9% vs 70.7%.

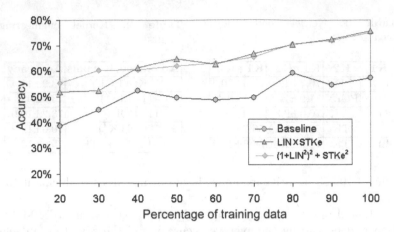

Fig. 6. Learning curves for GEOQUERIES250 corpora

Also, POLY^2 is better than STK^2 since it includes individual terms/words, which are not included by STK.

Next, the above point suggests that syntactic models can improve BOW although too many syntactic features (generated by STK) make the model unstable as suggested by the lower accuracy (70.1%) provided by STK$_e$×STK$_e$=STK$_e^2$. This consideration leads us to experiment with the model LIN × STK$_e$, which combines stems of the questions with syntactic constructs of SQL queries. This produces the highest result, i.e. 75.6%, suggesting that the syntactic parse tree of the SQL query is very reliable (it is obtained with 100% of accuracy) while the natural language parse tree, although accurate, introduces noise that degrades the overall feature representation. As a consequence it is more effective to use words only in the representation of the first member of the pairs.

Moreover, we experimented with very advanced kernels built on top of feature pair spaces as shown in Table 2. For example, we sum different pair spaces, STK$_e^2$ and POLY^2, and we apply the polynomial kernel on top of pair spaces by creating conjunctions, over feature pairs. This operation tends to increase too much the cardinality of the space and makes it ineffective. However, using the simplest initial space, i.e. LIN, to build pair conjunctions, i.e. $(1+LIN^2)^2$, we obtain a very interesting and high result, i.e. 73.6%[7]. This suggests that kernel methods have the potentiality to describe relational problems using simple building blocks although new theory describing the degradation of kernels when the space is too complex is required.

Finally, to study the stability of our complex kernels, we compared the learning curve of the baseline model, i.e. LIN+LIN, with the those of best models, i.e. LIN×STK$_e$ and STK^2+(1+LIN^2)^2. Figure 6 shows that complex kernels are

---

[7] Although the Std. Dev. associated with the model accuracy is high, the one associated with the distribution of difference between the model accuracy is much lower. Considering that we used 10 folds, it is easy to verify that $(1+LIN^2)^2$ is better than LIN^2 at a 90% confidence limit.

not only more accurate but also more stable, i.e. their accuracy grows smoothly according to the increase of training data.

5.3 Related Work

As the literature suggest, NLIDBs can be classified according to the approach employed in deriving an SQL query that retrieves the answer of a given NL question against a database. In this section we review three systems based on different approaches and that were also tested on the GEOQUERIES250. For a complete review of many NLIDB refer to Chandra and Mihalcea [15].

Systems based on authoring interface rely on semantic grammar specified by an expert user to interpret question over the database. CatchPhrase [16] is an authoring tool where the author is asked to name database elements, tailor entries and define additional concepts. This tool achieves 80% Recall and 86% Precision.

Another approach is based on enriching the information contained in the pairs. An example is given by Precise system [12]. Reducing the problem of finding a semantic interpretation of ambiguous phrases to a graph matching problem, Precise achieves 100% Precision on a subset of semantically tractable questions (77,5% Recall).

The machine learning approach, that induces semantic grammar from a corpus of correct pairs of questions and queries, has been used in Krisp [17]. Krisp performs semantic parsing mapping sentences into their computer-executable meaning representations. For each production in the meaning representation language it trains an SVM classifier based on string subsequence kernels. Then it uses these classifiers to compositionally represent a natural language sentence in their meaning representations. Krisp achieves approximatively 94% Precision and 78% Recall. Our system, also based on the machine learning approach, does not decline to answer any questions and shows an accuracy of 76% when the SQL query of the pair with the highest rank is executed to retrieve the answer of the paired question.

Finally, another approach for mapping natural language sentences to artificial language using lambda calculus has been proposed in [18]. Such method showed 96.3% in Precision and 79.3% in Recall on the extended corpus GeoQueries880.

Regarding the use of tree kernels for natural language tasks several models have been proposed and experimented [4,19,20,21,22,23,24,25,26,27].

6 Conclusions

In this paper, we approach the problem of mapping natural language semantics into programming languages by automatically learning a model based on lexical and syntactic description of the training examples. In our study, these are pairs of NL questions and SQL queries, which we annotated by means of our semi-automatic algorithm based on the initial annotation available in the GEOQUERIES250 corpus.

To represent syntactic/semantic relationships expressed by the pairs above, we largely adopted kernel methods along with SVMs. We designed innovative combinations between different kernels for structured data applied to pairs of objects. To our knowledge, the functions that we propose for relational semantics description are novel. The experiments of the automatic question translation system show a satisfactory accuracy, i.e. 76%, although large improvement are still possible.

The main contributions of our study are: (i) we show that our automatic mapping between question and SQL queries is viable, (ii) in at least one task we have proved that kernel products are effective, (iii) syntax is important to map natural language into programming languages and (iv) we have generated a corpus for future studies, which we are going to make publically available.

In the future we would like to extend this research by focusing on advanced shallow semantic approaches such as predicate argument structures [28,29].

Acknowledgments

This research has been partially supported by the EC Project#231126, Living-Knowledge.

References

1. Basili, R., Moschitti, A., Pazienza, M.: A text classifier based on linguistic process-ing. In: Proceedings of IJCAI 1999, Machine Learning for Information Filtering (1999)
2. Charniak, E.: A maximum-entropy-inspired parser. In: Proceedings of NAACL 2000 (2000)
3. Shawe-Taylor, J., Cristianini, N.: Kernel Methods for Pattern Analysis. Cambridge University Press, Cambridge (2004)
4. Collins, M., Duffy, N.: New ranking algorithms for parsing and tagging: Kernels over discrete structures, and the voted perceptron. In: Proceedings of ACL 2002 (2002)
5. Zhang, D., Lee, W.S.: Question classification using support vector machines. In: Proceedings of the 26th annual international ACM SIGIR conference on Research and development in informaion retrieval, pp. 26–32. ACM Press, New York (2003)
6. Moschitti, A., Bejan, C.: A semantic kernel for predicate argument classification. In: CoNLL 2004, Boston, MA, USA (2004)
7. Moschitti, A., Coppola, B., Pighin, D., Basili, R.: Engineering of syntactic features for shallow semantic parsing. In: Proceedings of ACL 2005 Workshop on Feature Engineering for Machine Learning in NLP, USA (2005)
8. Moschitti, A., Pighin, D., Basili, R.: Tree kernels for semantic role labeling. Com-putational Linguistics 34(2), 193–224 (2008)
9. Moschitti, A., Quarteroni, S., Basili, R., Manandhar, S.: Exploiting syntactic and shallow semantic kernels for question/answer classification. In: Proceedings of ACL 2007, Prague, Czech Republic (2007)
10. Moschitti, A.: Kernel methods, syntax and semantics for relational text categoriza-tion. In: CIKM 2008: Proceeding of the 17th ACM conference on Information and knowledge management, pp. 253–262. ACM, New York (2008)

11. Moschitti, A., Zanzotto, F.: Fast and effective kernels for relational learning from texts. In: Ghahramani, Z. (ed.) Proceedings of the 24th Annual International Conference on Machine Learning, ICML 2007 (2007)
12. Popescu, A.M., Etzioni, O., Kautz, H.: Towards a theory of natural language interfaces to databases. In: Proceedings of the 2003 International Conference on Intelligent User Interfaces, Miami, pp. 149–157. Association for Computational Linguistics (2003)
13. Moschitti, A., Pighin, D., Basili, R.: Semantic role labeling via tree kernel joint inference. In: Proceedings of CoNLL-X, New York City (2006)
14. Joachims, T.: Making large-scale SVM learning practical. In: Schölkopf, B., Burges, C., Smola, A. (eds.) Advances in Kernel Methods (1999)
15. Chandra, Y., Mihalcea, R.: Natural language interfaces to databases, University of North Texas, Thesis, M.S. (2006)
16. Minock, M., Olofsson, P., Näslund, A.: Towards building robust natural language interfaces to databases. In: Kapetanios, E., Sugumaran, V., Spiliopoulou, M. (eds.) NLDB 2008. LNCS, vol. 5039, pp. 187–198. Springer, Heidelberg (2008)
17. Kate, R.J., Mooney, R.J.: Using string-kernels for learning semantic parsers. In: Proceedings of the 21st ICCL and 44th Annual Meeting of the ACL, Sydney, Australia, July 2006, pp. 913–920. Association for Computational Linguistics (2006)
18. Zettlemoyer, L.S., Collins, M.: Learning to map sentences to logical form: Structured classification with probabilistic categorial grammars. In: UAI, pp. 658–666 (2005)
19. Kudo, T., Matsumoto, Y.: Fast Methods for Kernel-Based Text Analysis. In: Hinrichs, E., Roth, D. (eds.) Proceedings of ACL, pp. 24–31 (2003)
20. Cumby, C., Roth, D.: Kernel Methods for Relational Learning. In: Proceedings of ICML 2003, Washington, DC, USA, pp. 107–114 (2003)
21. Culotta, A., Sorensen, J.: Dependency Tree Kernels for Relation Extraction. In: ACL 2004, Barcelona, Spain, pp. 423–429 (2004)
22. Kudo, T., Suzuki, J., Isozaki, H.: Boosting-based parse reranking with subtree features. In: Proceedings of ACL 2005, US (2005)
23. Toutanova, K., Markova, P., Manning, C.: The Leaf Path Projection View of Parse Trees: Exploring String Kernels for HPSG Parse Selection. In: Proceedings of EMNLP 2004, Barcelona, Spain (2004)
24. Kazama, J., Torisawa, K.: Speeding up Training with Tree Kernels for Node Relation Labeling. In: Proceedings of EMNLP 2005, Toronto, Canada, pp. 137–144 (2005)
25. Shen, L., Sarkar, A., Joshi, A.k.: Using LTAG Based Features in Parse Reranking. In: EMNLP, Sapporo, Japan (2003)
26. Zhang, M., Zhang, J., Su, J.: Exploring Syntactic Features for Relation Extraction using a Convolution tree kernel. In: Proceedings of NAACL, New York City, USA, pp. 288–295 (2006)
27. Zhang, D., Lee, W.: Question classification using support vector machines. In: Proceedings of SIGIR 2003, Toronto, Canada. ACM, New York (2003)
28. Giuglea, A.M., Moschitti, A.: Knowledge discovery using framenet, verbnet and propbank. In: Meyers, A. (ed.) Workshop on Ontology and Knowledge Discovering at ECML 2004, Pisa, Italy (2004)
29. Giuglea, A.M., Moschitti, A.: Semantic Role Labeling via Framenet, Verbnet and Propbank. In: Proceedings of ACL 2006, Sydney, Australia (2006)

Search Interface to a Mayan Glyph Database Based on Visual Characteristics[*]

Grigori Sidorov[1], Obdulia Pichardo-Lagunas[1], and Liliana Chanona-Hernandez[2]

[1] Natural Language and Text Processing Laboratory,
Center for Research in Computer Science (CIC), National Polytechnic Institute (IPN),
Av. Juan Dios Batiz, s/n, Zacatenco, 07738, Mexico City, Mexico
sidorov@cic.ipn.mx, ayilina@hotmail.com
[2] Superior School of Mechanical and Electric Engineering (ESIME),
National Polytechnic Institute (IPN),
Av. Juan Dios Batiz, s/n, Zacatenco, 07738, Mexico City, Mexico

Abstract. During years, the community of Mayan researchers was not open to the usage of computer tools. Still, the progress of the computer science and the current state of Mayan research proves the necessity of this type of software. We present the project related to the development of Mayan script database, which is the first necessary step in development of computer representation of Mayan script data. The database contains several tables and allows for various queries. The main idea of the project is the development of the system that would allow managing Mayan script data for specialists and as well for persons without any previous knowledge of Maya. The system includes structural visual description of glyph images and search facilities based on the structural description, along with traditional database search facilities. The user can define any set of visual characteristics. The glyphs are characterized using a "naive" feature set. Fully working prototype is presented. Preliminary evaluation of the efficiency is carried out. In fact, the system can be applied to any set of images that can be assigned structural features.

Keywords: Search interface, Mayan glyph database, visual structural description.

1 Introduction

Mayan hieroglyphic writing contains more than thousand of different glyph signs. Many of them are variations of one sign (allographs), other glyphs are different signs with the same reading (homophones). Some signs are glyph variants that were used during a certain period of time or in a certain area.

Mayan writing system can be described as a logosyllabic system ([1, 3, 4, 5, 7]), based on signs that represent words (logograms) and syllables that also can be used as

---

[*] Work done under partial support of Mexican Government (CONACYT projects 50206-H and 83270, SNI) and National Polytechnic Institute, Mexico (projects SIP 20080787, 20091587 and 20090772; COFAA, PIFI).

H. Horacek et al. (Eds.): NLDB 2009, LNCS 5723, pp. 222–229, 2010.
© Springer-Verlag Berlin Heidelberg 2010

phonetic signs. There are approximately 200 different syllabic (i.e., purely phonetics) signs, of which about a 60 percent are homophones. Namely, there are about 80 syllables in the classic Mayan language according to its phonetics; still more than 200 glyphs signs were used in the phonetic writing. The Mayans used a system of writing capable to register complete oral manifestations of their language.

The discoveries of the last decades in Mayan epigraphy field allowed deciphering of almost all documents and known inscriptions, according to the dominant theories of deciphering. Though, the deciphering is different according to the adopted theory. Nevertheless, there is no kind of somewhat sophisticated computational tool that can be used by the Mayanists in their investigations. These investigations are normally carried out manually on the basis of facsimiles of documents and Mayan inscriptions ([6, 8, 9]).

On the other hand, the public in general should also have possibilities of using some specific computer tools if they are interested in reading Mayan inscriptions. Though they are not specialists in the field, the subject is interesting from the general cultural perspective.

Even though the specialists are aware of the existence of various dictionaries in the Internet, they are not frequently used. Even more: during years, the community of Mayan researchers was very skeptic as far as the usage of any type of computer tools. Still, the progress of computer science and the current state of Mayan research proves the necessity of this type of software, because it makes the work of a researcher more fast, reliable and productive.

We present the project related to the development of the interface to the Mayan glyph database, which is the first necessary step in development of computer representation of Mayan script data.

We believe that developing of this application would be impossible without support of investigators of Maya and without considering their needs as end users. Our purpose is development of the application in which investigators of Maya would participate in design and implementation. Also, this application should be able to resolve specific problems that can solely be raised by a specialist.

This software will help to diminish the burden of many rather complex procedures that are performed manually by now, like context comparison or glyph identification. In addition, this software offers to the investigators a tool that facilitated the process of search and classification of Mayan glyphs and can serve as a didactic tool that helps in learning of the glyph signs that compose Mayan writing system.

The paper has the following structure. First, we describe the interface and its usage (Section 2), then in Section 3 we present the preliminary evaluation of the interface based on the comparison with the purely manual search, and finally the conclusions are drawn.

2 Interface Description

We based our development on John Montgomery's dictionary [2]. It contains hieroglyphs of Mayan classic writing, organized alphabetically by phonetic writing of words, phrases or syllables, and it also includes appendices that list signs using additional classification categories.

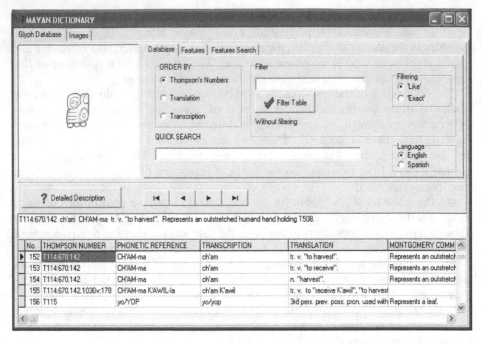

Fig. 1. Interface for the database (main dictionary view)

The developed application is a relational database – and the corresponding interface – that stores general information about glyphs: glyph image, translation, transcription, Thompson numbers (these can be considered as glyphs identifiers, ID; they were assigned in arbitrary manner by E. Thompson, an outstanding Mayan investigator), phonetic reference and descriptive notes corresponding to each glyph that forms part of John Montgomery's dictionary. Total size of the dictionary is 1,241 entries. The database is ordered according to Thompson numbers, but the system offers two more options for its ordering using the fields containing transcription or translation.

Due to normalization of the database, in the cases that a glyph has more than one meaning according to the dictionary, these are separated and marked with ID that corresponds to the sign and as well with the IDs that correspond to the number of meaning (translation); thus, different meanings are stored as separate records in the database.

The database is also capable of storing information provided by a user related to the structural visual description of the graphical image. This is based on a feature set created by each user if he wants to describe the glyphs using his own feature set, see below. The system has the predefined feature set. The user has a possibility to choose a desired feature set: the predefined one or his own set.

The information about glyphs is visualized as records in a standard tabular way, i.e., each visual field value corresponds to a specific database field. The following fields are used: glyph image, Thompson number, phonetic reading, meaning (translation from Maya), and comments.

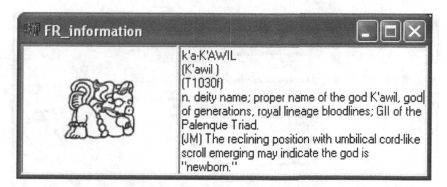

Fig. 2. Complete information about a glyph

Different meanings are stored as separate records (Fig. 1), thus, duplicating other field values. We chose this option for enabling searches in the field "meaning". Still, we have only one glyph image visible at a time and we present additionally the concatenated information of values referring the meaning (from several records) just below the glyph. This is justified by the fact that otherwise this information is presented in separate records and, thus, it is not clearly related to the same glyph.

The complete information about glyphs can be visualized within separate dialog window, see Fig. 2.

The user has an option to choose the ordering of the dictionary records by marking the corresponding field in the Button with multiple choices "Order by": *Thompson number, translation* or *transcription.*

This interface also allows for two types of searches: fast search and filtering. Both of them are performed in accordance with the selected ordering. The fast search moves the table cursor to the desired value according to the ordering in the corresponding field. It starts working with partial matching, thus, the cursor is moved to the first value that matches the current search field (though partially). The filtering restricts the records that are shown to the user according to the data present in the field "Filter" and the current ordering.

For the advanced search using structural features, the user can use predefined feature sets or create his set and describe glyphs according to this set.

The systems shows the window, where it provides the general information about a glyph. Also, a list of possible graphical features (characteristics) is presented, see Fig. 3.

The system allows for structural description of glyph images using various feature sets. This facility permits a user who is not familiar with Maya glyphs to make searches and identify the glyphs. The user can add and remove these features while describing an individual glyph. This makes possible assigning characteristics that correspond to each one of the glyphs. Each user (professional or amateur) can use his judgments for this characterizing.

The system stores each one of assigned feature values and allows visualizing of the characteristics that have been assigned previously to the glyph. Each set of features has its own name, thus, changing the set is just picking up a different set name. This allows for different users applying different feature sets prepared by them or by other users.

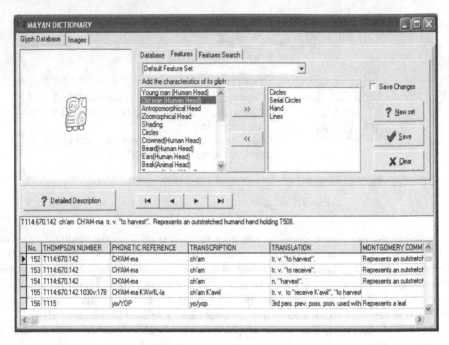

Fig. 3. Feature set assignment

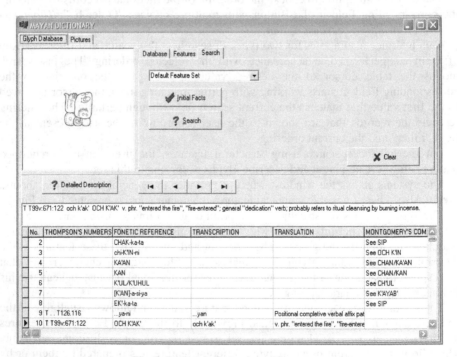

Fig. 4. Dialog for selection of feature set

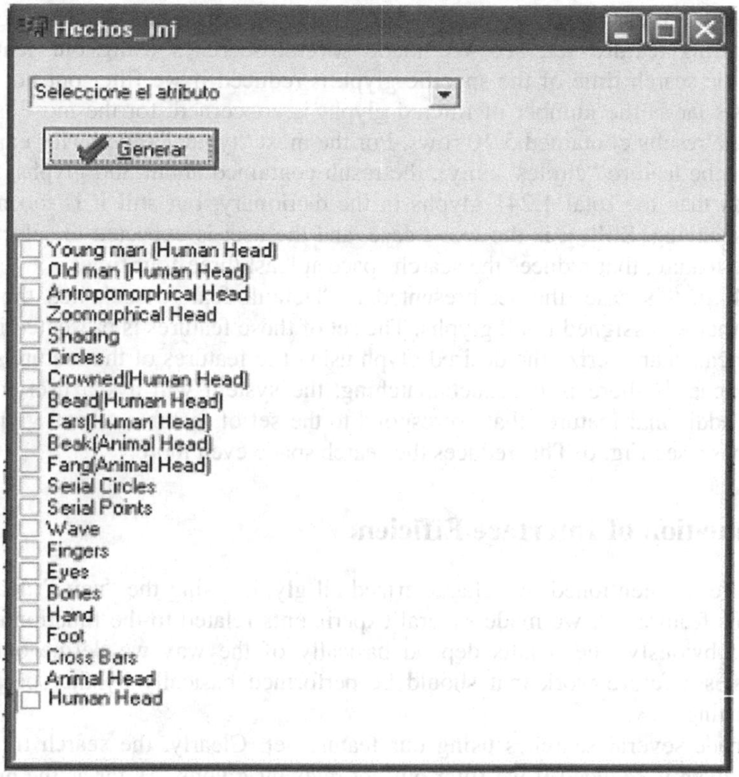

Fig. 5. Selection of the initial feature values

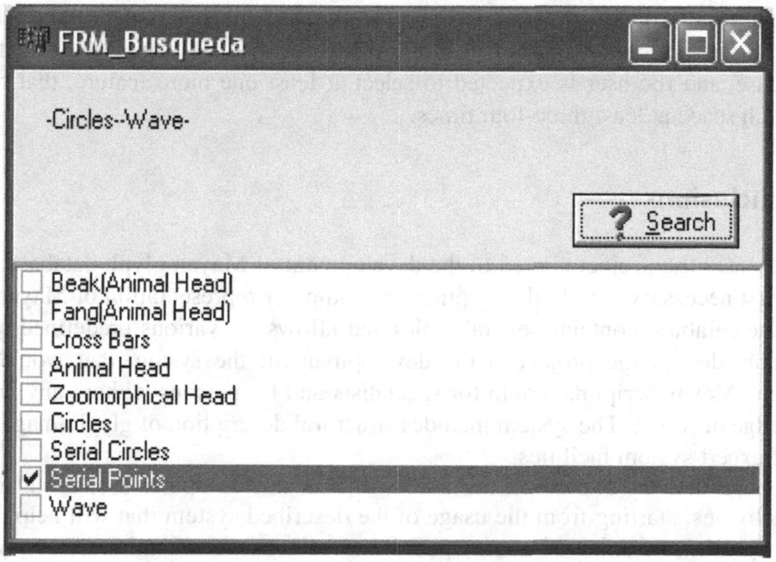

Fig. 6. Selection of the additional features

When the user wants to search for a glyph, first he should choose the corresponding feature set, see We made several searches using our feature set. Clearly, the search time of the specific glyph is reduced if we filter out not relevant glyphs. As far as the number of filtered glyphs is concerned, for the most "specific" glyphs, the results contained 5-10 rows. For the most "typical" glyph (for example, if we select the feature "circles" only), the result contained about 300 glyphs, which is much less than the total 1,241 glyphs in the dictionary, but still it is too many for manual scanning. Still, it is the worst case, and the user is expected to select at least one more feature, that reduces the search space at least three-four times.

Fig. 4. In this case, the set presented is "Default feature set", i.e., the "naive" features that we assigned to all glyphs. The set of these features is presented in Fig. 5. The user can characterize the desired glyph using the features of the feature set using this dialogue. If there is no exact matching, the system will ask a user to choose possible additional features that correspond to the set of features of real glyphs from the database, see Fig. 6. This reduces the search space even more.

3   Evaluation of Interface Efficiency

As we already mentioned, we characterized all glyphs using the "naive" feature set. Using this feature set, we made several experiments related to the functioning of the system. Obviously, the results depend basically of the way we define our feature set. This is a future work that should be performed basically by the specialists in Maya writing.

We made several searches using our feature set. Clearly, the search time of the specific glyph is reduced if we filter out not relevant glyphs. As far as the number of filtered glyphs is concerned, for the most "specific" glyphs, the results contained 5-10 rows. For the most "typical" glyph (for example, if we select the feature "circles" only), the result contained about 300 glyphs, which is much less than the total 1,241 glyphs in the dictionary, but still it is too many for manual scanning. Still, it is the worst case, and the user is expected to select at least one more feature, that reduces the search space at least three-four times.

4   Conclusions

We presented the project related to the development of Mayan glyph database, which is the first necessary step in development of computer representation of Mayan script data. The database contains several tables and allows for various predefined queries. The main idea of the project is the development of the system that would allow managing Mayan script data both for specialists and for persons without any previous knowledge of Maya. The system includes structural description of glyph images and a kind of expert system facilities.

We hope that the experts in the field will find this computer tool useful in their daily activities, starting from the usage of the described system that will help them to identify glyphs and reduce the time spent for search and classification of the glyphs.

The developed application will serve also as a didactic tool that helps not only the professional investigators and students, but it will also serve to any person interested in the Mayan writing system.

We plan to implement a determinist expert system based on the structural characteristics of glyph images. This will allow for performing automatic glyph classification, so that when the database contains the characteristics that correspond to each glyph according to the selected feature set, it will be possible to make reasoning procedures for quick glyph searches.

If the glyphs are classified, it will be easier to identify them within documents and Mayan inscriptions. As further investigation, we plan to develop a corpus where the glyphs are represented by their identifiers, for example, Thompson numbers. Note that the majority of existing corpus of Maya is just the digitalized images. Next interesting step in the future research is related to applying statistical methods of corpus linguistics to confirm the results of the deciphering procedures.

Speaking more generally, the presented system can be applied to any set of images that can be assigned a set of structural features. In this perspective, we are considering the implementation not only of the binary features (the feature is present or it is absent), but the features that allow non binary values.

References

1. Foundation for the advancement of Mesoamerican Studies Inc.,
   http://www.famsi.org
2. John Montgomery and Christophe Helmke. Dictionary of Mayan hieroglyphs. Hippocrene Books Inc. (2007)
3. Coe, M.D., Van Stone, M.: Reading the Maya Glyphs. Thames & Hudson (2001)
4. Calvin, I.E.: Maya Hieroglyphs Study Guide,
   http://www.famsi.org/mayawriting/calvin/index.html
5. Montgomery, J.: How to Read Maya Hieroglyphs. Hippocrene Books Inc. (2003)
6. Voss, A.W., Kremer, H.J.: Estudio epigráfico sobre las inscripciones jeroglíficas y estudio iconográfico de la Fachada del Palacio de los Estucos en Acanceh Yucatán, México. Instituto de Americanística y Etnología, Universidad de Bonn, Alemania (Noviembre 2000)
7. Knorosov, Y.V.: La antigua escritura de los pueblos de America Central. Biblioteca Obrera, Mexico (1954)
8. Shele, L.: Notebook for the Maya Hieroglyphic Writing. In: Proc. of Workshop at Texas Austin, Institute of Latin American Studies. University of Texas (1978)
9. Proskourniakoff, T.: Historical data in the inscriptions of Yaxchilan. In: Part 1. The reign of Shield-Jaguar. Estudios de Cultura Maya (3), Centro de Estudios Mayas, IIFL, UNAM, Mexico (1963)

Temporal Expression Identification Based on Semantic Roles*

Hector Llorens, Estela Saquete, and Borja Navarro

Natural Language Processing Research Group
University of Alicante, Spain
{hllorens,stela,borja}@dlsi.ua.es

Abstract. Following TimeML (TIMEX3) specifications, we present a study analyzing to what extent are semantic roles useful in temporal expression identification task, as well as, a list of the potential applications of this combination. For that purpose, two approaches of a temporal expression identification system based on semantic roles have been developed: (1) Baseline and (2) TIPSem-Full. The first approach is a direct conversion from a temporal semantic role to a temporal expression. The second one processes and converts all temporal roles into correct TIMEX3, using a set of transformation rules defined in this paper. These two approaches have been evaluated using TimeBank corpus. The evaluation results confirm that the application of semantic roles to temporal expression identification task can be valuable, obtaining, for TIPSem-Full, an 73.4% in F1 for relaxed span and a 65.9% in F1 for strict span.

Keywords: TE identification, Semantic Roles, TimeML, TimeBank.

1 Introduction

Recently, there has been an intensive research on automatic treatment of temporal expressions (TEs), events and their relations over natural language (NL) text [1]. This interest has been motivated by the wide range of natural language processing (NLP) areas that take advantage of considering temporal information, such as summarization or question answering (QA) [2]. All this research effort has been reflected in recent specialized workshops and conferences [3,4,5,2], as well as in evaluation forums [6,7].

The main goal of automatic treatment of temporal information is making it explicit in order to be easily extracted and used by other applications. For this reason, different annotation schemes have been created [8,9,10]. Concretely, TimeML [10] has been recently adopted as *de facto* standard by a large number of researchers [2].

Semantic role labeling (SRL), another NLP area, has achieved important results in the last years [11]. SRL consists of determining basic event structures

---

* This paper has been supported by the Spanish Government, project TIN-2006-15265-C06-01 where Hector Llorens is funded under a FPI grant (BES-2007-16256).

H. Horacek et al. (Eds.): NLDB 2009, LNCS 5723, pp. 230–242, 2010.

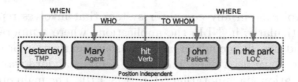

Fig. 1. SR example illustration

in a sentence, detecting semantic relations among entities and events. All semantic predicates are identified around the central event of each sentence. Once identified, they are classified as an argument role (agent, patient, etc.) or an adjunct role (locative, temporal, etc.). The temporal role represents *"WHEN"* an event takes place, so that semantic roles (SR) information could be useful on TE identification task (see fig. 1).

Semantic roles could help on recognize TEs that are ambiguous at lower language analysis levels. That is to say, expressions like *"April"* which can be interpreted as a TE (month) or person named entity (i.e., the female proper name) are equal at lexical level, morphological level, and even at syntactic level in some sentences. The following example illustrates this ambiguity:

Example 1.
He blamed the failure on April.   (female proper name)
I went to Canada on April.   (month, TE)

As shown in this example, both sentences are very similar at lower language analysis levels. Both *"April"* expressions are nouns, proper names, and part of a prepositional phrase together with the *"on"* preposition. However, in the first sentence *"on April"* is a person named entity and in the second one it is a TE. At semantic roles level, the expression *"on April"* represents a numbered argument role in first sentence while it represents a temporal role in the second one. Therefore, in this case, semantic roles provide a valuable information to identify TEs.

Taking these statements as starting point, the objective of this work is determining to what extent it is effective the application of SR in TE identification. For this purpose, a TE identification system based on SR has been implemented following TIMEX3 TimeML guidelines. In order to test its validity, this system has been evaluated using TimeBank corpus [12].

The paper is structured as follows: Section 2 focuses on the background of temporal information processing and SRL fields, Section 3 provides detailed information about our proposal and implementation of a TE identification system based on SR, and Section 4 includes its evaluation. Finally, conclusions and some further work proposals are presented.

2   Background

The fact that temporal aspects of NL are important is not a new issue in artificial intelligence (AI) [13]. The representation of time was pointed out as one of the most crucial problems in any computer system that involves representing the world [14]. Particularly, in NLP, the applications where temporal information processing would benefit include areas such as summarization or QA [2].

Several efforts have been done in order to define standard ways to represent temporal information via annotating NL text [8,9,10]. The main objective of this annotation is making temporal information explicit through annotation schemes. In computational linguistics, the construction of annotated corpora has been useful for developing and testing different approaches, specially in the application of machine learning (ML) techniques. Since temporal information extraction was included in Message Understanding Conference (MUC) context, there have been three important annotation schemes for temporal information:

- **STAG:** It was developed by Setzer and Gaizauskas [8] with the aim of identifying events in news and the relationship of these events with points in a temporal line. This scheme defines: TIMEX (inherit from MUC), SIGNAL and EVENT tags.
- **TIDES:** Under the support of the DARPA and ACE, TIDES [9] arose with the purpose of annotating only TEs and its normalized values (ISO 8601) in text using TIMEX2 tags. This annotation standard was used in TERN-2004 evaluation forum [6]. Examples of systems using this scheme are TempEx [3], Chronos [15] and TERSEO [16].
- **TimeML:** After STAG and TIDES, TimeML [10] was developed under the sponsor of ARDA. TimeML is a rich specification language for events and TEs in NL text. It combines and extends features of both preceding schemes, which makes it a more powerful annotation scheme.

TimeML was developed in the context of three AQUAINT workshops and projects: TERQAS-2002 workshop [4], TANGO-2003 workshop [5] and TARSQI project [17]. Under this framework, TimeML was set out to enhance QA systems to answer temporally-based questions about the events and entities in news articles. TimeML defines TIMEX3 tag for TEs, SIGNAL tag for temporal signals, EVENT tag for events, and TLINK, ALINK and SLINK tags for different kind of temporal relations.

An English corpus illustrating an example TimeML annotation was created together with the first version of this annotation scheme. It was TimeBank 1.1 [12]. The last version of the corpus, TimeBank 1.2[1], was compiled following the TimeML 1.2.1 [18] specifications. This version is now considered a gold standard for the annotation scheme and has been published by Linguistic Data Consortium (LDC). Moreover, TimeBank 1.2 was used in TempEval-2007 [7] evaluation forum as reference corpus for identifying temporal relations in NL text. An in-depth analysis of TimeBank corpus can be found in [19,20].

There have been different works on developing algorithms for automatically tagging NL text following TIMEX3 specifications.

---

[1] http://www.ldc.upenn.edu/Catalog/CatalogEntry.jsp?catalogId=LDC2006T08

On the one hand, TTK, developed in context of the TARSQI project [17], accomplishes this task using the GUTime module. This module was developed at Georgetown University and extends the capabilities of the MITRE TempEx tagger [3]. TempEx was developed using a set of hand-crafted rules to tag TEs, following the TIDES standard. This solution was benchmarked on training data from TERN 2004 [6] at 85% and 78% in F-measure for TIMEX2 identification and exact bounding respectively.

On the other hand, the work of Boguraev and Ando [20] presents an evaluation on automatic TimeML annotation over TimeBank using machine learning (ML) techniques. The results obtained on recognition of TIMEX3 over TimeBank using 5-fold cross validation were 81.7% in F for strict span and 89.6% for relaxed span.

In the last years, automatic semantic role labeling has generated a great research interest in NLP [11]. Several SRL tools were developed and evaluated in SensEval-3 workshop and CoNLL-2005 Shared Task evaluation forum. Since then, many works on the application of SR to other NLP fields have appeared [21,22,23].

Only one reference about using SR for temporal information processing has been found in literature [24]. This work was presented in TempEval-2007 and the usage of SR was related to help in the task of identifying temporal relations, but not TEs. Therefore, to our knowledge, our work represents the first time that SR have been applied to TE identification using TimeML TIMEX3 standard.

3 Proposal and Implementation

With the main objective of studying the application of SR to TE identification task, we developed a system accomplishing this task, following the indications of TimeML TIMEX3.

Our system uses the SRL tool developed by University of Illinois CCG group[2] [25] to tag SR over English text. This tool was presented in CoNLL-2004 and obtained a 77.44% F1 in temporal role labeling. It uses the PropBank role set [26] in which the temporal semantic role is represented by the AM-TMP label, hereafter referred to as TSR. Apart from SRL, this tool provides part-of-speech (PoS) tagging and syntactic tree structure.

The implemented system is called TIPSem: Temporal Information Processing based on Semantic Roles. The implementation includes two different approaches: Baseline and TIPSem-Full.

3.1 Baseline

This implementation tags each TSR identified by the SRL tool as a TIMEX3 temporal expression. It has been developed only as a baseline to measure how accurate are PropBank temporal roles by their own on representing TIMEX3. The following example shows this transformation.

---

[2] http://l2r.cs.uiuc.edu/~cogcomp/srl-demo.php

Example 2.
```
SR:     [I A0] [saw V] [you A1] [yesterday AM-TMP].
TIMEX3:  I saw you <TIMEX3>yesterday</TIMEX3>.
```

3.2 TIPSem-Full

First tests of Baseline showed that a temporal role is not defined exactly as a TIMEX3. For that reason, the cases where TSR differs from TIMEX3 have been identified. Once identified, a set of transformation rules between TSR and TIMEX3 has been defined and implemented in the TIPSem-Full system. The rules developed are the following.

TSR to TIMEX3 transformation rules

1. **Removing subordination of TSR**: If a TSR corresponds to a subordination clause it does not correspond to a TIMEX3. A TSR extent is a whole semantic predicate. However, according to TimeML specifications, the full extent of a TIMEX3 tag must correspond to one of the following categories: noun phrase (*"yesterday"* NP), adjective phrase (*"3-day"* ADJP) or adverbial phrase (*"3 days ago"* ADVP). The information of temporal subordination is treated by TimeML using other tags, but not TIMEX3.

   Example 3.
   ```
   SR:     She was happy [when I saw her AM-TMP].
   TIMEX3: She was happy when I saw her.
   ```

   Our system detects subordination using the syntactic tree and removes it.

2. **Removing TSR overlapping**: Due to the fact that each verb in a sentence has its own role annotation, it is possible to find overlapped TSRs.

   Example 4.
   ```
   SR: It took effect [[Aug. 9 AM-TMP] AM-TMP] pending ratification.
   TIMEX3: It took effect <TIMEX3>Aug. 9</TIMEX3> pending ratification.
   ```

   If two TSRs are overlapped, our system chooses the one that produces the shortest annotation because TIMEX3 corresponds to the minimum syntactic unit as explained in the previous rule.

3. **Splitting TSR**: A TSR composed of more than one NP can contain two or more related TIMEX3. TSRs with NPs linked by a temporal preposition or a coordination conjunction are denoting an anchoring relation or a conjunction relation respectively. They are generally marked with independent TIMEX3 tags. There are two exceptions for this rule. Firstly, times *"[ten minutes to four]"*, where the "to" preposition is denoting an specification relation. And secondly, the preposition *"of"* (*"at the end of 1999"*), which is usually part of the expression.

Example 5.

```
SR: I will go [for a week from Sunday AM-TMP].
TIMEX3: I will go for <TIMEX3>a week</TIMEX3> from <TIMEX3>Sunday</TIMEX3>.
SR: I was in Paris [in 1999 and 2000 AM-TMP].
TIMEX3: I was in Paris in <TIMEX3>1999</TIMEX3> and <TIMEX3>2000</TIMEX3>.
```

Our system looks for prepositions or coordination conjunctions in every TSR containing more than one NP. If they are found and do not represent an exception, the TSR is split in n TIMEX3 corresponding to each NP, excluding either prepositions or coordination conjunctions.

4. **Removing prepositions on single-NP TSR**: As described above, a TSR generally differs from a TIMEX3 on its boundaries. If a TSR has any preposition out of the minimum syntactic unit, the preposition is not included as part of the TIMEX3.

Example 6.

```
SR: She was born [in 1981 AM-TMP].
TIMEX3: She was born in <TIMEX3>1981</TIMEX3>.
```

This rule scans all TSR containing only one NP and strips out any preposition, transforming this kind of TSR into correct TIMEX3.

5. **Tagging resulting TSR as TIMEX3**: Finally, after the application of all the previous rules, resulting TSR are directly tagged as TIMEX3.

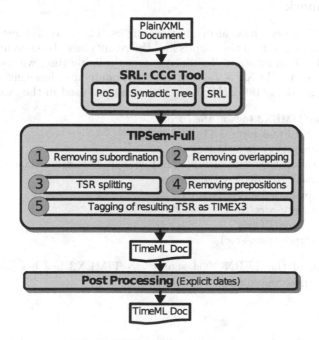

Fig. 2. TIPSem-Full Architecture

Post-processing. Due to the fact that SRL relies on verbs, nominal sentences can not be labeled. These sentences are commonly found in titles, brackets, notes, etc. Hence, as a post-processing step, a TE tagger capable of identifying basic explicit TEs (such as dates) is needed.

Example 7.
```
The attack in June 1999.
(Barcelona 07-1992).
```

The implemented system executes a simple TE tagger that only detects explicit dates and unambiguous single-word TEs according to TIMEX3 guidelines.

TIPSem-Full system architecture. The architecture is shown in Figure 2. As described above, the system implements 5 transformation rules, each one of them treating one aspect of temporal semantic roles in order to transform them into valid TimeML TIMEX3 tags.

4   Evaluation

The implemented approaches, Baseline and TIPSem-Full, have been evaluated and analyzed using the following benchmark.

4.1   Benchmark

TimeBank 1.2 has been used as evaluation corpus. The systems have been tested in TE identification within the corpus and the results have been compared to the original TIMEX3 annotation. To make a more reliable test, we used the 1228 TIMEX3 between <TEXT> tags in corpus documents, ignoring the explicit dates of their headers (186 TIMEX3). The measures used in the evaluation are:

- **POS**: Total TIMEX3 tags in the corpus.
- **ACT**: TIMEX3 tags returned by the system.
- Correct (**corr**): Correct instances
- Incorrect (**inco**): Correctly identified but wrongly bounded instances
- Missing (**miss**): Not detected instances
- Spurious (**spur**): False positives
- Precision (**prec**): corr/ACT
- Recall (**rec**): corr/POS
- **F1**: (2*prec*rec)/(prec+rec)

An adaptation of the TERN-2004 scorer[3] to TIMEX3 has been used for the calculation of these measures.

4.2   Results

Table 1 shows the obtained results. For each system, span strict S and span relaxed R results are indicated. S refers to strict match of both boundaries of a

Table 1. TimeBank TIMEX3 identification results

System		POS	ACT	corr	inco	miss	spur	prec	rec	F
Baseline	R	1228	1410	764	0	464	646	0.542	0.622	**0.579**
	S	1228	1410	368	396	464	646	0.261	0.300	**0.279**
TIPSem-part	R	1228	1180	727	0	501	453	0.616	0.592	0.604
(NoSub)	S	1228	1180	366	361	501	453	**0.310**	0.298	0.304
TIPSem-part	R	1228	1139	727	0	501	412	0.638	0.592	0.614
(NoOverl)	S	1228	1139	366	361	501	412	**0.321**	0.298	0.309
TIPSem-part	R	1228	1196	736	0	492	460	0.615	0.599	0.607
(Splitting)	S	1228	1196	384	352	492	460	0.321	**0.313**	0.317
TIPSem-part	R	1228	1181	736	0	492	445	0.623	0.599	0.611
(NoPrep)	S	1228	1181	669	67	492	445	**0.566**	**0.545**	0.555
TIPSem-Full	R	1228	1244	907	0	321	337	0.729	0.739	**0.734**
(PostProc)	S	1228	1244	814	93	321	337	0.654	0.663	**0.659**

TIMEX3 expression (exact extent) while R results consider as correct every tag identifying a TIMEX3 even if it is wrongly bounded.

The Baseline approach obtains 57.9% in F1 for span relaxed TIMEX3 identification, but it falls to 27.9% in the span strict case.

The TIPSem-Full evaluation has been made incrementally, applying, one by one, the rules defined in section 3, in order to analyze in detail to what extent each transformation rule of the full approach has improved the results of the Baseline approach. The first two rules are focused on omitting TSR that do not represent TIMEX3, and therefore, focused on improving the system precision. If we include the removing subordination rule the precision rises from 26.1% to 31.0% for S case. In addition, if we apply the removing overlapping rule the precision increases to 32.1%. The third rule, splitting TSR, divides roles in order to get correct TIMEX3. Hence, its aim is to improve both precision and recall. Results show that, for S case, recall improves from 29.8% to 31.3% but precision does not improve. The reasons for that fact will be explained in the error analysis subsection. The application of removing prepositions from single-NPs rule improves both precision and recall of S case to 56.6% and 54.5% respectively. This is because the inclusion of preceding prepositions as part of a role is an intrinsic characteristic of SR and many TSR in the corpus include a preposition that must be removed to represent a correct TIMEX3. Thus, the application of this rule produces a great improvement in the system. Finally, the post-processing step improves the recall to 66.3%, reaching a 73.4% and a 65.9% in F1 for R and S case respectively. This improvement in recall is due to the fact that many explicit TEs in the corpus appear out of the scope of TSR. Some of them appear in nominal sentences, between brackets, etc. And the rest of them appear tagged with numbered argument roles such as A1 or A2.

There are no strictly comparable results in the literature. The closest evaluation is the one done by Boguraev and Ando [20] using a ML approach. That evaluation described in section 2 performed a test for English TIMEX3 using a 5-fold cross validation over TimeBank. The results obtained were an 81.7% and an 89.6% F1 for strict and relaxed span respectively.

---

[3] http://fofoca.mitre.org/tern.html#scorer

Comparing our systems, we notice that the TIPSem-Full approach doubles Baseline results for strict case. It indicates that the improvements introduced to the TIPSem-Full approach through the transformation rules have resolved the differences between TSR and TIMEX3 described in section 3.

Results show that, taking temporal roles as TIMEX3 TEs without any post processing (Baseline), they are reasonably good in the span relaxed identification case, but not in the strict case. However, if we adapt this approach to TIMEX3 TimeML specifications (TIPSem-Full), we obtain much higher results.

In order to provide a deeper analysis of this results the following subsection present a detailed error analysis studying where does the presented system fail.

4.3   Detailed Error Analysis (TIPSem-Full)

In order to determine which aspects of our full approach are failing, we analyzed in depth the errors found in the evaluation. Table 2 illustrates some examples of the most frequent errors.

Table 2. Most frequent errors in the evaluation

Spurious	Missing	Incorrect (Correct)
14 now	36 third-quarter	3 earlier this month (this month)
4 a year	11 the third quarter	2 a year-earlier (year-earlier)
4 currently, recently	3 previously, recently	2 the year-ago period (year-ago period)

– **Spurious** (337/1244): The spurious value represents the amount of TIMEX3 tagged by our system that are not tagged in the TimeBank corpus. On the one hand, among most frequent errors, there are some correct TIMEX3 like *"now, a year,..."*. It is because some correct TIMEX3 that should be tagged, appear untagged in TimeBank. On the other hand, other spurious words like *"this stage, the TWA flight,..."* point out real errors and one of the most important problems of the presented system, as illustrated in the following example.

Example 8.
```
...defendant's assets [before the trial AM-TMP]          (single-NP)
...defendant's assets before <TIMEX3>the trial</TIMEX3>
```

Example 9.
```
...spent ten million dollars [for this house AM-TMP]          (single-NP)
...spent ten million dollars for <TIMEX3>this house</TIMEX3>
```

Example 10.
```
...[after a coup d'etat last year AM-TMP]          (2 NPs)
...after <TIMEX3>a coup d'etat</TIMEX3> <TIMEX3>last year</TIMEX3>
```

Example 11.
```
[the occupiers in the Gulf AM-TMP]                                    (2 NPs)
<TIMEX3>the occupiers</TIMEX3> in <TIMEX3>the Gulf</TIMEX3>
```

Different problems are causing these errors. In example 7, the preposition is correctly removed but the implementation is unable to detect that the NP is not representing a TE, but an event, and it can not be differentiated from correct cases like *"before [this year]"*. In example 8, a similar error is produced, but this time by a SRL tool failure. Example 9 presents the same situation but only one of the two NPs is incorrectly tagged. Furthermore, if we apply the splitting rule over a spurious expression, like in example 10, errors are multiplied. This fact produces that the splitting rule does not improve precision in this evaluation. In all cases, some extra information is required to determine whether a NP can be a TIMEX3 or not. One possible solution is the usage of temporal triggers but this is not the aim of this paper, which is focused in the influence of SR on TE identification.

- **Missing** (321/1228): Missing value points out which TIMEX3 are tagged in the corpus but missed by our system. On the one hand, frequently missed TIMEX3 like *"number-quarter"* style expressions have been analyzed. The analysis has revealed that most of these expressions appear in sentences in which the verb requires a numbered argument or in which these expressions are part of a prepositional phrase without verb. On the other hand, expressions like *"recently"*, found among spurious errors, are also found among missing errors, representing a TimeBank inconsistency.

Example 12.
```
Freightways Inc. reported [a 77% drop in third-quarter net income A1].
...the earning power they lost [to inflation in the last decade A2]
```

In general, found missing errors are produced by: SRL tool errors, syntactic analysis errors, and also by TEs appearing out of the scope of TSR, like the ones in the examples.

- **Incorrect** (93/1244): Span errors are mostly produced by left bounds *"the, a,..."*. Our system follows strictly TimeML specifications while TimeBank seems to have some inconsistent annotations. Apart from this, real span errors have been found in expressions like *"[10 p.m]. [Wednesday]"*. This expression, representing an specification relation, has been tagged by SRL tool using two TSR because the tool has interpreted the last period of *"10 p.m."* as a sentence separation mark. For that reason, the syntactic tree wraps *"10 p.m."* and *"Wednesday"* using different sentence constituents. A rule for joining such kind of expressions is needed. However, the automatic joining process will not be a trivial issue, due to the fact that both elements are annotated as separated constituents. None cases have been found in the corpus but joining is also useful for split hours like *"[at quarter] [to four]"*.

Summarizing, most important errors are the following: firstly, temporal roles not representing TIMEX3 but events, secondly, errors in the labeling produced by the SRL tool or TEs not annotated as TSR. Thirdly, syntactic analysis errors. And finally, corpus inconsistencies.

5   Conclusions and Further Work

This work presents an in-depth study on how do semantic roles perform for temporal expression identification. For this purpose, a TE identification system based on SR has been developed, following the TimeML TIMEX3 specifications. Two approaches of this system have been implemented: (1) Baseline and (2) TIPSem-Full. The Baseline approach is a direct application of temporal semantic role to TIMEX3 identification and the TIPSem-Full approach is an adaptation of Baseline to cope TIMEX3 specifications.

These two approaches have been evaluated using TimeBank as reference corpus. The TIPSem-Full approach has obtained an 73.4% in F1 for relaxed span and a 65.9% in strict span. Results show that Baseline approach perform considerably well in the span relaxed identification, but not in the strict case. However, if we adapt this Baseline system to TIMEX3 specifications, using the transformation rules defined in this work, results obtained are much higher. Hence, we can conclude that the application of SR to temporal expression identification task is valuable. Furthermore, the contribution of this paper has valuable applications, which are proposed next as further work.

On the one hand, due to the fact that this evaluation has been made using an automatic SRL tool over the TimeBank reference corpus the SRL errors have been propagated, decreasing the results of the presented approach. Our next experiment will be the application of this approach over corpora manually tagged with semantic roles to minimize this error. Moreover, we set out the hypothesis that the approach could be valid for other European languages that share several features at semantic roles level. The AnCora[4] corpus copes all the needs for these studies. AnCora is the largest corpus annotated manually with semantic roles for Spanish and Catalan languages. A potential application of this work is to help human annotators on tagging TimeML TEs in corpora manually annotated with SR in languages for which there is no TimeML corpora available (i.e., Spanish). This application is specially interesting for two reasons:

- If no TimeML TIMEX3 training data is available for the target corpora language, state of the art ML approaches are not applicable.
- If ML approaches are not applicable, the cost of developing a good knowledge-based TIMEX3 tagger is much higher than the application of the approach described in this work.

On the other hand, the error analysis showed that some TSR are not denoting TIMEX3 but events. Furthermore, studying some of the defined rules such as

---

[4] http://clic.ub.edu/ancora/

preposition removing, we find that what we are doing is separating temporal signals from TIMEX3. Taking into account these statements, it would be interesting to apply SR to tagging not only TIMEX3, but also TimeML SIGNAL and EVENT tags. Therefore, this will be done, also, as further work.

References

1. Mani, I., Pustejovsky, J., Gaizauskas, R.: The Language of Time: A Reader. Oxford University Press, Oxford (2005)
2. Schilder, F., Katz, G., Pustejovsky, J.: Annotating, Extracting and Reasoning About Time and Events. In: Schilder, F., Katz, G., Pustejovsky, J. (eds.) Annotating, Extracting and Reasoning about Time and Events. LNCS (LNAI), vol. 4795, pp. 1–6. Springer, Heidelberg (2007)
3. Mani, I., Wilson, G.: Robust temporal processing of news. In: ACL Annual Meeting, NJ, USA, ACL, pp. 69–76 (2000)
4. Pustejovsky, J.: TERQAS: Time and Event Recognition for Question Answering Systems. In: ARDA Workshop (2002)
5. Pustejovsky, J., Mani, I., Belanger, L., Boguraev, B., Knippen, B., Litman, J., Rumshisky, A., See, A., Symonen, S., van Guilder, J., van Guilder, L., Verhagen, M.: ARDA summer workshop on graphical annotation toolkit for TimeML. Technical report, MITRE (2003)
6. TERN-2004: Time Expression Recognition and Normalization Evaluation Workshop (2004), http://fofoca.mitre.org/tern.html
7. Verhagen, M., Gaizauskas, R.J., Hepple, M., Schilder, F., Katz, G., Pustejovsky, J.: Semeval-2007 task 15: Tempeval temporal relation identification. In: Proceedings of the 4th International Workshop on Semantic Evaluations, ACL, pp. 75–80 (2007)
8. Setzer, A., Gaizauskas, R.: Annotating Events and Temporal Information in Newswire Texts. In: LREC 2000, Athens, pp. 1287–1294 (2000)
9. Ferro, L., Gerber, L., Mani, I., Sundheim, B., Wilson, G.: TIDES 2005 Standard for the Annotation of Temporal Expressions. Technical report, MITRE (2005)
10. Pustejovsky, J., Castaño, J.M., Ingria, R., Saurí, R., Gaizauskas, R.J., Setzer, A., Katz, G.: TimeML: Robust Specification of Event and Temporal Expressions in Text. In: IWCS-5, 5th Int. Workshop on Computational Semantics (2003)
11. Gildea, D., Jurafsky, D.: Automatic labeling of semantic roles. Computational Linguistics 28(3), 245–288 (2002)
12. Pustejovsky, J., Hanks, P., Saurí, R., See, A., Gaizauskas, R.J., Setzer, A., Radev, D.R., Sundheim, B., Day, D., Ferro, L., Lazo, M.: The TIMEBANK Corpus. In: Corpus Linguistics, pp. 647–656 (2003)
13. Allen, J.F.: Maintaining knowledge about temporal intervals. Communications of ACM 26(11), 832–843 (1983)
14. Allen, J.F.: Time and time again: The many ways to represent time. International Journal of Intelligent Systems 6, 341–355 (1991)
15. Negri, M., Marseglia, L.: Recognition and Normalization of Time Expressions: ITC-irst at TERN 2004. Technical report, Information Society Technologies (2004)
16. Saquete, E., Muñoz, R., Martínez-Barco, P.: Event ordering using TERSEO system. Data Knowledge Engineering 58(1), 70–89 (2006)
17. Verhagen, M., Mani, I., Saurí, R., Knippen, R., Jang, S.B., Littman, J., Rumshisky, A., Phillips, J., Pustejovsky, J.: Automating temporal annotation with TARSQI. In: ACL, NJ, USA. ACL, pp. 81–84 (2005)

18. Saurí, R., Littman, J., Knippen, R., Gaizauskas, R.J., Setzer, A., Pustejovsky, J.:
    TimeML Annotation Guidelines 1.2.1 (2006), http://www.timeml.org/
19. Boguraev, B., Pustejovsky, J., Ando, R.K., Verhagen, M.: TimeBank evolution
    as a community resource for TimeML parsing. Language Resources and Evalua-
    tion 41(1), 91–115 (2007)
20. Boguraev, B., Ando, R.K.: Effective Use of TimeBank for TimeML Analysis. In:
    Schilder, F., Katz, G., Pustejovsky, J. (eds.) Annotating, Extracting and Reasoning
    about Time and Events. LNCS (LNAI), vol. 4795, pp. 41–58. Springer, Heidelberg
    (2007)
21. Moreda, P., Llorens, H., Saquete, E., Palomar, M.: Automatic Generalization of a
    QA Answer Extraction Module Based on Semantic Roles. In: Geffner, H., Prada,
    R., Machado Alexandre, I., David, N. (eds.) IBERAMIA 2008. LNCS (LNAI),
    vol. 5290, pp. 233–242. Springer, Heidelberg (2008)
22. Melli, G., Shi, Z., Wang, Y., Popowich, Y.L.,, F.: Description of SQUASH, the SFU
    Question Answering Summary Handler for the DUC-2006 Summarization Task. In:
    Document Understanding Conference, DUC (2006)
23. Narayanan, S., Harabagiu, S.: Question answering based on semantic structures.
    In: COLING, Morristown, NJ, USA, p. 693. ACL (2004)
24. Hagège, C., Tannier, X.: XRCE-T: XIP temporal module for TempEval campaign.
    In: TempEval (SemEval), Prague, Czech Republic, pp. 492–495. ACL (2007)
25. Punyakanok, V., Roth, D., Yih, W., Zimak, W.t., Zimak, D., Tu, Y.: Semantic role
    labeling via generalized inference over classifiers. In: HLT-NAACL (CoNLL 2004),
    MA, USA, pp. 130–133. ACL (2004)
26. Palmer, M., Gildea, D., Kingsbury, P.: The Proposition Bank: An Annotated Cor-
    pus of Semantic Roles. Computational Linguistics 31 (2005)

Evaluation of Arabic Machine Translation System Based on the Universal Networking Language

Noha Adly[1,2] and Sameh Al Ansary[1,3]

[1] Bibliotheca Alexandrina, Alexandria, Egypt
[2] Department of Computer and Systems Engineering, Faculty of Engineering,
Alexandria University, Alexandria, Egypt
[3] Department of Phonetics and Linguistics, Faculty of Arts, Alexandria University,
Alexandria, Egypt

Abstract. This paper evaluates a machine translation (MT) system based on the interlingua approach, the Universal Network Language (UNL) system, designed for Multilanguage translation. The study addresses evaluation of English-Arabic translation and aims at comparing the MT systems based on UNL against other systems. Also, it serves to analyze the development of the system understudy by comparing output at the sentence level. The evaluation is performed on the Encyclopedia of Life Support Systems (EOLSS), a wide range corpus covering multiple linguistic and cultural backgrounds. Three automated metrics are evaluated, namely BLEU, F_1 and F_{mean} after being adapted to the Arabic language. Results revealed that the UNL MT outperforms other systems for all metrics.

Keywords: Machine Translation, Natural Language Processing, Natural Language Generation, Evaluation of MT, Universal Networking Language, Encyclopedia of Life Support Systems, Interlingua.

1 Introduction

Research in machine translation (MT) has spanned several approaches. Statistical machine translation has been the approach most widely used, see [13] for a recent survey. The Interlingua approach relies on transforming the source language to a language-independent representation, which can then be transformed to the target language. When multilingual translation is of interest, the interlingua approach allows to build a system of N languages with a linear effort while the statistical approach would require a quadratic effort. The challenge with the interlingua approach is to design a language independent intermediate representation that captures the semantic structures of all languages while being unambiguous. The interlingua has been used on limited task-oriented domains such as speech translation for specific domains [8]. Few efforts studied machine translation based on Interlingua, but on a limited scale, for Indian languages [20], Korean language [10] and Arabic language [19].

The UNL System promises a representation of all information, data and knowledge that humans produce in their own natural languages, in a language independent way, with the purpose of overcoming the linguistic barrier in Internet. The UNL is an

H. Horacek et al. (Eds.): NLDB 2009, LNCS 5723, pp. 243–257, 2010.

artificial language that has lexical, syntactical and semantic components as does any natural language. This language has been proven tractable by computer systems, since it can be automatically transformed into any natural language by means of linguistic generation processes. It provides a suitable environment for computational linguists to formalize linguistic rules initiation from semantic layer.

The Encyclopedia of Life Support Systems (EOLSS) [6] is an Encyclopedia made of a collection of 20 online encyclopedias. It is a massive collection of documentation, under constant change, aiming at different categories of readers coming from multiple linguistic and cultural backgrounds. EOLSS is an unprecedented global effort over the last ten years, with contributions from more than 6000 scholars from over 100 countries, and edited by nearly 300 subject experts. The result is a virtual library equivalent to 200 volumes, or about 123,000 printed pages.

Availing EOLSS in multiple languages is a main goal of its initiators. However, translating EOLSS in every possible language is a daunting task that requires years of work and large amount of human and financial resources, if done in the conventional ways of translation. The UNDL Foundation proposed to use the UNL System for representing the content of EOLSS in terms of language independent semantic graphs, which in turn can be decoded into a target natural language, generating a translation of EOLSS documents into multiple languages. With the UNL System, this can be achieved in a relative shorter period of time, and at lower costs in comparison to costs of traditional translation. Work has actually started with the six official languages of the United Nations. 25 documents, forming around 15,000 sentences have been enconverted from EOLSS to UNL. The UNL version of EOLSS is sent to the UNL language centers for deconversion. It is a prototype for translating massive amount of text; done in anticipation to the deconversion in many other languages of the world.

The Arabic language center has completed the deconversion of the 25 documents of the prototype and automatically generated the equivalent Arabic language text. The purpose of this paper is to evaluate the quality of the translated text. The objective of the evaluation is twofold. First, it is desirable to evaluate the strength and weakness of the machine translation generated through the UNL system and compare it against other MT systems. Second, it is aimed to set up a framework of evaluation that can be applied on a frequent and ongoing basis during the system development, in order to guide the development of the system based on concrete performance improvements.

The rest of the paper is organized as follows. Section 2 gives a brief description of the UNL system and describes its usage for the automated translation of the EOLSS. Section 3 presents a brief description of the Arabic dictionary and generation rules. Section 4 describes the automated metrics used in the performance evaluation and introduces some adaptation of the metrics to suit the Arabic language. Section 5 gives an overview of the process of the data preparation and presents the experimental design. Section 6 presents the different conducted experiments and discusses the results. Finally, Section 7 concludes the paper.

2 The UNL System

The architecture of the UNL system (Fig. 1) comprises three sets of components [23]:

1. *Linguistic components*: dictionaries that include Universal Words (UWs) and their equivalents in natural languages, grammatical rules responsible for

producing a well formed sentence in the target natural language and knowledge base for representing a universal hierarchy of concepts in natural languages;

2. *Software components*: two software programs for converting content from natural languages to UNL (the EnConverter) and vice versa (the DeConverter). The EnConverter is a language independent parser that provides synchronously a framework for morphological, syntactic and semantic analysis. It is designed to achieve the task of transferring the natural language to the UNL format or UNL expressions. The DeConverter is a language independent generator that provides synchronously a framework for morphological and syntactic generation, and word selection for natural collocation. DeConverter can deconvert UNL expressions into a variety of native languages, using the Word Dictionary, formalized linguistic rules and Co-occurrence Dictionary of each language;

3. *System interfacing components*: protocols and tools enabling the flow of UNL documents throughout the web.

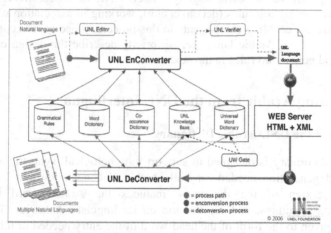

Fig. 1. The core architecture of the UNL system

2.1 UNL Language Components

The UNL consists of Universal Words (*UWs*), *Relations*, *Attributes*, and UNL Knowledge Base (UNL KB). The *UWs* constitute the vocabulary of the UNL, *Relations* and *Attributes* constitute the syntax of the UNL and the UNL KB constitutes the semantics of the UNL. Formally, a UNL expression can be viewed as a semantic network, whose nodes are the *UWs*, linked by arcs labeled with the UNL *Relations* which express the objective meaning of the speaker. *UWs* are modified by the so-called *Attributes* to convey the subjective meaning of the speaker (For more details see [23]). The UNL KB constitutes the semantic background of the UNL System. It is constituted by the binary direct relations between two UWs. With these links, a conceptual network can be shaped to form a lattice structure. The structure allows for implementing the principle of inheritance in the definition of concepts.

The UNL KB is meant to assure robustness and precision to the UNL System, both to the NL-UNL encoverting, and to the UNL-NL deconverting processes. In the former case, the UNL KB would be used as a sort of word sense disambiguation device. In the latter, the UNL KB would allow for the deconversion of UWs not enclosed in the target language dictionaries.

2.2 Using UNL in Machine Translation of EOLSS

Translation with the UNL system is a two-step process. The first step deals with Enconverting the content of the EOLSS from the source language (English) to UNL (the universal representation). This process is called the *UNLization* process; it is carried out with the use of the English-UNL Enconverter. Initially, some post-editing is needed, but as the performance of the English Enconverter and the technical dictionaries improve, human intervention will be gradually reduced, and productivity will be increased.

The second step deals with Deconverting EOLSS content from UNL to a target natural language [2, 3]. This Deconversion process is a task to be carried out by the UNL-Language Server of each language. Each UNL Language Server contains a dictionary and generation rules (deconversion), working in association with the UNL KB, which are the enabling components in this process. Since we are concerned with the generation of the Arabic language, we briefly describe the design of the Arabic dictionary and generation rules in the next section.

3 Generating Arabic from the UNL Interlingua

3.1 Design of the UNL-Arabic Dictionary

The Arabic dictionary is designed to support morphological, syntactic and semantic analysis and generation needed for both Arabic EnConversion and DeConversion rules. The design of the dictionary includes the Arabic word heading, its corresponding meanings, and information on its linguistic behavior. The focus of attention is given to the form of the head word of the entry needed to fulfill language analysis and generation tasks adequately. The entries are stem-based to avoid adding all possible inflectional and derivational paradigms of each lexical item to the dictionary, and to minimize the number of entries in the dictionary which will give more efficiency in the analysis and generation tasks and minimize the processing time (e.g. instead of storing أكاديمية، أكاديميتنا، أكاديميات etc., only أكاديمي will be stored). The Arabic UNL dictionary stores three types of linguistic information. First, morphological information which is responsible for correctness of the morphology of words; it describes the changes that occur within a word when it is attached to various suffixes and prefixes in different contexts. Second, syntactic information to generate well-formed Arabic sentence structure; it determines grammatical relations coded as the presence of adjuncts and arguments in isolation or as sub-categorization frames, and describes grammatical relations between words. Third, semantic information about the semantic classification of words that allows for correct mapping between semantic information in UNL-graphs and syntactic structure of the sentence under generation. The following examples represent full records of lexical entries:

[أوَى]{أوَى}"accommodate(agt>person,obj>person)"(ST,1.2,2V,3V,V1,1?) <A,0,0>;
[إنو] {إنو} "accommodate(agt>person,obj>person)" (ST,1.2,2V,3V,V6,1?) <A,0,0>;
[أوَي]{أوَيْتُ}"accommodate(agt>person,obj>person)"(ST,1.2,2V,3V,V2,1?)<A,0,0>;
[أو] {أو} "accommodate(agt>person,obj>person)" (ST,1.2,2V,3V,V5,1?) <A,0,0>;
[أوَي]{أوَي}"accommodate(agt>person,obj>person)"(ST,1.2,2V,3V,V3,1?) <A,0,0>;
[يؤوي]{يؤوي}"accommodate(agt>person,obj>person)"(ST,1.2,2V,3V,1?) <A,0,0>;
[أوَ] {أوَ} "accommodate(agt>person,obj>person)" (ST,1.2,2V,3V,V4,1?) <A,0,0>;
[أوَه]{أوَاه}"accommodate(agt>person,obj>person)" (ST,1.2,2V,3V,V7,1?) <A,0,0>;

The example above shows the different word forms of the verb "أوَى" that are stored in the Arabic dictionary with different linguistic information about each form to guide the grammar to pick the appropriate one according to the syntactic structure and the tense of the sentence.

3.2 Design of the Arabic Generation Grammar

The Arabic language is a morphologically and syntactically rich language and its generation is very complex. Hence, the technical design of the Arabic generation grammar is divided into several stages, namely the lexical mapping stage, the syntactic stage and the morphological stage. The lexical mapping stage deals with identifying the target lexical items. The syntactic stage deals with the order of words in the node list, and morphological stage specifies how to form words and deals with agreement gender, number, person and definiteness. The different stages are illustrated in Fig. 2.

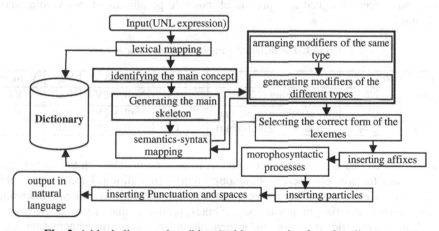

Fig. 2. A block diagram describing Arabic generation from interlingua

Lexical Mapping: The lexical mapping stage performs the mapping between the meaning conveyed by the concepts of the intermediate representation (UNL intelingua) and the lexical items of the target language. For example, the word "answer" can be translated in the Arabic language as "يجيب" or "إجابة" but it is expressed by two concepts "answer(agt>thing,obj>thing)" which is mapped with the corresponding Arabic verb "يجيب" and the concept "answer(icl>notion) which is mapped with the corresponding Arabic noun"إجابة".

Syntactic Stage: The syntactic stage is concerned with the order of words in the node list; it can be divided into two phases. The first phase is concerned with building the main skeleton of the sentence. The starting node in the UNL network is the 'entry' node that refers to the main concept of the sentence which is marked as "@entry". The phase continues to generate the arguments of the main concept concerning the suitable Arabic syntactic structure in either a nominal structure (Topic-Comment) or in a verbal structure (VSO). The second phase in the grammar deals with generating the modifiers. One of the challenges faced in this stage is when a given node in the semantic network is modified by more than one modifier of the same type. The Arabic generation grammar is designed to control the insertion of nodes in such a situation. The generation process highlights a basic point which is the type and number of syntactic arguments a predicate takes are determined by the type and number of semantics arguments that a predicate expresses. This actually reflects the interface between semantics and syntax in Natural Language Generation.

Morphological stage: The Morphological stage is concerned with two axes. First, inserting affixes (prefixes and suffixes) to the node list to generate the final form of the entries according to the linguistic features attached to each entry in the dictionary. The features are in turn based on the form of the dictionary entries selected to represent different paradigms representing lexemes. For example, the form of the defective verb "كان" 'be' changes according to subject pronouns. Therefore, three forms have been designed to represent all possible paradigms of this verb as shown in Table 1:

Table 1. The different paradigms of the same lexeme

Hw	Reading	Uw	Pattern	V1	V2	V3	V_form
[كان]	كانَ	be(aoj>thing,obj>thing)	1.1	Null	2V	Null	V1
[كن]	كن	be(aoj>thing,obj>thing)	1.1	Null	2V	Null	V3
[كون]	يكون	be(aoj>thing,obj>thing)	1.1	Null	2V	Null	V2

Each of the entries is given a different code, to be used in selecting the form required to represent the concept "be(aoj>thing,obj>thing)". In addition, based on the subject of the sentence a given affix will be added to the head word to generate the realized form. Second, inserting prepositions, attributes, pronouns that are needed because of the Arabic syntactic structure under generation and inserting punctuation marks. Spaces will be added at the end of the morphological phase after inserting all nodes from the node net. Spaces separate all nodes except nodes that represent affixes.

4 Performance Evaluation Metrics

Research in MT depends heavily on the evaluation of its results. Many automated measures have been proposed to facilitate fast and cheap evaluation of MT systems. Most efforts focus on devising metrics based on measuring the closeness of the output of MT systems to one or more human translation; the closer it is, the better it is. The challenge is to find a metric to be produced at low cost while correlating highly with human evaluation. The metric should be consistent and reliable. The most commonly used MT evaluation metric in recent years has been BLEU [15], an *n*-gram precision metric that demonstrated a high correlation with human judgment of system adequacy and fluency.

Various researchers have noted, however, some shortcomings in the metric due to being mainly a precision metric and its lack of consideration of the recall. Recall has been found to be extremely important for assessing the quality of MT output [9], as it reflects to what degree the candidate translation covers the entire content of the reference translation. Several metrics have been introduced recently that take precision and recall into account. GTM [14, 22] used a balanced harmonic mean of unigram precision and recall. METEOR [9] used a weighted harmonic mean placing more weight on recall than on precision and shown that this leads to better correlation. Recent development of METEOR [1, 4, 7] introduced unigram matching based on stemmed forms and synonyms matching using Wordnet. Other proposed methods for MT evaluation include TER [21], a metric based on the Levenshtein distance, but applied on the word level rather than the character level. It measures the number of edit operations needed to fix a candidate translation so that it semantically matches a reference translation. A related metric is CDER [11], which is based on the edit distance but accounts for an operation that allows for reordering of word blocks.

Several evaluations of the above metrics were conducted [12, 17] but there were no conclusions as to whether one of them supersedes the others. To achieve a balance in our evaluation, we chose BLEU, as it has been the primary metric used by most systems. But also we selected two metrics that incorporates recall, namely F_1 and F_{mean} which are based on GTM. These will be described in the following.

4.1 BLEU Metric

The main principle behind BLEU [15] is the measurement of the overlap in unigrams and higher order *n*-grams of words, between a *candidate* translation being evaluated and a set of one or more *reference* translations. The main component of BLEU is *n*-gram precision: the proportion of the matched *n*-grams out of the total number of *n*-grams in the candidate translation.

To avoid exceeding the counts of a word in the candidate with respect to its occurrence in any single reference, they introduced the *modified n-gram precision*. All candidate *n*-gram counts and their corresponding maximum reference counts are computed. The candidate counts are clipped by their corresponding reference maximum value, summed, and divided by the total number (unclipped) of candidate *n*-grams. The precision p_n for each *n*-gram order is computed separately, and the precisions are combined via a geometric averaging.

Recall, which is the proportion of the matched *n*-grams out of the total number of *n*-grams in the reference translation, is not taken into account directly by BLEU. Instead, BLEU introduces a Brevity Penalty, which penalizes translations for being "too short". The brevity penalty is computed over the entire corpus and was chosen to be a decaying exponential in r/c, where c is the length of the candidate corpus and r is the effective length of the reference corpus. Therefore

$$BLEU = BP.exp \ (\Sigma_{n=1}^{N} w_n \log p_n) \ , \ \text{where } BP = \begin{cases} 1 & if \ c > r \\ e^{(1-\frac{r}{c})} & if \ c \leq r \end{cases}$$

The BLEU metric captures two aspects of translation: *adequacy* and *fluency*. Adequacy accounts for setting the words right, which is measured by BLEU with small n. Fluency captures the word order, the construction of the sentence and its well-formedness. It has been shown in [12,16] that shorter n-grams correlates better with adequacy with 1-gram being the best predictor, while longer n-grams has better fluency correlation. Typical values used by most systems is BLEU-4 [12]. The Smoothed technique proposed in [12] has been implemented in order to account for reliable score at the sentence level.

4.2 F_1 and F_{mean} Metrics

Both F_1 and F_{mean} metrics take into account Precision P and recall R and are based on unigram matching. F_1 is the harmonic mean[18] of the precision and recall, $F_1 = \frac{2PR}{P+R}$. F_{mean} [9], is similar to F_1, but recall is weighted more heavily than precision. $F_{mean}= \frac{10PR}{9P+R}$ where the weights are chosen to maximize the correlation with human judgment.

The definition of precision P and recall R, are adopted from [14,22]: given a set of candidates Y and a set of references X, precision$(Y|X)=\frac{|X \cap Y|}{|Y|}$ and recall$(Y|X)=\frac{|X \cap Y|}{|X|}$. Both are proportional to $|X \cap Y|$, the size of the set intersection of the pair of texts. The definition of the intersection is introduced by the aid of a grid, where every cell in the grid is the coordinate of some word in the candidate text with some word in the reference text. Whenever a cell in the grid coordinates two words that are identical is called a *hit*. Computing the match size as the number of hits in the grid would result in double counting. Therefore, the definition is based on the concept of "maximum matching" from graph theory [5]. A *matching* is a subset of the hits in the grid, such that no two hits are in the same row or column. The *match size* of a matching is the number of hits in the subset. A *maximum matching* is a matching of maximum possible size for a particular grid. The *maximum match size* (MMS) is the size of any maximum matching. The MMS is divided by the length of the candidate text (C) or the length of the reference text (F) to obtain the precision or the recall, respectively: precision$(C|F)=\frac{MMS(C,F)}{|C|}$ and recall$(C|F)=\frac{MMS(C,F)}{|F|}$.

In order to reward longer matches, a generalized definition of the match size is adopted; size(M)= $\sqrt[e]{\Sigma_{r \in M} length(r)^e}$, where r is a run, defined as a contiguous sequences of matching words appearing in the grid as diagonally adjacent hits running in parallel to the main diagonal. For $e>1$ computing MMS is NP-hard, therefore it is obtained using a greedy approximation that builds a matching by iteratively adding the largest non-conflicting aligned blocks. The parameter e is adjusted to weighting matching longer runs. A typical value of e is 2. To account for multiple references, the references are concatenated in arbitrary order. Then the maximum matching is computed, with a barrier between adjacent references preventing runs to cross the barriers. Finally, the MMS is normalized with respect to the length of the input texts.

4.3 Adaptation of the Metrics to the Arabic Language

The described metrics have been primarily applied and customized for the English language. For instance, they provide the option to account for case sensitivity. While the Arabic language does not have case sensitivity, but it does have some other features that need to be accounted for. The evaluation metrics have been modified such that they can adapt to some peculiarities in the Arabic language, which are tolerated by human being. For instance, we consider the following cases:

- ا, أ, إ: It is quite common for people to write the letter ا, instead of أ or إ. Since this error is tolerated by human, we modified the evaluation metrics such that they take this into consideration as follows: if the candidate token includes an ا, while the corresponding token in the reference translation is with a hamza (أ or إ) for all references, the token is given a score α, where 0≤α≤1. If on the other hand, the candidate token includes a hamza (أ or إ) then it must match one reference token with the hamza in the same position, otherwise it is given a zero score.
- ى and ي: since mixing ى with ي is a common error that could be tolerated by humans, the modification entails giving a score 0≤α≤1 for a candidate token not matching a token in any reference because a ى mixed with ي or vice versa.
- ة and ه / ا and أ: mixing ة with ه or mixing ا with أ are considered errors that are not tolerated in the algorithm and are given a score of zero.

It should be noted that we do not account for all possible cases. Rather, we introduce the methodology that other special cases could follow to tune the metrics to suit the different levels of tolerance needed. The above cases are used only as examples implemented in our evaluation.

5 Datasets and Experimental Design

The experiments reported in this paper are conducted on datasets prepared from the EOLSS. Preparing our own test datasets stemmed from the desire to evaluate the UNL MT systems on real data sets and real applications. Further, there are no publically available datasets for the language pair English-Arabic as the ones available from NIST or Linguistic Data Consortium (LDC).

Experiments are conducted using data drawn from the EOLSS encyclopedia, which is used as the English corpus. The test dataset contains around 500 sentences, composed of 8220 words, drawn randomly from 25 documents containing around 15,000 sentences. The length of the test sentences varied; with a mean 16.44 and standard deviation. The random selection ensured that the dataset covers the whole range of the 25 documents.

The output of the UNL system is evaluated and compared to other available systems supporting translation from English to Arabic. Three systems are considered: Google, Tarjim of Sakhr and Babylon.

Four reference translations have been prepared for the test dataset. Four professional translators were provided with the English sentences and they were requested to generate the Arabic translation, without being exposed to any output from MT system. The dataset was not split among different translators, that is each translator processed the entire dataset to ensure the same style within a reference.

Post edited versions of the UNL output have been prepared using human annotators and used in different experiments with dual purpose. One, is to evaluate the improvements introduced by post editing a machine output. Second, is to measure how far the UNL output is from the closest acceptable translation that is fluent and correct. A similar idea has been adopted in [21], where they create a *human-targeted reference*, a single reference targeted for this system, and shown to result in higher correlation with human judgment. Four post edited translations were prepared:

- PE-UNL: This form of post editing was performed by providing monolingual Arabic annotators with the Arabic output from the UNL MT and the UNL representation resulting from the encoding. The annotator was requested to correct the output by fixing any errors resulting from the lack of a grammar rule or the lack of semantic representation in a UNL expression.
- PE-En1 and PE-En2: This post editing was performed by providing bilingual annotators with the original English sentences, the UNL MT output and they were requested to perform the minimum changes needed to correct the output generated by the UNL such that the final output is fluent, grammatically correct and has the correct meaning relative to the English version. However, they were not requested to generate the best translation.
- PE-Pub: This post editing was conducted by an expert in the Arabic language, who was given the PE-En1 and was asked to render the sentence qualified for publishing quality, that is, making the article cohesive and ensuring that the sentence is written in a typical Arabic style.

Basic preprocessing was applied to all datasets. The sentences were tokenized, with removal of punctuation and diacritization. These preprocessing seemed essential as some systems use diacritization while others do not. Also some translators include different punctuation and diacritization, while others do not, or do with different degrees. Therefore, it seemed that removing them would result in a fairer comparison.

In the conducted experiments, the different parameters have been set as follows. For the BLEU, we use uniform weight $w_n = 1/N$ and we vary the n-gram; although in some experiments we only show BLEU-1 and BLEU-4 reflecting adequacy and fluency respectively. For the F1 and F_{mean} metrics, the exponent e has been set to 2, which is the typical value used for weighting longer runs. For the adaptation introduced to the Arabic language, the parameter α has been set to 0.7.

6   Results

6.1   Evaluation Using Professional Reference Translations

The dataset is evaluated for the three metrics using the four references obtained from the professional translators. One feature of BLEU not captured by unigram based metrics is the notion of word order and grammatical coherence achieved by the use of the higher level of n-grams. The n-grams are varied from 1 to 10 and BLEU has been computed for the four MT systems as shown in Figure 3(a). It is observed that UNL results in the best score, followed by Google, Sakhr, then Babylon. These results are statistically significant at 95% confidence. For $n=1$, which accounts for adequacy, it shows that all of the systems, except Babylon, provide reasonable adequacy, with UNL being the best. For higher n-grams, which captures fluency and the level of grammatical well formedness, as expected BLEU decreases as n increases. It is noted though that UNL provides better fluency than others. While on adequacy ($n=1$), UNL shows an improvement of 28% and 12% over Sakhr and Google respectively, for $4 \leq n \leq 10$, the improvement ranges from 42% to 144% over Google and 94% to 406% over Sakhr. For Babylon, it is observed that the decay is very fast, indicating the lack of fluency in its output.

When recall is taken into account, represented in F_1 and F_{mean}, Figure 3(b), it is noticed that UNL still outperforms all other, with significant improvement over Sakhr and Babylon, but with only marginal improvements over Google.

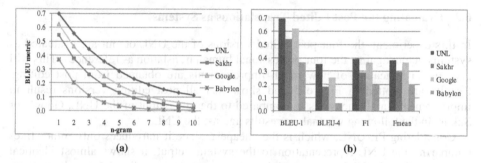

(a)   (b)

Fig. 3. (a) BLEU metric for MT systems, varying n-grams (b) BLEU, F_1 and F_{mean} for MT systems. Results are obtained with professional human translation references.

6.2 Evaluation Using Post Edited References

In this experiment, we present the evaluation of the dataset making use of the post edited versions of the UNL output as references. They are sought to be good choices for references, since they are considered acceptable translations, yet, they are cheap to obtain. Also, they can be considered a possible substitute for subjective human judgment of MT quality. A similar approach has been adopted in [21].

From Figure 4, it is observed that the UNL is better than the three other systems; Google and Sakhr show similar performance while Babylon shows the poorest results. Although results are expected to be biased towards the UNL, it is observed that results follow the same trend as the ones obtained from the professional human translations. Hence, the post edited versions could be considered a cheap and quick way of obtaining the tendency of the systems behavior.

It is worth mentioning that, analyzing the UNL performance with respect to its post edited versions gives an indication of how far it is from the closest acceptable translation. It is noted that the large values of BLEU, F_1 and F_{mean} for UNL is a good indicator that the output is not far off.

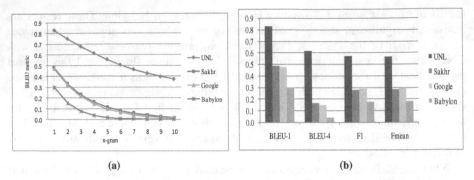

(a)   (b)

Fig. 4. (a) BLEU metric for MT systems, varying n-grams (b) BLEU, F_1 and F_{mean} for MT systems. Results are obtained with post edited references.

6.3 Evaluating the Post Edited Translations as Systems

In this experiment, the four post edited versions of the UNL output are evaluated as systems output against the four professional human translation as references. This will give us an indication of how much improvements are obtained from post editing. Results are plotted in Figure 5 and show that all post editing versions result in improvements in all measures as compared to the raw output of the UNL, Google or Sakhr. In the following we analyze results against the UNL.

Considering PE-UNL, which is the cheapest, since it introduces only minor fixes comparing the UNL representation to the system output; it shows almost identical performance to the UNL output with improvements not exceeding 3% for all metrics.

Examining a more expensive post editing, namely PE-En1 and PE-En2, both of them yield an improvement. PE-En1 improves BLEU with a range from 12% to 41%, with higher improvements achieved for higher n-grams. Also it results in improvements around 15% for both F_1 and F_{mean}. PE-En2 on the other hand gives much higher improvements, ranging from 19% to 250% on BLEU and 42% on F_1 and F_{mean}. It should be noted that the qualifications of the two persons who performed the post editing were the same, so the degree of improvement obtained is subjective and needs to be weighed against its cost.

Turning to PE-Pub, which is the most expensive post editing, results are disappointingly low, especially in comparison to PE-En1 which was the source PE-Pub departed from. Since PE-Pub is a publishing quality; it ensures cohesion and typical Arabic style, which will result in removing structural interference such as cataphora, inappropriate nominal chunking or inappropriate coordination. For example, the English sentence "The management of freshwater resources" is translated by all systems and translators as "إدارة موارد المياه العذبة" which is a correct translation. However, the Arabic editor changed it to "موارد المياه العذبة وإدارتها" to remove nominal chunks resulting from three successive nouns. This results in mismatch of PE-Pub with all references, hence, a low score. This implies that features such as cohesion and typical Arabic style are not captured by any of the MT metrics and more efforts needs to be exerted to devise metrics that account for these features.

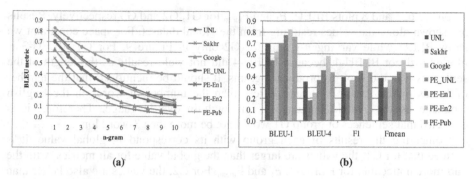

(a)   (b)

Fig. 5. (a) BLEU metric, varying *n*-grams (b) BLEU, F_1 and F_{mean}. Evaluating post edited versions as systems output.

6.4 Responsiveness of the Systems to the Complexity of the Corpus

The test dataset has been categorized into three groups according to the difficulty of the sentences. Difficulty is judged by linguists based on the complexity of the structure of the sentence as well as its length. The first group G1 contains simple sentences, group G2 contains moderate sentences while group G3 contains complex sentences. The categorization by the linguists resulted in G1, G2 and G3 containing 50, 215 and 235 sentences respectively.

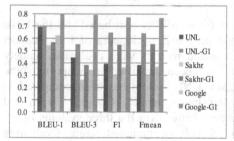

Fig. 6. BLEU, F_1 and F_{mean} for group G1   **Fig. 7.** BLEU, F_1 and F_{mean} for group G2

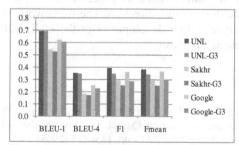   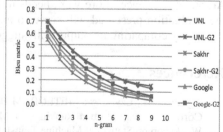

Fig. 8. BLEU, F_1 and F_{mean} for group G3   **Fig. 9.** BLEU metric, varying *n*-grams, for G2

Figure 6, 7 and 8 plots BLEU, F_1 and F_{mean} for G1, G2 and G3 respectively. Results are shown along the values resulting from the global dataset. For space constraint we show BLEU while varying n-grams for G2 only in Figure 9. For G1, BLEU-3 is plotted and not BLEU-4, because BLEU-4 did not yield results as the length of the sentences was too short to produce 4-grams. It should be mentioned that the number of words in the Arabic language is less than the number of words in the English language as the Arabic language is an agglutinative language. Therefore, it is expected that sentences of group G1 would not be more than 3-word long.

Comparing the results of each group with its corresponding global value, it is noticed that for G1, the values are larger than the global value for all metrics, with the gap more noticeable for BLEU-3, F_1 and F_{mean}. For G2, the values are also larger than the global values for all metrics, with smaller differences than in the case of G1. However, for G3, the values are constantly lower than the global values. This implies that simple and moderate sentences yield high values for all metrics, while complex statements are the ones which results in low values.

Comparing the results of the 3 MT systems within the same group, it is observed that Google results in the best score for G1, with improvements reaching 42%. However, UNL shows higher values for G2 and G3 on all metrics reaching improvements of 23% and 52% over Google; 66% and 107% over Sakhr respectively. This implies that UNL outperforms Sakhr and Google in generation of sentences with complex structure.

7 Conclusions

In this research, we presented an evaluation for a MT system based on the UNL system. The evaluation has been conducted on the Encyclopedia of Life Support Systems (EOLSS). Three widely used automated metrics were evaluated, namely BLEU, F_1 and F_{mean}. The three metrics have been modified to adapt to some peculiarities in the Arabic language. The MT UNL system has been compared to other systems supporting English-Arabic translation, namely Google, Tarjim and Babylon. Results revealed that UNL performed better than the three systems on all metrics, especially when generating sentences with a complex structure. Evaluating annotated versions of the UNL output shown that they can be used as cheap references in order to highlight the tendency of the systems behavior. Results also revealed that current metrics do not capture features such as cohesion and typical Arabic style; hence, more work needs to be done in this direction. The framework of the evaluation presented will serve to analyze further development of the UNL MT system by comparing its output with suggested changes.

References

1. Agrawal, A., Lavie, A.: METEOR, M-BLEU and M-TER: Evaluation Metrics For High Correlation with Human Rankings of Machine Translation Output. In: Proc. of the 3rd Workshop on Statistical Machine Translation, Ohio, June 2008, pp. 115–118 (2008)
2. Alansary, S., Nagi, M., Adly, N.: A Semantic Based Approach for Multilingual Translation of Massive Documents. In: The 7th International Symposium on Natural Language Processing (SNLP), Pattaya, Thailand (2007)

3. Alansary, S., Nagi, M., Adly, N.: Generating Arabic text: The Decoding Component of an Interlingual System for Man-Machine Communication in Natural Language. In: The 6th International Conference on Language Engineering, Cairo, Egypt, December 6-7 (2006)
4. Banerjee, S., Lavie, A.: METEOR: An Automatic Metric for MT Evaluation with Improved Correlation with Human Judgments. In: Proc. of ACL Workshop on Intrinsic and Extrinsic Evaluation Measures for Machine Translation and Summarization, Ann Arbor (2005)
5. Cormen, T., Leiserson, C., Rivest, R., Stein, C.: Introduction to Algorithms, 2nd edn. MIT Press, Cambridge (2001)
6. Encyclopedia of Life Support Systems, http://www.eolss.net/
7. Lavie, A., Agarwal, A.: METEOR: An Automatic Metric for MT Evaluation with High Levels of Correlation with Human Judgments. In: Proceedings of the Second ACL Workshop on Statistical Machine Translation, Prague, June 2007, pp. 228–231 (2007)
8. Lavie, A., Pianesi, F., Levin, L.: The NESPOLE! System for Multilingual Speech Communication over the Internet. IEEE Transactions on Audio, Speech and Language Processing 14(5) (September 2006)
9. Lavie, A., Sagae, K., Jayaraman, S.: The Significance of Recall in Automatic Metrics for MT Evaluation. In: Frederking, R.E., Taylor, K.B. (eds.) AMTA 2004. LNCS (LNAI), vol. 3265, pp. 134–143. Springer, Heidelberg (2004)
10. Lee, Y., Yi, W., Seneff, S., Weinstein, C.: Interlingua-based Broad-coverage Korean-to-English Translation in CCLINC. In: Proceedings of the 1st International Conference on Human Language Technology Research, san Diego (2001)
11. Leusch, G., Ueffing, N., Ney, H.: CDER: Efficient MT Evaluation Using Block Movements. In: Proceedings of the 13th Conference of the European Chapter of the Association for Computational Linguistics (2006)
12. Lin, C., Och, J.: ORANGE: a Method for Evaluation Automatic Evaluation Metrics for Machine Translation. In: COLING 2004, Switzerland, August 2004, pp. 501–507 (2004)
13. Lopez, A.: Statistical Machine Translation. In: ACM Comp. Surveys, vol. 40 (August 2008)
14. Melamed, I., Green, R., Turian, J.: Precision and Recall of Machine Translation. In: Proceedings of the HLTNAACL 2003, Canada, pp. 61–63 (2003)
15. Papineni, K., Roukos, S., Ward, T., Zhu, W.: BLEU: a Method for Automatic Evaluation of Machine Translation. In: 40th Annual Meeting of the Association for Computational Linguistics (ACL), Philadelphia, pp. 311–318 (2002)
16. Papineni, K., Roukos, S., Ward, T., Henderson, J., Reeder, F.: Corpus-based Comprehensive and diagnostic MT evaluation: Initial Arabic, Chinese, French, and Spanish results. In: Proceedings of Human Language Technology 2002, San Diego, CA (2002)
17. Przybocki, M., Sanders, G., Le, A.: Edit Distance: a Metric for Machine Translation Evaluation. In: LREC (2006)
18. Rijsbergen, C.: Information Retrieval, 2nd edn., Butterworths, London (1979)
19. Shaalan, K., Monem, A., Rafea, A., Baraka, H.: Mapping Interlingua Representations to Feature Structures of Arabic Sentences. In: The Challenge of Arabic for NLP/MT. International Conference at the British Computer Society, London, pp. 149–159 (2006)
20. Sinha, R., et al.: ANGLABHARTI: a Multilingual Machine Aided Translation from English to Indian Languages. In: IEEE Intl. Conference on Systems, Man and Cybernetics (1995)
21. Snover, M., Dorr, B., Schwartz, R., Micciulla, L.: A Study of Translation Edit Rate with Targeted Human Annotation. In: Proceedings of AMTA, Boston (2006)
22. Turian, J., Shen, L., Melamed, I.: Evaluation of Machine Translation and its Evaluation. In: Proceedings of MT Summit IX (2003)
23. Uchida, H., Zhu, M., Della Senta, T.: The Universal Networking Language. In: UNDL Foundation (2005)

Quality of Subsumption Hierarchies in Ontologies

Geir Solskinnsbakk[1], Jon Atle Gulla[1], Veronika Haderlein[2],
Per Myrseth[2], and Olga Cerrato[2]

[1] Department of Computer and Information Science
Norwegian University of Science and Technology
Trondheim, Norway
{geirsols,jag}@idi.ntnu.no
[2] Det Norske Veritas (DNV)
Oslo, Norway
{Veronika.Haderlein,Per.Myrseth,Olga.Cerrato}@dnv.com

Abstract. Ontologies are becoming increasingly more popular tools for many tasks, such as information integration, information retrieval, knowledge management and extraction etc. The cost and complexity of developing good ontologies is high, and therefore it is important to be able to verify the ontology and detect flaws early. In this paper we propose an approach to expose desirable properties of ontological structures. The approach is based on an ontological profile which is an ontology extended with a vector of weighted terms describing the semantics of each concept of the ontology. We describe four hypotheses for the relations among the classes of the ontology and perform experiments to verify them. Our initial findings are that the experiments support the hypotheses.

1 Introduction

With the emergence of the Semantic Web [2] and Semantic Web related technologies, ontologies are becoming steadily more important. This also means that the quality aspect of ontologies is becoming even more important. Ontologies are formal specifications of shared conceptualizations [3] and are used for a wide range of tasks, such as information integration, information retrieval, knowledge management and extraction etc. Designing ontologies is an expensive and time consuming task, and adding to the complexity is the problem of modelers not being appropriately familiar with the domain, requiring domain experts which are not expert modelers to chip in. On the other hand, there has been quite a lot of work done on automatic ontology building. However, the task of ontology development is complex, and automatic ontology building is a hard task. Therefore it is important to be able to verify the consistency of the ontology and find mistakes as early as possible. We thus present in this paper a method for exposing desirable properties in the hierarchical structure of the ontology with respect to subsumption. The proposed approach is based on the concept of ontological profile, which is an ontology extended with term vectors for each

H. Horacek et al. (Eds.): NLDB 2009, LNCS 5723, pp. 258–268, 2010.

concept. The term vector describes the semantics of the concept with respect to an underlying text collection. The objective of our approach is to be able to specify and verify desirable properties of ontological structures. Our initial experiments show encouraging results towards the objective. The rest of the paper is structured as follows; Section 2 gives a short introduction to some of the most important related work on the field of ontology taxonomy verification, while Section 3 gives a short introduction to the concept of ontological profiles. Section 4 gives an overview of the approach, and specifies hypotheses about structure properties. Details about the experiment are found in Section 5, while the results are reported and commented in Section 6. Finally the paper is concluded in Section 7.

2 Related Work

Guarino et al. presents in [4] the methodology OntoClean for validating taxonomic relations in ontologies. The methodology is based on philosophical conceptions like essence, identity, and unity. These conceptions are used as metaproperties on the taxonomic structure describing the classes ([4] uses the term property about classes). The metaproperties described by the authors are Rigidity (R), Unity (U), Dependence (D), and Identity (I). The taxonomic structure of the ontology is verified by first tagging the concepts with metaproperties, and next the tags in the subsumption hierarchy are analyzed. The methodology imposes constraints on the subsumption of concepts with certain combinations of metaproperties. By removing subsumption relations that violate these constraints, relationships that were ill-defined from the designers part are avoided. Highly related to the work of Guarino et al. is the work of Völker et al. [9], which describes a tool, AEON, to automatically tag a RDF/OWL ontology with metaproperties. The authors use positive and negative evidence found in a large corpus (the Web) as basis for tagging classes with metaproperties. For each metaproperty the authors use a set of patterns to gather positive and negative evidence by querying the Google API. The patterns are in the form of a natural language string with a variable that is replaced by the concept in question. One example of such a pattern is *"is no longer (a|an)? x"*[9]. By replacing the variable x with the concept in question (e.g. "student") this specific pattern provides negative evidence for the Rigid metaproperty (a class is considered Rigid when an instance can not stop being an instance of that class, e.g. a human can not stop being a human, on the other hand, a student can stop being a student). The results from Google are analyzed by part-of-speech tagging the result to remove results that give false match for the pattern. Finally they use a classifier to decide whether or not a metaproperty is applicable to the concept based on the evidence.

3 Ontological Profiles

An ontological profile is an ontology extended with vectors of terms for each concept of the ontology. These terms are regarded as semantic descriptions of the concept at hand, based on an underlying document collection. Moreover, the terms are weighted to reflect the strength of the relation between the concept and the term. By using a document collection as the basis for the ontological profile, both the terms and their weights are reflections of the real world usage of the ontology concepts. For a definition of concept vector see Definition 1.

Definition 1. *Concept Vector. The definition given here is adapted from Su[8]. Let T be the set of n terms in the document collection used for construction of the ontological profile. $t_i \in T$ denotes term i in the set of terms. Then the concept vector for concept j is defined as the vector $C_j = [w_1, w_2, \ldots, w_n]$ where each w_i denotes the semantic relatedness weight for each term t_i with respect to concept C_j.*

The method of constructing an ontological profile is systematic, in that we have defined a way of assigning documents to and creating vectors from these documents to represent concepts of the ontology. For a more detailed description of ontological profiles, see [7].

We have done (and are still doing) research on using ontological profiles for information retrieval (IR) purposes, for an introduction to the use of ontological profiles for IR see for example [7]. Further, ontological profiles have been used for ontology alignment purposes [8].

4   Approach

Our approach of using vectors of weighted terms associated with ontological concepts, gives us a sort of semantic description of the notions used in real world texts describing the concept. The terms contained in the vectors are restricted to stemmed versions of certain word classes (part-of-speech tagging), with certain frequent words removed (details described in Section 5). The main objective of our approach is to compare these text based descriptions of concepts, concept vectors, with each other. The comparison is used as a basis to define and verify desirable properties in the subsumption hierarchy of the ontology.

To give the reader a real example of the vectors created from the DNV ontology, we can take a look at the concept *careers* which has the following *phrase vector* definition (only top 5 terms are shown with weights in subscript):

$$C_{careers} = [\text{"high ambitions"}_{5.9}, \text{"extra dimension"}_{5.9}, \text{"profiling film"}_{5.9},$$
$$\text{"vacant positions"}_{5.2}, \text{"dnv uk"}_{3.3}]$$

As we see from the vector these terms are quite reasonable for what you would expect from a recruitment page. Another example from the ontology is the concept *maritime* which is an import business area for DNV. The *phrase vector* is represented as follows:

$$C_{maritime} = [\text{"ship classification"}_{3.5}, \text{"maritime industry"}_{3.1},$$
$$\text{"strong base"}_{2.9}, \text{"printed editions"}_{2.9}, \text{"ships life"}_{2.9}].$$

Since the vector representation carries a semantic description of the concept, we should be able to find interesting properties that should hold for hierarchical relations of good quality by comparing the vectors of super classes and sub classes. We use two different approaches for the comparison of the vectors. The first is cosine similarity [1], and the second is reducing the vectors to sets (by disregarding the weights of the vectors) and performing set operations on the resulting sets. The cosine similarity is calculated as in Equation 1, where C_i and C_j are the concept vectors for concept i and j respectively, $w_{n,l}$ is the weight for term n in concept vector l, t the total number of terms, and $sim(C_i, C_j)$ is the cosine similarity between C_i and C_j.

$$sim(C_i, C_j) = \frac{\sum_{n=1}^{t} w_{n,i} \times w_{n,j}}{\sqrt{\sum_{n=1}^{t} w_{n,i}^2} \times \sqrt{\sum_{n=1}^{t} w_{n,j}^2}} \tag{1}$$

The approach does not consider the semantics of separate words other than removing words that are stop words, and words that are not tagged as nouns using part-of-speech-tagging. There is one exception to the last point, we include adjectives that are parts of noun phrases (described in Section 5). Moreover, we do not consider the implicit relations between words, only how the words sum up to describe the concepts of the ontology.

We have proposed four different hypotheses about properties of the hierarchical relations in the ontologies based on the two measures just described (cosine similarity and set relations).

4.1 Hypotheses

We will in this subsection describe our four hypotheses and argue for why they seem reasonable. For the sake of making the explanation of the hypothesis simpler, we will use the notation described below. A concept C of the ontology has n sub classes, C_i. All concepts C_i (under C) are then said to be siblings at the same abstraction level of the ontology. Further, the super concept's vector representation is given by S, and its corresponding set representation (by disregarding the weights of S) is given by S'. Likewise, for the sub classes, the vector representation of sub class C_i is given by U_i and the corresponding set representation is given by U_i'.

Hypothesis 1. *The relationship between super and sub class is stronger than between the sub classes.*

In other words, Hypothesis 1 states that we expect to find that the relationship between a class and its subsumer is stronger than the relationship between the class and its siblings on the same level of the hierarchy. Put in terms of the ontology relations, evidence supporting this hypothesis should be for each S and all its corresponding sub vectors U_i, U_j: $sim(S, U_i) > sim(U_i, U_j)$ and $sim(S, U_j) > sim(U_i, U_j)$. The argument for this is that the sub-super relation carries some commonalities that we should be able to observe. On the other hand the commonalities of the sibling relations are mainly carried by the relationship with the super class relation, and should thus not be as prominent.

Hypothesis 2. *Characterizations of super class and sub class overlap semantically, but refer to different levels of abstraction*

Hypothesis 2 expresses that although there is a relation between the super class and its sub class, the set representation of the two should be different. The motivation for this is that while the super class has a broader definition, touching some of the relevant aspects of the sub class, the sub class should have its own, narrower description, fleshing out on the more detailed aspects. Evidence supporting this hypothesis should be found in the form $S' \setminus U_i' \neq \emptyset$ and $U_i' \setminus S' \neq \emptyset$ thus signaling that neither set contains fully of the other set.

Hypothesis 3. *Commonalities among subclasses are defined by their super class.*

By Hypothesis 3 we mean that the super class, being the more general class in the hierarchy, defines some least common set that should be found amongst the sub classes. In other words, we expect that the super class has a partitioned terminology, one describing abstract features, and one describing more specific features of the concept. For this hypothesis to be true, we would expect to find a common terminology amongst the sub classes that also would be found in the super class' terminology. Specifically, we would expect to find that the specific part of the super class' description is shared with the intersection of the sub classes. In terms of set relation we can say that evidence supporting this hypothesis should be $(\bigcap_{i=1}^{n} U_i') \setminus S' = \emptyset$.

Hypothesis 4. *There are abstract features of a super class that are not shared by any subclass.*

Hypothesis 4 states that the super class is defined at a higher level of abstraction using terminology that is not directly applicable to the lower level of abstraction of the sub class. Using the same argumentation as in Hypothesis 3, we can say that the terminology is partitioned. Whereas Hypothesis 3 tests the commonalities between the super class and sub classes (specific features of the super class), we are here interested in the abstract features of the super class. The super class should describe the concept at a higher level, and thus contain terminology that is not interesting to deal with on the more detailed level of the sub class. Evidence supporting this hypothesis should thus be $S' \setminus (\bigcup_{i=1}^{n} U_i') \neq \emptyset$ and $|S' \setminus (\bigcup_{i=1}^{n} U_i')| < |S'|$.

5   Experiment

The data set we have based our experiments on is the web site of a large corporation, DNV[1], which has activities spanning globally. The web site for this

---

[1] http://www.dnv.com

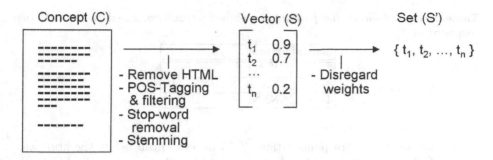

Fig. 1. Overview of vector construction

company has been downloaded for two separate temporal snapshots, namely 2004 and 2008. We made a simple parser application that uses the site map of the site as a basis, all pages listed in the site map were attempted downloaded. The site map for 2004 contained 207 files of which we successfully downloaded all, while the site map for 2008 contained 424 files, which we also managed to successfully download. However, only 369 of the files from 2008 contained content (the rest were subject to "http 404 file not found" or internal server errors at the company). As the actual data source for the files we used the web site of the Internet Archive [2] and their Wayback Machine.

Further we have used this site and its hierarchy as a ontology, using the structure of the site to create a subclass hierarchy. This is not considered an actual ontology by many, but our main purpose in this experiment is to look at the hierarchical relationships between concepts, so for our use this will suffice. Furthermore, we assume that the ontology of DNV is good with respect to subsumption, as it has been developed over many years and is continuously updated. Thus we deem that the results we get from our experimental results should be valid for concluding about how a good ontology is structured.

Figure 1 shows the overall techniques used during construction of the concept vectors, discussed in more detail below. For each node in the ontology (web page) we created a concept vector. As we regard each of the web pages in the hierarchy as a single ontology concept, it is quite straight forward to assign documents to each concept. We simply view the single document found at the concept as its textual description.

Preprocessing and cleaning of the html documents is the first step for creating a concept vector for each concept in the ontology. First, we remove any html tags, script tags, common structures (such as menus) etc. from the documents, leaving us with clean text files. Since html tagging is mainly a layout formatting, authors are not always very good at writing grammatically correct pages with respect to the textual content, which hampers the results of the next step in the process, part-of-speech (POS) tagging . This lead us to apply the following solution. Assuming that any text within certain types of tags is either a sentence/paragraph or a stand-alone collection of words (e.g. the entry in a bullet

---

[2] http://www.archive.org

Table 1. Part-of-speech tag patterns for phrase recognition, based on the patterns suggested in [5]

Length	Pattern
2	NN, NJ
3	NNN, NJN , JNN, JJN
4	NNNN

point), we inserted extra punctuations ("."·) into the html before the html was removed. By not applying this solution we could in the worst case end up with a whole table as a full sentence.

For POS-tagging we used the Stanford Tagger v1.6 [3]. The POS tags were further used to remove words of unwanted word-classes, and recognize phrases in the text. As a basis for the phrase extraction we use a set of POS tag patterns slightly different than the ones suggested by Justeson and Katz [5]. The tag patterns we use are shown in Table 1, where N is a noun (we have not differentiated between words within the noun class), and J is an adjective (also here we do not differentiate between words within the adjective class).

All phrases found by the tag patterns were added to the concept vector without any frequency filtering. This will of course result in some noise, but we found that filtering on a frequency of 2 would disqualify a large number of good phrases found in the text. Thus the benefit of adding more good phrases was considered to be higher than the disadvantage of the small amount of noise that was added. In addition to the phrases found based on the tag patterns in Table 1, we also added single nouns (NN, NNS, NNP, NNPS) that appear in more than two documents to the concept vectors.

Next the phrases were split into its sub terms, and all these were added to the concept vectors (using the frequency of the phrase). We note that in the step of breaking up the phrases, there are still some terms that can be adjectives, although we specified that only nouns (single terms) should be added to the vector. The argumentation we use is that the adjectives present in a noun phrase are more important for the semantic description of the concept than other "stand-alone" adjectives, and thus are added to the vector.

To make sure that the vectors are containing only the most meaning bearing terms, we next remove the stop words (if any, since the POS-filtering should have removed a substantial portion). The final processing of the terms is to apply stemming using the Porter stemming algorithm [6]. This ensures that terms having the same general meaning are collapsed to a single term. Finally, the weight of the terms are calculated based on the familiar $tf \times idf$ score[1] depicted in Equation 2, where $w_{i,j}$ is the weight of term i in concept vector j, $freq_{i,j}$ is the frequency of term i in the concept vector j, $max_l freq_{l,j}$ is the frequency of the most frequent term l in concept vector j, N is the number of concept vectors, and n_i is the number of concept vectors containing term i.

---

[3] http://nlp.stanford.edu/software/tagger.shtml

$$w_{i,j} = \frac{freq_{i,j}}{max_l freq_{l,j}} \times log\frac{N}{n_i} \qquad (2)$$

To compute the set operations needed for Hypothesis 2-4, we disregard the weight information for the terms, and regard each vector as a set, adding access methods to perform set operations on the resulting set. The cosine similarity calculations were performed according to Equation 1.

6   Results

We will in this section present and comment upon the results obtained from our experiments. The experiment consisted of running two tests for each of the four hypotheses. For all the tests we located all the classes that had two or more sub classes. For the 2004 collection this left us with 26 super classes, and for the 2008 collection we were left with 60 super classes. Please refer to the example from Section 4.1 for the notation used to describe the classes/vectors/sets.

The first part of the experiment was concerning Hypothesis 1 (which deals with the strength of the sub-super relation in contrast to the sibling relations). We specified the success criteria for the hypothesis to be $sim(S, U_i) > sim(U_i, U_j)$ and $sim(S, U_j) > sim(U_i, U_j)$. We had a total of 26 (2004) and 60 (2008) classes with 2 or more sub classes, and for the sub-super relation analysis we examined the cosine similarity between the super class and each of its sub classes (totaling 172 relations for 2004 and 344 for 2008). For each of the super classes we examined the sibling relation (according to how sibling relations were defined in Section 4.1) for each pair of sub classes. The number of sibling relations analyzed totaled 744 for the 2004 collection, and for the 2008 collection 1482. Looking at the results from Table 2 we see that the mean cosine similarity between super and sub concepts is higher than the similarity between siblings. This seems to support the hypothesis, and we see that the results seem to agree for both collections.

Next, we look at the part of the experiment concerning Hypothesis 2 (super class and sub class are different). Recall that we specified the success criteria for the hypothesis to be $S' \setminus U_i' \neq \emptyset$ and $U_i' \setminus S' \neq \emptyset$. Looking at the results in Table 3 we can see that this indeed seems to be the case. We see that both result sets are quite large, meaning that there is some partial overlap between the two, and that both concepts seem to have their own specificities. The super class should have some general more abstract terms which are more appropriate

Table 2. The results of the experiment for Hypothesis 1

Variable	2004	2008
Mean sub-super similarity	0.347	0.348
Mean sibling similarity	0.197	0.219
Number of concepts having a mean sibling similarity larger than mean sub-super similarity	5	6

Table 3. The results of the experiment for Hypothesis 2

Variable	2004	2008
Mean number of terms in S'	65.5	71.3
Mean number of terms in U_i'	65.8	71.4
Mean number of terms in $S' \setminus U_i'$	43.3	44.8
Mean number of terms in $U_i' \setminus S'$	41.7	46.6

at a more abstract level, while the sub class should contain some terms that are more specialized and appropriate at a more detailed level. It is however not an easy task to analyze what an optimal result for this test is. Very large result sets would indicate little overlap, while small result sets indicates a high degree of overlap. The optimal size of the result set remains an open issue.

The third part of the experiment concerns whether commonalities among the sub classes are defined by their super class (refer to Hypothesis 3). Further we specified the evidence criterion for supporting this hypothesis as being $(\bigcap_{i=1}^{n} U_i') \setminus S' = \emptyset$. From the results in Table 4 we can see that this indeed seems to be the case. For 2004 we have 18 (out of 26) empty result sets, while we have for 2008 28 (out of 60) empty result sets. Further we note that the mean size of the result sets is quite low, 0.7 and 3.7 for 2004 and 2008, respectively. The interpretation of this result is that there is some commonality between the sub classes that is also defined by the super class. The result in our opinion is quite clear, as supported by both collections.

Table 4. The results of the experiment for Hypothesis 3

Variable	2004	2008
Mean number of terms in S'	65.5	71.3
Mean number of terms in $(\bigcap_{i=1}^{n} U_i')$	13.7	18.4
Mean number of terms in $(\bigcap_{i=1}^{n} U_i') \setminus S'$	0.7	3.7
Empty result sets	18	28

The last part of the experiment was concerning Hypothesis 4. The test run was based on subtracting the super class from the union of the sub classes $(S' \setminus (\bigcup_{i=1}^{n} U_i'))$, and the evidence criteria were specified as $S' \setminus (\bigcup_{i=1}^{n} U_i') \neq \emptyset$ and $|S' \setminus (\bigcup_{i=1}^{n} U_i')| < |S'|$. The results are depicted in Table 5 and we can see that for each of the collections there is a single empty result set, meaning that the entire super vector is contained in the union of the sub vectors. This result shows that the super set does not contain any specific semantics not carried in the sub sets. On the other hand, for the remaining sets we see that the result set contains on average 15.3 and 23 terms for 2004 and 2008, respectively. We see that the results (both for 2004 and for 2008) agree quite well with the evidence criterion given. We interpret this as that the semantics of the super concept can be split in two; one general (abstract) part most appropriate for the higher

Table 5. The results of the experiment for Hypothesis 4

Variable	2004	2008
Mean number of terms in S'	65.5	71.3
Mean number of terms in $(\bigcup_{i=1}^{n} U_i')$	222.2	223.7
Mean number of terms in $S' \setminus (\bigcup_{i=1}^{n} U_i')$	15.3	23.0
Empty result sets	1	1

level, and one more specific part that is also shared with the lower level (the sub classes).

7 Conclusion

In this paper we have presented an approach for describing properties in a subsumption hierarchy of an ontology, and run tests that verify that our hypotheses seem reasonable. We have used as basis the DNV ontology defining the structure of the www.dnv.com web site. Since this ontology has been developed for several years and is subject to continuous update, we deem it as a good ontology for our evaluation. The first thing we can note from the results is that there seems to be a stronger relation between a class and its subsumer, than between a class and its siblings on the same level of abstraction. This is a nice result pointing out how the relations should be between sub/super class and siblings in the ontology. Further, we found that there is indeed a difference between the super class and the sub class, while they still retain some similarity. We interpret this as the super class having a broader definition, and carrying a vocabulary more suited for the higher level of abstraction, while the sub class has a more detailed vocabulary geared towards the lower level of abstraction. This may point out that classes in a subsumption hierarchy that overlap to a high degree possibly should not have the subsumption relation.

We also found that the commonalities between the sub classes (represented as the intersection of the sub classes) defines a common set that also can be found in the super class. If the commonalities were not found in the super class this could possibly be a indication that the relation is inappropriate, or that we really are dealing with concepts that should be even further down the hierarchy (a missing class between the super and sub classes). Finally we found that the super class and the sub classes overlap partially semantically. The super class contains one part specifying its more abstract semantics (not shared by the sub classes), while the sub classes contain one part specifying their more detailed nature.

All of these findings are supported both in the 2004 and the 2008 version of the ontology, even though the ontology itself has evolved over the time period. This indicates that the ontology has been updated by sound principles.

There are however some weaknesses in our approach. First, we do not look at the semantic relations between the words in the sets/vectors. Second, and this is more of a concern towards the validity of our experiments, the amount of text used to construct the sets/vectors is limited. It would thus be an interesting point for further work to do a more thorough analysis of the approach with a larger data set. Lastly, our approach does not specify what the classes mean, rather how they are defined by words. Guarino [4] using the OntoClean methodology tries in contrast to define the semantics of the classes.

Acknowledgment. This research was carried out as part of the LongRec project, project no. 176818/I40, funded by the Norwegian Research Council.

References

1. Baeza-Yates, R., Ribeiro-Neto, B.: Modern Information Retrieval. ACM Press, New York (1999)
2. Berners-Lee, T., Hendler, J., Lassila, O.: The Semantic Web. Scientific American (May 2001)
3. Gruber, T.R.: A translational approach to portable ontologies. Knowledge Acquisition 5(2) (1993)
4. Guarino, N., Welty, C.A.: An overview of OntoClean. In: Staab, S., Studer, R. (eds.) Handbook on ontologies. International Handbooks on Information Systems. Springer, Heidelberg (2004)
5. Justeson, J., Katz, S.: Technical terminology: some linguistic properties and an algorithm for indentification in text. Natural Language Engineering 1(1), 9–27 (1995)
6. Porter, M.F.: An algorithm for suffix stripping. Program 14(3), 130–137 (1980)
7. Solskinnsbakk, G., Gulla, J.A.: Ontological Profiles as Semantic Domain Representations. In: Kapetanios, E., Sugumaran, V., Spiliopoulou, M. (eds.) NLDB 2008. LNCS, vol. 5039, pp. 67–78. Springer, Heidelberg (2008)
8. Su, X.: Semantic Enrichment for Ontology Mapping. PhD Thesis, Norwegian University of Science and Technology, Trondheim, Norway (2004)
9. Völker, J., Vrandecic, D., Sure, Y.: Automatic Evaluation of Ontologies (AEON). In: Gil, Y., Motta, E., Benjamins, V.R., Musen, M.A. (eds.) ISWC 2005. LNCS, vol. 3729, pp. 716–731. Springer, Heidelberg (2005)

Evaluating Semantic Relations by Exploring Ontologies on the Semantic Web

Marta Sabou, Miriam Fernandez, and Enrico Motta

Knowledge Media Institute (KMi)
The Open University, Milton Keynes, United Kingdom
{R.M.Sabou,M.Fernandez,E.Motta}@open.ac.uk

Abstract. We investigate the problem of evaluating the correctness of a semantic relation and propose two methods which explore the increasing number of online ontologies as a source of evidence for predicting correctness. We obtain encouraging results, with some of our measures reaching average precision values of 75%.

1 Introduction

The problem of understanding how two concepts relate to each other has been investigated in various fields and from different points of view. Firstly, the level of relatedness between two terms is a core input for several Natural Language Processing (NLP) tasks, such as word sense disambiguation, text summarization, annotation or correction of spelling errors in text. As a result, a wide range of approaches to this problem have been proposed which mainly explore two paradigms. On the one hand, corpora-based methods measure co-occurrence in a given context (usually characterized by means of linguistic patterns) across large-scale text collections [4,14]. On the other hand, knowledge rich methods use world knowledge explicitly declared in ontologies or thesauri (usually, WordNet) as a source of evidence for relatedness [3].

Secondly, from the beginnings of the Semantic Web (SW), where semantic relations are the core components of ontologies, the task of identifying the actual semantic relation that holds between two concepts has received attention in the context of the ontology learning field [5]. Finally, recent years have seen an evolution of Semantic Web technologies, which lead both to an increased number of online ontologies and to a set of mature technologies for accessing them[1]. These changes have facilitated the appearance of a new generation of applications which are based on the paradigm of reusing this online knowledge [6]. These applications differ substantially from the typical knowledge-based AI applications (as well as some of the early SW applications) whose knowledge base is provided a-priory rather than being acquired through re-use during runtime. They also reform the notion of knowledge reuse, from an ontology-centered view,

---

[1] http://esw.w3.org/topic/TaskForces/CommunityProjects/LinkingOpenData/
SemanticWebSearchEngines

H. Horacek et al. (Eds.): NLDB 2009, LNCS 5723, pp. 269–280, 2010.

to a more fine-grained perspective where individual knowledge statements (i.e., semantic relations) are reused rather than entire ontologies. In the case of these applications, it is therefore important to estimate the correctness of a relation, especially when it originates from a pool of ontologies with varying quality.

The problem we investigate in this paper is evaluating the correctness of a semantic relation. Our hypothesis is that the Semantic Web is not just a motivation for investigating this problem, but can actually be used as part of the solution. We base this hypothesis on the observation that the Semantic Web is a large collection of knowledge-rich resources, and, as such it exhibits core characteristics of both data source types used in NLP for investigating relatedness: knowledge resources (structured knowledge) and corpora (large scale, federated). Earlier research has showed that although contributed by heterogeneous sources, online ontologies provide a good enough quality to support a variety of tasks [17]. It is therefore potentially promising to explore this novel source and to investigate how NLP paradigms can be adapted to a source with hybrid characteristics such as the SW. We phrase the above considerations into two research questions:

1. *Can the SW be used as a source for predicting the correctness of a relation?*
2. *Can we adapt existing NLP paradigms to the SW?*

To answer these questions we present two methods that explore online ontologies to estimate the correctness of a relation and which are inspired from two core paradigms used for assessing semantic relatedness. We perform an extensive experimental evaluation involving 5 datasets from two topic domains and covering more than 1400 relations of various types. We obtain encouraging results, with one of our measures reaching average precision values of 75%.

We start by describing some motivating scenarios where the evaluation of semantic relations is needed (Section 2). Then, we describe two measures designed for this purpose and give details over their implementation (Sections 3 and 4). In Section 5 we detail and discuss our experimental investigation and results. An overview of related work and our conclusions finalize the paper.

2   Motivating Scenarios

In this section we describe two motivating scenarios that would benefit from measures to evaluate the correctness of a semantic relation.

Embedded into the NeOn Toolkit's ontology editor, the Watson plugin[2] supports the ontology editing process by allowing the user to reuse a set of relevant ontology statements (equivalent to semantic relations) drawn from online ontologies. Concretely, for a given concept selected by the user, the plugin retrieves all the relations in online ontologies that contain this concept (i.e., concepts having the same label). The user can then integrate any of these relations into his ontology through a mouse click. For example, for the concept *Book* the plugin would suggest relations such as:

---

[2] http://watson.kmi.open.ac.uk/editor_plugins.html

- $Book \subseteq Publication$
- $Chapter \subseteq Book$
- $Book - containsChapter - Chapter$

The relations are presented in an arbitrary order. Because of the typically large number of retrieved relations it would be desirable to rank them according to their correctness. To date, however, no such methods exist thus hampering the user in finding the correct relations first, or indeed preventing him from reusing incorrect ones (e.g., $Chapter \subseteq Book$ where subsumption has been used incorrectly to model a meronymy relation).

As a second scenario we consider ontology matching [7], a core Semantic Web task. This task leads to establishing a set of mappings between the concepts of two input ontologies (i.e., an alignment). While these mappings take the form of semantic relations of various types, the focus of the community has primarily been in deriving and evaluating equivalence relations by comparing them against a-priory, manually-built, gold-standard alignments. However, as more and more matchers are capable of identifying other mappings than equivalence, the current gold-standard based evaluations need to be revised as it is impossible to manually predict all types of relations that would hold between the elements of two ontologies [16]. We hope that the methods described in this paper could be used as a way to automatically assess the correctness of alignments containing more than just equivalence mappings.

3   Evaluating the Correctness of Semantic Relations

To formally define our problem, let us denote a semantic relation as a triple $< s, R, t >$, where s is the source concept (or domain), t is the target concept (or range) and R denotes the relation that holds between the two concepts. R can define a wide range of relation types, such as hyponymy, disjointness, meronymy or simply any associative relation. Our aim is to derive a set of methods that can predict the level of correctness of such a relation, i.e., whether it is likely to be correct or incorrect. For the purposes of this work, we distinguish between a relation being *generically correct* and *correct or relevant in a given context*. When we decide on generic correctness we estimate the generic consensus on a relation independently of an interpretation context, while contextual-correctness or relevance should also take into account a given interpretation context. In this work we focus on generic correctness and leave contextual issues for future work.

In this section we propose two measures that exploit the large amount of online ontologies for predicting the correctness of a semantic relation. The measures are based on two different paradigms. The first measure explores the knowledge declared in online ontologies to predict correctness and as such it resembles the knowledge-rich methods reported in [3]. The second measure treats the Semantic Web as a corpus of ontologies for measuring the likely relatedness of the concepts involved in the relation and the popularity of that relation. As such, it is inspired from corpora-based methods similar to those described in [4,14].

3.1 Exploring Ontologies as Knowledge Artifacts

The measures in this section explore ontologies as knowledge artifacts and are based on the intuition that explicitly declared relations are more likely to be correct than implicit ones (i.e., those which are derived through reasoning).

Let $< s, R, t >$ be a relation which we wish to evaluate. Let n be the number of online ontologies such that each ontology O_i contains concepts similar to s and t ($s'_i = s$ and $t'_i = t$) and that a relation equivalent to R ($R'_i = R$) is declared explicitly (or can be inferred) between s'_i and t'_i. For example, for the statement $aircraft \supseteq helicopter$ there are three ontologies (shown in Table 1) that explicitly (or implicitly) declare such a relation.

Table 1. Examples of derivation paths for $aircraft \supseteq helicopter$

Derivation Path and Ontology	Path Length
$O_1 : Aircraft \supseteq O_1 : Helicopter$ $O_1 =$`http://reliant.teknowledge.com/DAML/Mid-level-ontology.owl`	1
$O_2 : Aircraft \supseteq O_2 : Helicopter$ $O_2 =$`http://reliant.teknowledge.com/DAML/Transportation.owl`	1
$O_3 : Aircraft \supseteq O_3 : HeavierThanAirCraft \supseteq O_3 : Rotorcraft$ $\supseteq O_3 : Helicopter$ $O_3 =$`http://www.interq.or.jp/japan/koi_san/trash/aircraft3.rdf`	3

Our measure relies on the hypothesis that there is a correlation between the length of the derivation path and the correctness of the relation. In particular, we think that longer paths probably lead to the derivation of less obvious relations, which are therefore less likely to be correct. To verify this hypothesis we compute three values: $AveragePathLength_R$ is the average of the lengths of all derivation paths for relation R (e.g., in our case $(1 + 1 + 3)/3 = 1.66$), $minLength_R$ is the length of the shortest derivation path that lead to R (in our case, 1), and $maxLength_R$ is the length of the longest derivation path associated to R (in our case, 3). Formally:

$$AveragePathLength_R = \frac{\sum_i PathLength_{R,O_i}}{n}$$

$$minLength_R = min_i(PathLength_{R,O_i}); maxLength_R = max_i(PathLength_{R,O_i})$$

3.2 Exploring Online Ontologies as a Corpus

Unlike in the previous section, the focus of the measures presented here is on exploring the Semantic Web as a corpus of ontologies for computing concept relatedness and relation popularity.

For a relation $< s, R, t >$ to be evaluated, we define $RelatednessStrength_{s,t}$ as the ratio between the number of ontologies from which a relation can be deduced between s and t (i.e., $|O_{s,r,t}|$) and the number of all ontologies where these concepts are mentioned but not necessarily related (i.e., $|O_{s,t}|$). This measure is an indication of how likely it is that the two concepts are related. Indeed, if all the ontologies that mention s and t also lead to deriving a relation between them, then s and t are likely to be related. This measure takes its values in the interval $(0,1]$, with low values corresponding to terms that are weakly related, and 1 to those that are related in all ontologies that they are mentioned. For example, *Rodents* and *Animals* appear in 5 ontologies and each of these ontologies leads to a relation between them. While this measure does not inform about the correctness of a particular relation R, we assume that a relation established between terms that are not likely to be related is less likely to be correct than a relation established between closely related terms. Formally:

$$RelatednessStrength_{s,t} = \frac{|O_{s,r,t}|}{|O_{s,t}|}$$

Table 2. Examples of relations between *honey* and *food*

Relation	Derivation Path and Ontology
sibling	$O_1 : Honey \subseteq O_1 : Sweetener \subseteq O_1 : SweetTaste \subseteq$ $O_1 : PartiallyTangible$ $O_1 : Food \subseteq O_1 : FoodOrDrink \subseteq O_1 : HumanScaleObject \subseteq$ $O_1 : PartiallyTangible$ $O_1 =$http://secse.atosorigin.es:10000/ontologies/cyc.owl
$\subseteq$	$O_2 : Honey \subseteq O_2 : Food$ $O_2 =$http://sweet.jpl.nasa.gov/ontology/substance.owl
$\subseteq$	$O_3 : honey \subseteq O_3 : sweetener \subseteq O_3 : flavoring \subseteq$ $O_3 : plant - derived - foodstuff \subseteq O_3 : foodstuff \subseteq O_3 : food$ $O_3 =$http://morpheus.cs.umbc.edu/aks1/ontosem.owl

We then define $StrengthRelation_R$ for measuring the popularity of a relation R over any type of relations that can be derived between s and t. This measure also takes its values from $(0,1]$, with the lowest values indicating that R has a low popularity (and therefore it is likely to be incorrect) and a value of 1 being obtained when R is the only relation derivable between these concepts (and therefore it is likely to be correct). For example, as shown in Table 2, because it is more popular amongst online ontologies, the $\subseteq$ relation between *honey* and *food* will have a higher value for this measure (i.e., 0.66) than the *sibling* relation between the same concepts (i.e., 0.33). Formally:

$$StrengthRelation_R = \frac{freq(R)}{allRels_{s,t}}$$

Note that we have also experimented with various ways of normalizing these measures, however, we do not present them because experimental evaluation has showed a less optimal behavior than for the original measures.

4 Implementation

We implemented our measures using the services of the Watson[3] semantic web gateway. Watson crawls and indexes a large number of online ontologies[4] and provides a comprehensive API which allows exploring these ontologies.

We have also built an algorithm that, using Watson, extracts relations between two given terms from online ontologies. The algorithm is highly parameterized[5]. For the purposes of this study we have configured it so that for each pair (A,B) of terms it identifies all ontologies containing the concepts A' and B' corresponding to A and B from which a relation can be derived between these terms. Correspondence is established if the labels of the concepts are lexical variations of the same term. For a given ontology (O_i) the following derivation rules are used:

- if $A'_i \equiv B'_i$ then derive $A \xrightarrow{\equiv} B$;
- if $A'_i \sqsubseteq B'_i$ then derive $A \xrightarrow{\sqsubseteq} B$;
- if $A'_i \sqsupseteq B'_i$ then derive $A \xrightarrow{\sqsupseteq} B$;
- if $A'_i \perp B'_i$ then derive $A \xrightarrow{\perp} B$;
- if $R(A'_i, B'_i)$ then derive $A \xrightarrow{R} B$;
- if $\exists P_i$ such that $A'_i \sqsubseteq P_i$ and $B'_i \sqsubseteq P_i$ then derive $A \xrightarrow{sibling} B$.

Note that in the above rules, the relations between A'_i and B'_i represent both explicit and implicit relations (i.e., relations inherited through reasoning) in O_i. For example, in the case of two concepts labeled $DrinkingWater$ and tap_water, the algorithm deduces the relation $DrinkingWater \xrightarrow{\sqsubseteq} tap\_water$ by virtue of the following subsumption chain in the TAP ontology: $DrinkingWater \sqsubseteq Flat\text{-}DrinkingWater \sqsubseteq TapWater$.

5 Experimental Evaluation

In this section we describe the experimental evaluation of the measures detailed in Section 3. We have used the implementation presented in Section 4 over the datasets described in Section 5.1. We then further explore and analyze the results for both measure types (Sections 5.2 and 5.3).

5.1 Data Sets

As experimental data we have used datasets from the domain of ontology matching, in the form of alignments obtained in two different test-cases put forward by the Ontology Alignment Evaluation Initiative[6](OAEI), an international body that coordinates evaluation campaigns for this task.

---

[3] http://watson.kmi.open.ac.uk
[4] Estimated to 250.000 during the writing of this paper.
[5] A demo of some of these parameters and an earlier version of the algorithm are available at http://scarlet.open.ac.uk/
[6] http://oaei.ontologymatching.org/

Table 3. Overview of the experimental data sets and their characteristics

Data Set	Nr. of Relations	Type of Relations	Domain
AGROVOC/NALT	380	$\subseteq, \supseteq, \bot$	Agriculture
OAEI'08 301	112	$\subseteq, \supseteq, \bot$, named relations	Academia
OAEI'08 302	116	$\subseteq, \supseteq, \bot$, named relations	Academia
OAEI'08 303	458	$\subseteq, \supseteq, \bot$, named relations	Academia
OAEI'08 304	386	$\subseteq, \supseteq, \bot$, named relations	Academia
Total	1452		

The AGROVOC/NALT data set has been obtained by performing an alignment between the United Nations' Food and Agriculture Organization (FAO)'s AGROVOC ontology and its US equivalent, NALT. The relations established between the concepts of the two ontologies are of three types: $\subseteq$, $\supseteq$ and $\bot$. Each relation has been evaluated by two experts, as described in more detail in [15].

The OAEI'08 dataset represents the alignments obtained by the Spider system on the 3** benchmark datasets and their evaluation [16]. This dataset contains four distinct datasets representing the alignment between the benchmark ontology and the MIT (301), UMBC(302), KARLSRUHE(303) and INRIA(304) ontologies respectively. Besides the $\subseteq$, $\supseteq$ and $\bot$ relation types, this data set also contains named relations (e.g., $inJournal(Article, Journal)$). Table 3 provides a summary of these datasets and their characteristics.

5.2   Results for the Derivation Path Based Measures

To investigate the correlation between the characteristics of the derivation path and the correctness of a relation, we computed the $AveragePathLength_R$, $minLength_R$ and $maxLength_R$ values for all relations in our five datasets. Then, for each dataset we computed the mean value for $AveragePathLength_R$ for relations judged to be false (F-Mean) and those judged to be true (T-Mean). We also repeated these calculations for the dataset obtained by merging the relations in all datasets. The values of these computations are shown in columns two and three of Table 4. We notice that for all datasets there is a clear difference between the mean path length of true and false relations, where false relations, on average, have a longer derivation path (always over 2) than the true ones (always under 2). This is already a good indication that this measure captures a valid hypothesis.

We continued our investigations by computing a threshold value for which the assignment of correctness values correlates best with that of the human judgement. This was measured in terms of a precision value computed as the ratio of correctly assessed relations with that threshold over all relations in the dataset. Columns four and five of Table 4 show our results. We note that there is considerable variation in the values of the optimal threshold between datasets and that some are very close to the extreme values (e.g., in the case of AGROVOC/NALT, and OAEI'08 304 the best threshold is close to F-Mean, while for OAEI'08 302 the threshold is almost identical with T-Mean). Given this situation we tried to

Table 4. Correlation between the derivation path characteristics and correctness

Data Set	$AveragePathLength_R$		Best Threshold	Prec.	Best Threshold'	Prec.'
	F-Mean	T-Mean				
AGROVOC/NALT	2.07	1.77	2.08	65%	2.00	71%
OAEI'08 301	2.58	1.29	1.66	86%	1.29	**94%**
OAEI'08 302	2.50	1.70	1.71	74%	1.71	80%
OAEI'08 303	2.83	1.76	2.60	76%	2.00	78%
OAEI'08 304	2.31	1.81	2.25	69%	2.00	73%
Merged Datasets	2.46	1.73	2.33	71%	**2.00**	**75%**

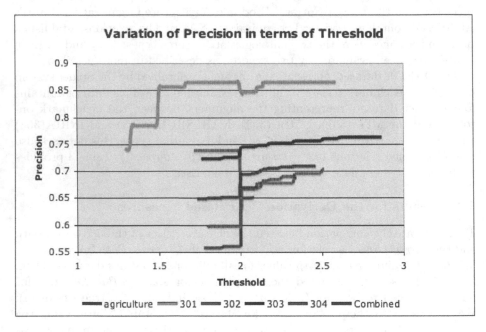

Fig. 1. Precision variation in terms of threshold values set for the length of the derivation path

approximate a global optimal threshold by computing it on the merged dataset. This yielded the value 2.33. The precision values per dataset vary from a minimum of 65% to a maximum of 86%, and we obtained an average precision for the merged dataset of 71%. Figure 1 graphically depicts the variation of precision in terms of threshold for all the five datasets and the merged datasets.

When examining the values of the $minLength_R$ and $maxLength_R$ measures, we observed that the overwhelming majority of relations that were deduced with paths of different lengths (i.e., their min and max path values were different) were correct relations. A good example is that of $aircraft \supseteq helicopter$ which is explicitly declared in two ontologies, while in another ontology this relation is defined in terms of a chain of more fine-grained relations (see Table 1). Another

Table 5. Average values for True and False relations, best threshold and precision values for *RelatednessStrength* and *StrengthRelation*

Data Set	Relatedness Strength		Best Thresh.	Prec.	Strength Relation		Best Thresh.	Prec.
	T	F			T	F		
AGROVOC/NALT	0.91	0.88	0.89	45%	0.34	0.34	0.34	36%
OAEI'08 301	0.81	0.75	0.75	41%	0.36	0.04	0.33	42%
OAEI'08 302	0.80	0.75	0.80	46%	0.38	0.11	0.11	38%
OAEI'08 303	0.58	0.50	0.55	43%	0.15	0.11	0.12	53%
OAEI'08 304	0.63	0.55	0.59	46%	0.23	0.15	0.16	56%

example relates to relations involving plants or animals, such as $goat \subseteq animal$. Some ontologies contain these relations explicitly (i.e., with a path length of 1), while others contain a more fine-grained path between these concepts, e.g., $goat \subseteq ungulate \subseteq mammal \subseteq vertebrate \subseteq animal$[7]. We have incorporated this observation in the calculation of the best threshold as follows: any relation which has the $AveragePathLength_R$ over the threshold but whose values for $minLength_R$ and $maxLength_R$ differ, is considered to be a True relation. The recomputed values for the best threshold and the corresponding precision are shown in the last two columns of Table 4. Remarkably, in the case of most datasets this observation has lowered the threshold and for all datasets it increased the precision to values ranging now from 71% to 94%. On the combined dataset this lead to a threshold of 2% and a precision value of 75%. We regard these values as illustrative for our derivation path based measures.

5.3   Results for the Corpora Inspired Measures

In columns two and three of Table 5 we present the average values of the *RelatednessStrength* measure for True and False relations respectively. Our hypothesis for this measure was that correct relations will most likely be declared between highly related terms (i.e., where the value of this measure is high), while the inverse will hold for false relations. Indeed, this hypothesis is verified by the obtained numbers as, for all datasets, on average, True relations are established between terms with higher *RelatednessStrength* than False ones. We note however, that the difference between the average values of this measure for True and False relations is rather small thus potentially decreasing its discriminative power. Indeed, this is verified when computing the best threshold and the corresponding precisions (columns four and five), as the precision values are quite low, not even reaching 50%.

In the second half of Table 5 we present the results of our experiments for the *StrengthRelation* measure. Our hypothesis was that high values of this measure, corresponding to popular relations, will mostly characterize True relations, while False relations will be associated with lower values. This hypothesis has been

---

[7] http://morpheus.cs.umbc.edu/aks1/ontosem.owl

verified in four out of five datasets, where the average value of the measure is lower for False relations than for True relations. The AGROVOC/NALT dataset is an exception, where both values are the same. We also notice that the difference between these values is higher than for the previous measure. This has a positive effect on the discriminative value of the measure, and indeed, we obtain higher precision values than for *RelatednessStrength* (up to 56%).

We conclude that, overall, the *StrengthRelation* measure has a better behavior than *RelatednessStrength*, although both are clearly inferior to the derivation path based measures discussed before. We think this is primarily due to the fact that, despite its increasing size, the Semantic Web is still rather sparse and as such negatively affects any corpus based measures. These measures could potentially be strengthened when combined with path based measures.

6   Related Work

An overview of related work suggests that various approaches are used to evaluate relatedness or semantic relations. The output of measures that provide a relatedness (or similarity) coefficient [3,14] has been evaluated through theoretical examination of the desirable mathematical properties [10], by assessing their effect on the performance of other tasks [3], and mainly by comparison against human judgement by relying on gold-standards such as the Miller Charles data set [13] or WordSim353[8]. The field of ontology learning has focused on learning taxonomic structures (consisting of hyponymy relations) and other types of relations [5]. For example, Hearst pattern based techniques have been successfully scaled up to the Web in order to identify certain types of relations such as hyponymy, meronymy [18] or complex qualia structures [5]. The evaluation measures used to assess the correctness of the learned relations either rely on comparison to a conceptual structure that plays the role of a gold-standard (mostly using the measures described in [12]) or on expert evaluation. Note that the techniques that use Hearst patterns on the Web can implicitly be used to verify whether a relation is of a given type. As such, these techniques are the most similar to the presented work, with the difference that they explore the Web (a large body of *unstructured* knowledge) rather than the Semantic Web (a collection of *structured* knowledge).

Another important body of work exists in the context of ontology evaluation (see two recent surveys for an overview [2], [9]), where existing approaches are unevenly distributed in two major categories. On the one hand, a few principled approaches define a set of well-studied, high level ontology criteria to be manually assessed (e.g., OntoClean [8], Ontometric [11]). On the other hand, *automatic* approaches cover different evaluation perspectives (coverage of a corpus, similarity to a gold standard ontology) and levels (e.g., labels, conceptual structure). Common to these approaches is that they focus on evaluating an

---

[8] http://www.cs.technion.ac.il/~gabr/resources/data/wordsim353/
wordsim353.html

ontology as a whole rather than on assessing the correctness of a given relation as we do in this work.

7   Conclusions and Future Work

In this paper we investigated the problem of predicting the correctness of a semantic relation. Our hypothesis was that the Semantic Web can be used as a source of knowledge for this task and that existing NLP paradigms can be adapted to explore online ontologies.

Based on our experimental results, we can conclude that the Semantic Web is a promising source of information for addressing the relation evaluation problem. Indeed, a combination of our measures which explore ontologies as knowledge artifacts lead to an average precision value of 75% (with an individual result of 94% for one of the datasets). Our results have also shown that the measures inspired from different paradigms had varying performance. The measures that explored the knowledge provided by ontologies outperformed those that regarded the Semantic Web as a corpus. A simple explanation could be the still sparse nature of the Semantic Web which hampers its meaningful use as a corpus. Our future work will focus in trying to enhance and combine the methods from these two paradigms, as well as complementing them with other sources than the SW.

Additionally to our conclusions, we observe a potential of using the proposed measures for evaluating ontology characteristics. For example, in the case of a relation that is derived from paths of different lengths, we can conclude that the ontology which leads to the shorter path is more concise (less detailed) than the one which leads to a longer derivation path. While valuable, such estimations of conceptual complexity have been difficult to capture with current ontology evaluation measures such as those described in [1].

In this work we have taken some simplifying assumptions which will be revisited during future work. Firstly, we gave a broad definition of correctness without distinguishing between different types of correct or incorrect relations. In future work we plan to identify and individually investigate different types of correct/incorrect relations. Secondly, when counting named relations we have assumed that a relation can have a single lexicalization. This assumption is however not verified in a minimal number of cases when a given semantic relation is present with different labels. Finally, in the case of path based measures we have given the same weight to each relation within a path, although, it is well-known from NLP, that even within the same ontology, different relations often cover different conceptual distances and should be weighted differently [3]. Our ongoing work explores ontology structure characteristics (e.g., depth, breadth) as a way to predict the granularity of the conceptual space covered by relations.

Acknowledgements

Work funded by the NeOn IST-FF6-027595 and X-Media IST-FP6-026978 projects.

References

1. Alani, H., Brewster, C.: Ontology Ranking based on the Analysis of Concept Structures. In: Proc. of the Third Int. Conf. on Knowledge Capture. ACM, New York (2005)
2. Brank, J., Grobelnik, M., Mladenic, D.: A survey of ontology evaluation techniques. In: Proc. of the Conf. on Data Mining and Data Warehouses (2005)
3. Budanitsky, A., Hirst, G.: Evaluating WordNet-based measures of semantic distance. Computational Linguistics 32(1), 13–47 (2006)
4. Calibrasi, R.L., Vitanyi, P.M.: The Google Similarity Distance. IEEE Transactions on Knowledge and Data Engineering 19(3), 370–383 (2007)
5. Cimiano, P.: Ontology Learning and Population from Text: Algorithms, Evaluation and Applications. Springer, Heidelberg (2006)
6. d'Aquin, M., Motta, E., Sabou, M., Angeletou, S., Gridinoc, L., Lopez, V., Guidi, D.: Towards a New Generation of Semantic Web Applications. IEEE Intelligent Systems 23(3), 20–28 (2008)
7. Euzenat, J., Shvaiko, P.: Ontology Matching. Springer, Heidelberg (2007)
8. Guarino, N., Welty, C.A.: An Overview of OntoClean. In: Staab, S., Studer, R. (eds.) Handbook on Ontologies. Springer, Heidelberg (2004)
9. Hartmann, J., Sure, Y., Giboin, A., Maynard, D., Suarez-Figueroa, M.C., Cuel, R.: Methods for ontology evaluation. Knowledge Web Deliverable D1.2.3 (2005)
10. Lin, D.: An information-theoretic definition of similarity. In: Proc. of the 15th Int. Conf. on Machine Learning (1998)
11. Lozano-Tello, A., Gomez-Perez, A.: ONTOMETRIC: A Method to Choose the Appropriate Ontology. Journal of Database Management 15(2), 1–18 (2004)
12. Madche, A., Staab, S.: Measuring similarity between ontologies. In: Proc. of the European Conf. on Knowledge Acquisition and Management (2002)
13. Miller, G.A., Charles, W.G.: Contextual Correlates of Semantic Similarity. Language and Cognitive Processes 6(1), 1–28 (1991)
14. Mohammad, S., Hirst, G.: Distributional Measures as Proxies for Semantic Relatedness. Submitted for peer review
15. Sabou, M., d'Aquin, M., Motta, E.: Exploring the Semantic Web as Background Knowledge for Ontology Matching. Journal on Data Semantics XI (2008)
16. Sabou, M., Gracia, J.: Spider: Bringing Non-Equivalence Mappings to OAEI. In: Proc. of the Third International Workshop on Ontology Matching (2008)
17. Sabou, M., Gracia, J., Angeletou, S., d'Aquin, M., Motta, E.: Evaluating the Semantic Web: A Task-based Approach. In: Proc. of ISWC/ASWC (2007)
18. van Hage, W., Kolb, H., Schreiber, G.: A Method for Learning Part-Whole Relations. In: Proc. of the 5th Int. Semantic Web Conf. (2006)

On the Assessment of Text Corpora[*]

David Pinto[1], Paolo Rosso[2], and Héctor Jiménez-Salazar[3]

[1] Faculty of Computer Science,
B. Autonomous University of Puebla, Mexico
dpinto@cs.buap.mx
[2] Natural Language Engineering Lab. - ELiRF
Universidad Politécnica de Valencia, Spain
prosso@dsic.upv.es
[3] Department of Information Technologies,
Autonomous Metropolitan University, Mexico
hgimenezs@gmail.com

Abstract. Classifier-independent measures are important to assess the quality of corpora. In this paper we present supervised and unsupervised measures in order to analyse several data collections for studying the following features: *domain broadness*, *shortness*, *class imbalance*, and *stylometry*. We found that the investigated assessment measures may allow to evaluate the quality of gold standards. Moreover, they could also be useful for classification systems in order to take strategical decisions when tackling some specific text collections.

1 Introduction

Many algorithms devoted to document categorization have been tested on classical corpora such as Reuters and 20 Newsgroups in order to determine their quality. However, up to now the relative hardness of those corpora has not been completely determined.

The relative clustering hardness of a given corpus may be of high interest, since it would be helpful to determine whether or not the usual corpora used to benchmark the clustering algorithms are hard enough.

Moreover, when dealing with raw text corpora, if it is possible to find a set of features involved in the hardness of the clustering task itself, ad-hoc clustering methods may be used in order to improve the quality of the obtained clusters. Therefore, we believe that this study would be of high benefit.

In [1], the authors attempted to determine the relative hardness of different Reuters-21578 subsets by executing various supervised classifiers. However, in their research work it is not defined any measure for determining the hardness of these corpora, neither the possible set of features that could be involved in the process of calculating the relative hardness of some corpus.

---

[*] This research work was partially supported by the CICYT TIN2006-15265-C06 project.

H. Horacek et al. (Eds.): NLDB 2009, LNCS 5723, pp. 281–290, 2010.

The aim of our proposal is to evaluate classifier-independent features which could help on determining the hardness of a given corpus. As far as we know, research work in this field nearly have been carried out in literature.

For the purpose of our investigation, we took into account four different corpus features: *domain broadness, shortness, class imbalance,* and *stylometry*. We consider that these features will be sufficient to evaluate the relative hardness of a document collection. The aim is, for instance, to agree on whether the quality of the gold standard is good enough or not.

The description of the features to be investigated together with the corresponding assessing measures is given as follows.

Domain broadness. The goal is to evaluate the broadness of a given corpus. We assume (see for instance [2]) that it is easier to classify documents belonging to very different categories, for instance "sports" and "seeds", than those belonging to very similar ones, e.g. "barley" and "corn" (Reuters-21578). The attempt is to indicate the *domain broadness* degree of a given corpus. A binary classifier would assign, respectively, the tags *wide* to the former "sports-seeds" collection and *narrow* to the latter "barley-corn" one.

Shortness. The term frequency is crucial for the majority of the similarity measures. When dealing with very short texts, the frequency of their vocabulary is very low and, therefore, the clustering algorithms have the problem that the similarity matrix has very low values. Therefore, we believe that independently of the clustering method used, the average text length of the corpus to be clustered is an important feature that must be considered when evaluating its relative difficulty. The formula introduced by Herdan [3] has extensively been used for measuring lexical richness of documents [4] such as, vocabulary richness for authorship attribution [5].

Class imbalance. The document distribution across the corpus is another feature that we consider important to take into account. There may exist different levels of difficulties depending on whether the corpus is balanced or not. This feature is even more relevant when the corpus is used with the purpose of benchmarking different classifiers, for instance in the different tasks of an international competition such as SemEval[1]. Let us suppose that the corpus is totally unbalanced and, that for some reason exists a clue of that. If so, then some participants would "wisely" force their system to obtain the least possible number of clusters in order to get the best performance (unfair for the rest of the teams). The *imbalance* degree of a given corpus is also closely-related to the external corpus validation measure used (e.g. *F*-Measure) and, therefore, the obtention of a single value for measuring it will clearly be of high benefit. Two research works that deal with the problem of class imbalance are the ones presented in [6] and [7]. Particularly, in the former paper it is claimed that class (category) imbalances hinder the performance of standard classifiers.

Stylometry. It refers to the linguistic style of a writer. The goal is to determine the authorship of a set of documents. Even if in our case, the aim

---

[1] http://nlp.cs.swarthmore.edu/semeval/

is not to attribute the authorship but to distinguish between scientific and other kind of texts. Due to the specific writing style of researchers, when the collection to be clustered is scientific then a new level of difficulty arises. This observation has its basis in domain-dependent vocabulary terms that are not considered in the pre-processing step (for example, in the elimination of stopwords phase). There have been carried out several approaches on the statistical study of writing style (stylometry) field [8]. Morover, up to now, it stills an active research area [9,10].

The rest of this paper is structured as follows. The following section describes the measures we have used in the study of assessment of the quality of text corpora. Section 3 presents the evaluation of standard corpora used in the task of categorization. Finally, the conclusions are given.

2   Corpus Assessment Measures

The supervised vs. unsupervised nature way of measuring each of the mentioned corpus features is very important. Some measures evaluate the gold standard of the target corpus and, therefore, they are devoted to evaluate the classification given by the "experts". The other evaluations are meant to be obtained without any knowledge of the distribution of the documents and, therefore, they may be used to either evaluate general features of the collection or to improve, for instance, clustering results from an unsupervised viewpoint.

In the following sub-sections we present both, the supervised and unsupervised versions of the previously introduced corpus features.

2.1   Domain Broadness Evaluation Measures

In this approach, we assume (see for instance [2]) that it is easier to classify documents belonging to very different categories, for instance "sports" and "seeds", than those belonging to very similar ones, e.g. "barley" and "corn" (Reuters-21578). The attempt is to indicate the *domain broadness* degree of a given corpus. A binary classifier would assign, respectively, the tags *wide* to the former "sports-seeds" collection and *narrow* to the latter "barley-corn" one.

Using statistical language modeling. The first approach presented for the assessment of gold standards makes use of Statistical Language Modeling (SLM) in order to calculate probabilities of sequences of words in the different classes of a gold standard and, thereafter, to determine the domain broadness degree of the corresponding corpus by using two different variants, namely supervised and unsupervised.

SLM is commonly used in different natural language application areas such as machine translation, part-of-speech tagging, information retrieval, etc [11,12,13]. However, it has been originally known by its use in speech recognition (see for instance [14]) which stills the most important application area.

Informally speaking, the goal of SLM consists in building a statistical language model in order to estimate the distribution of words/strings of natural language. The calculated probability distribution over strings S of length n, also called n-grams, attempts to reflect the relative frequency in which S occurs as a sentence. In this way, from a text-based perspective, such a model tries to capture the writing features of a language in order to predict the next word given a sequence of them.

In our particular case, we have considered that every hand-tagged category of a given corpus would have a language model. Therefore, if this model is very similar to the rest of them which were calculated from the remaining categories, then we could affirm that the corpus is narrow domain. Our proposal approaches also in an unsupervised way the problem of determining the domain broadness of a given corpus by calculating language models for v partitions of the corpus without any knowledge about the expert document categorization. However, due to the fact that the perplexity is by definition dependent on the text itself, we should make sure that the text chosen is representative of the entire corpus [15].

Given a corpus made up of k categories $C = \{C_1, C_2, \cdots, C_k\}$, we obtain the language model of all the categories except C_i ($\bar{C}_i$) and, thereafter, we compute the perplexity (PP) of the obtained language model with respect to the model of C_i. That is, we use the category C_i as a test corpus and the remaining ones as a training corpus in a leave one out process. Formally, the *Supervised* Language Modeling Based (SLMB) approach for determining the domain broadness degree of the corpus C may be obtained as shown in Eq. (1).

$$SLMB(C) = \sqrt{\frac{1}{k} \sum_{i=1}^{k} \left(PP(C_i|\bar{C}_i) - \mu(PP(C|\bar{C}_i)) \right)^2} \qquad (1)$$

where

$$\mu(PP(C|\bar{C}_i)) = \frac{\sum_{i=1}^{k} PP(C_i|\bar{C}_i)}{k} \qquad (2)$$

The *Unsupervised* Language Modeling Based (ULMB) approach for assessing the domain broadness of a text corpus is computed as follows. Given a corpus C splitted into subsets C_i' of l documents, we calculate the perplexity of the language model of C_i' with respect to the model of a training corpus composed of all the documents not contained in C_i' ($\bar{C}_i'$). Formally, given $\bar{C}_i' \bigcup C_i' = C$ such as $\bar{C}_i' \bigcap C_i' = \emptyset$ and $k =$ Integer($\frac{|C|}{|C_i'|}$) with $|C_i'| \approx l$, the *unsupervised* broadness degree of a text corpus C may be obtained as shown in Eq. (3).

$$ULMB(C) = \sqrt{\frac{1}{k} \sum_{i=1}^{k} \left(PP(C_i'|\bar{C}_i') - \mu(PP(C|\bar{C}_i')) \right)^2} \qquad (3)$$

where

$$\mu(PP(C|\bar{C}_i')) = \frac{\sum_{i=1}^{k} PP(C_i'|\bar{C}_i')}{k} \qquad (4)$$

Using vocabulary dimensionality. This measure of calculating the domain broadness of a corpus assumes that those subsets belonging to a narrow domain will share the maximum number of vocabulary terms compared with the subsets which do not. In case of a wide domain corpus, it is expected that the standard deviation of vocabularies obtained from subsets of this corpus is greater than the one of a narrow domain corpus. We formalise the above mentioned idea as follows.

Given a corpus C (with vocabulary $V(C)$) which is made up of k categories C_i, the *Supervised* Vocabulary Based (SVB) measure for the domain broadness of C may be written as shown in Eq. (5).

$$SVB(C) = \sqrt{\frac{1}{k} \sum_{i=1}^{k} \left(\frac{|V(C_i)| - |V(C)|}{|C|} \right)^2} \qquad (5)$$

The Unsupervised version of the Vocabulary-Based (UVB) domain broadness evaluation measure would be useful when the gold standard is not available. Since the categories are unknown, we could then use each document (n) instead of the corpus categories (k). The *unsupervised* broadness evaluation measure (based on vocabulary dimensionality) of a corpus C made of n documents ($D_1, ..., D_n$) may be written as shown in Eq. (6).

$$UVB(C) = \sqrt{\frac{1}{n} \sum_{i=1}^{n} \left(\frac{|V(D_i)| - |V(C)|}{|C|} \right)^2} \qquad (6)$$

2.2 Shortness-Based Evaluation Measures

These evaluation measures assess features derived from the length of a text. Given a corpus C made up of n documents D_i, we present two *unsupervised* text length-based evaluation measures which take into account the level of shortness [3]. We directly calculated the arithmetic mean of Document Lengths (DL) and Vocabulary Lengths (VL) as shown in Eq. (7) and (8), respectively.

$$DL(C) = \frac{1}{n} \sum_{i=1}^{n} |D_i| \qquad (7)$$

$$VL(C) = \frac{1}{n} \sum_{i=1}^{n} |V(D_i)| \qquad (8)$$

2.3 Class Imbalance Degree Evaluation Measure

The class *imbalance* degree is an important feature that must be considered when corpora are categorized, since according to the imbalance degree there could exist different levels of difficulty [6]. This feature is even more relevant when the corpus is used for benchmarking different classifiers. Let us suppose that the corpus is totally unbalanced and, that for some reason there exist some clue of that. This fact could lead some participants to force their system to obtain the least possible number of clusters in order to get the best performance. In these conditions it would be quite difficult to carry out a fair evaluation and, therefore, to determine which is(are) the best system(s).

Given a corpus C (made of n documents) with a pre-defined gold standard composed of k classes (C_i), the Expected Number of Documents per Class is assumed to be: $ENDC(C) = \frac{n}{k}$.

The *supervised* Class Imbalance (CI) evaluation measure is calculated as the standard deviation of C with respect to the expected number of documents per class in the gold standard as shown in Eq. (9).

$$CI(C) = \sqrt{\frac{1}{k} \sum_{i=1}^{k} \left(|C_i| - ENDC(C) \right)^2} \tag{9}$$

2.4 Stylometric-Based Evaluation Measure

The aim of this measure is to determine whether a corpus is written with the same linguistic style or not.

For the analysis of slylometry introduced here, we make use of the Zipf law [16]. Formally, given a corpus C with vocabulary $V(C)$, we may calculate the probability of each term t_i in $V(C)$ as shown in Eq. (10) and the expected Zipfian distribution of terms as shown in Eq. (11). We used the classic version of the Zipf's law and, therefore, s was set to 1.

$$P(t_i) = \frac{freq(t_i, C)}{\sum_{t_i \in V(C)} freq(t_i, C)} \tag{10}$$

$$Q(t_i) = \frac{1/i^s}{\sum_{r=1}^{|V(C)|} 1/r^s} \tag{11}$$

The *unsupervised* Stylometric Evaluation Measure (SEM) of C is obtained by calculating the asymmetrical Kullback-Leibler distance of the term frequency distribution of C with respect to its Zipfian distribution, as shown in Eq. (12).

$$SEM(C) = \sum_{t_i \in V(C)} P(t_i) log \frac{P(t_i)}{Q(t_i)} \tag{12}$$

A summary of the presented assessment corpus measures is given in Table 1.

Table 1. Text corpora assessing measures

Short name	Description	Category	Approach
$SLMB(C)$	Language model perplexity	Broadness	Supervised
$ULMB(C)$	Language model perplexity	Broadness	Unsupervised
$SVB(C)$	Vocabulary of categories	Broadness	Supervised
$UVB(C)$	Vocabulary of document	Broadness	Unsupervised
$DL(C)$	Document length	Shortness	Unsupervised
$VL(C)$	Vocabulary size	Shortness	Unsupervised
$CI(C)$	Document distribution	Imbalance	Supervised
$SEM(C)$	Zipfian based distribution	Stylometric	Unsupervised

3 Computational Study of Testbed Corpora

The experiments were carried out over the following corpora: *WebKB* [17], *CICLing-2002* [18], *hep-ex* [7], *20 Newsgroups* (20NG) and the *R8* and *R52* subsets of the Reuters-21578 text categorization collections. Moreover, the 100 corpora which compose the *WSI-SemEval* collection were also used [19].

In order to assess how well each assessment measure performs, we have correlated the automatic ranking of the corpora (according to each measure) with respect to an expert manual ranking.

We have considered that there are no tied ranks and we have not made any assumptions about the frequency distribution of the evaluation measures. Moreover, the equi-distance between the different corpora evaluation value cannot be justified and, therefore, the correlation was calculated by means of the *Kendall tau* (τ) rank correlation coefficient [20].

$$\tau = \frac{2 \cdot P}{(e \cdot (e-1))/2} - 1 \tag{13}$$

where e is the number of items, and P is the number of concordant pairs obtained as the sum, over all the items, of those items ranked after the given item by both rankings.

This coefficient value lies between -1 and 1, and high values imply a high agreement between the two rankings. Therefore, if the agreement (disagreement) between the two rankings is perfect, then the coefficient will have the value of 1 (-1).

Tables 2 and 3 illustrate all the evaluation measures with the corresponding obtained value for each one of the evaluated corpus. Aside of each measure we may also see the associated manual ranking which is used to evaluate their performance.

The correlation results (see Table 4) show a high agreement between the automatic and manual corpus rankings for each one of the analysed measures (over 109 corpora). The lowest value (0.56) was obtained for two unsupervised measures ($ULMB(C)$ and $UVB(C)$). Therefore, we consider to have a good

Table 2. Corpus assessment measures for domain broadness

Corpus	$SLMB(C)$	$ULMB(C)$	$SVB(C)$	$UVB(C)$
CICLing-2002	38.9 / 1	63.6 / 1	1.73 / 1	2.70 / 1
hep-ex	298.2 / 2	93.8 / 2	2.75 / 2	3.07 / 2
WSI-SemEval	195.0 / 3	130.6 / 3	1.80 / 3	3.06 / 3
WebKb-Training	262.3 / 5	628.6 / 5	0.50 / 5	1.77 / 5
WebKb-Test	337.4 / 4	218.9 / 4	0.44 / 4	1.60 / 4
R52-Training	627.6 / 9	143.1 / 9	4.38 / 9	4.62 / 9
R52-Test	565.8 / 8	177.5 / 8	4.58 / 8	4.82 / 8
R8-Training	603.9 / 7	135.9 / 7	3.67 / 7	4.76 / 7
R8-Test	545.7 / 6	134.6 / 6	3.84 / 6	4.89 / 6
20NG-Training	694.4 / 11	400.2 / 11	5.23 / 11	6.08 / 11
20NG-Test	786.0 / 10	455.4 / 10	5.21 / 10	6.05 / 10

Table 3. Corpus assessment measures (stylometry, shortness and class imbalance)

Corpus	$SEM(C)$	$DL(C)$	$VL(C)$	$CI(C)$
CICLing-2002	0.301 / 11	70.5 / 7	48.4 / 7	0.036 / 3
hep-ex	0.271 / 10	46.5 / 1	36.8 / 1	0.280 / 11
WSI-SemEval	0.448 / 9	59.6 / 2	50.3 / 2	0.226 / 10
WebKb-Training	0.231 / 8	133.7 / 9	77.1 / 9	0.096 / 6
WebKb-Test	0.227 / 7	136.2 / 8	79.4 / 8	0.097 / 7
R52-Training	0.159 / 5	70.3 / 6	43.1 / 6	0.067 / 4
R52-Test	0.120 / 2	64.3 / 4	39.7 / 4	0.068 / 5
R8-Training	0.142 / 4	66.3 / 5	41.2 / 5	0.171 / 9
R8-Test	0.098 / 1	60.1 / 3	37.3 / 3	0.169 / 8
20NG-Training	0.154 / 6	142.7 / 11	84.3 / 11	0.004 / 1
20NG-Test	0.144 / 3	138.7 / 10	83.2 / 10	0.005 / 2

trade-off between the unsupervised characteristic and the relatively low Kendall tau value obtained by $ULMB$ and UVB. In particular, the time needed for calculating the $ULMB$ measure in huge collections may be prohibitive, but this issue may be alleviated by using sampling over the complete data set.

All the assessment measures were compiled in a on-line system which we have made available for all interested researchers[2]. Therefore, the assessment measures may be used not only to evaluate other corpora but to compare the results with the standard corpora already evaluated and presented in this research work.

Table 4. Correlation between the automatically and manually obtained ranking

Assessment measure	τ value
$SLMB(C)$	0.82
$ULMB(C)$	0.56
$SVB(C)$	0.67
$UVB(C)$	0.56
$SEM(C)$	0.86
$DL(C)$	0.96
$VL(C)$	0.78
$CI(C)$	1.00

4   Conclusions

We have presented a set of corpus evaluation measures that may be used to either, evaluate gold standards or to make decisions a priori when, for instance, clustering particular kinds of text collections such as, narrow domain short-text corpora [18].

All the proposed measures were executed over several corpora in order to determine their evaluation capability. We ranked each corpus according to the evaluation value given by the corresponding measure and, thereafter, we calculated the Kendall tau correlation coefficient in order to determine the correlation

---

[2] http://nlp.dsic.upv.es:8080/watermarker; http://nlp.cs.buap.mx/watermarker

degree between the automatically and the manually obtained ranking. Our findings indicate a strong agreement of all the evaluation measures with respect to the manual ranking.

The developed quality corpus analysis system would allow researches in different fields of linguistics and computational linguistics to easily assess their corpora with respect to the aforementioned corpus features.

References

1. Debole, F., Sebastiani, F.: An analysis of the relative hardness of Reuters-21578 subsets. Journal of the American Society for Information Science and Technology 56(6), 584–596 (2005)
2. Wibowo, W., Williams, H.: On using hierarchies for document classification. In: Proc. of the Australian Document Computing Symposium, pp. 31–37 (1999)
3. Herdan, G.: Type-Token Mathematics: A Textbook of Mathematical Linguistics. Mouton & Co., The Hague (1960)
4. Tweedie, F.J., Baayen, R.H.: How variable may a constant be?: Measures of lexical richness in perspective. Computers and the Humanities 32(5), 323–352 (1998)
5. Hoover, D.L.: Another perspective on vocabulary richness. Computers and the Humanities 37(2), 151–178 (2004)
6. Japkowicz, N.: The class imbalance problem: Significance and strategies. In: Proc. of the 2000 International Conference on Artificial Intelligence (IC-AI 2000), vol. 1, pp. 111–117 (2000)
7. Montejo-Ráez, A.: Automatic text categorization of documents in the High Energy Physics domain. Phd thesis, Granada University, Spain (2006)
8. Diederich, J., Kindermann, J., Leopold, E., Paass, G.: Authorship attribution with support vector machines. Applied Intelligence 19(1-2), 109–123 (2004)
9. Can, F., Patton, J.M.: Change of writing style with time. Computers and the Humanities 38(1), 61–82 (2004)
10. Hoover, D.L.: Corpus stylistics, stylometry, and the styles of henry james. Style 41(2), 174–203 (2007)
11. Brants, T., Popat, A.C., Xu, P., Och, F.J., Dean, J.: Large language models in machine translation. In: Proc. of the 2007 Joint Conference on Empirical Methods in Natural Language Processing and Computational Natural Language Learning (EMNLP-CoNLL), pp. 858–867 (2007)
12. Màrquez, L., Padró, L.: A flexible pos tagger using an automatically acquired language model. In: Proc. of the 35th annual meeting on Association for Computational Linguistics, pp. 238–245 (1997)
13. Ponte, J.M., Croft, W.B.: A language modeling approach to information retrieval. In: Research and Development in Information Retrieval, pp. 275–281 (1998)
14. Bahl, L.R., Jelinek, E., Mercer, R.L.: A maximum likelihood approach to continuous speech recognition. IEEE Transactions on Pattern Analysis and Machine Intelligence 5(2), 179–190 (1983)
15. Brown, P.F., Pietra, V.J.D., de Souza, P.V., Lai, J.C., Mercer, R.L.: Class-based n-gram models of natural language. Computational Linguistics 18(4), 467–479 (1992)
16. Zipf, G.K.: Human behaviour and the principle of least effort. Addison-Wesley, Reading (1949)
17. Cardoso-Cachopo, A., Oliveira, A.: Combining LSI with other classifiers to improve accuracy of single-label text categorization. In: First European Workshop on Latent Semantic Analysis in Technology Enhanced Learning - EWLSATEL 2007 (2007)

18. Pinto, D., Benedí, J.M., Rosso, P.: Clustering narrow-domain short texts by using the Kullback-Leibler distance. In: Gelbukh, A. (ed.) CICLing 2007. LNCS, vol. 4394, pp. 611–622. Springer, Heidelberg (2007)
19. Agirre, E., Soroa, A.: Semeval-2007 task 2: Evaluating word sense induction and discrimination systems. In: Proc. of the 4th International Workshop on Semantic Evaluations - SemEval 2007, pp. 7–12. Association for Computational Linguistics (2007)
20. Kendall, M.: A new measure of rank correlation. Biometrika 30, 81–89 (1938)

Improving Semantic Web Applications with Navigational Semantics

Jesús M. Hermida, Andrés Montoyo, and Jaime Gómez

Department of Software and Computing Systems, University of Alicante, Spain
{jhermida,montoyo,jgomez}@dlsi.ua.es

Abstract. This paper presents a new approach that includes semantic annotations of navigational information in Semantic Web Applications using as core resource a new ontology, NavOntology. Our purpose is using these annotations to retrieve just the information you are going to use from the Web. Finally, conclusions have been obtained from an initial case study.

Motivation. The SHOE project [1] designed one of the first methods of semantic annotation, which used its own ontologies to annotate static Web pages by means of an HTML extension. Since the architecture of the Semantic Web [2,3] standarised Web Ontology Language (OWL), the task of semantic annotation has not evolved drastically even though the use of ontologies opens new opportunities. Therefore, this paper introduces a new proposal of semantic annotation for Web applications in the Semantic Web, which uses different ontologies to describe their domain and their hypermedial structure.

Managing Navigational Semantics in Semantic Web Applications. To the best of our knowledge, the concept of semantic Web application (SWA) is ambiguous in bibliography. In this work, a SWA is a Web application able to automatically generate semantic annotations of its own information in a computer-understandable way. From our viewpoint, SWAs provide two benefits: *(i)* the improvement of the results in retrieving web information and *(ii)* providing its contents as a resource for other tasks, for instance, NLP tasks.

However, the use of ontologies in these Web pages is poorly exploited, other intrinsic aspects of a Web application can be represented thus improving the usability of the Web. At present, search agents do not understand where they are or how many Web pages they have left until finishing a search in one specific domain. In order to simplify the navigation of the agents in Web applications, we introduce *NavOntology*, an OWL ontology that captures the semantics of the Web navigation in a user-oriented point of view.

Organising in our concept of SWA, each Web page contains: *(i)* HTML code, the visualisation of the Web page; *(ii)* domain semantics (OWL code), the description of its data using instances of a domain ontology; *(iii)* navigational semantics (instances of NavOntology, OWL code), information about itself, its homepage, and those Web pages and links inside a *navigational window* of

H. Horacek et al. (Eds.): NLDB 2009, LNCS 5723, pp. 291–292, 2010.

certain depth (a new concept which determines the accessibility to the semantics of linked Web pages).

Case Study: DLSI Semantic Web Application. Applying this proposal in a case study is essential to extract valid conclusions. The selected case study was a Web application for the management of the information of our department (Spanish acronym, DLSI). Its domain was restricted to those lecturers, degrees and subjects related with this department.

Once the Web application was manually implemented, we checked its performance by simulating the behaviour of an agent when looking for information. With this purpose, we created a collection of specific questions, such as, *information about a subject whose name is "Analysis and Specification of Information Systems"*.

Discussion. From this initial case study, we can carry out a first analysis of the advantages and shortcomings of the proposal. The main benefits of the proposal are: *(i)* ontology-driven web navigation without processing HTML; *(ii)* domain information anticipation through the instances of NavOntology; and *(iii)* its scalability.

However, there are some shortcomings: *(i)* partial vision of the Web container from a single Web page; *(ii)* resources needed to generate the semantic representation of the page; and *(iii)* semantic generation becomes more complex (it would be necessary to create or adapt model-based design methods to facilitate its development).

Conclusions and Future Work. This paper proposes a new method for the inclusion of semantics in Web applications taking into consideration aspects of the navigational context where this information is shown. NavOntology provides new mechanisms for retrieving and extracting information in a "intelligent approach". It is necessary to apply it in wider scenarios to have an accurate measure of scalability and performance.

Our future work is focused on two aspects: *(i)* refining the current proposal and improving the shortcomings with real case studies; and *(ii)* studying the adaptation of a design method (e.g. OO-H[4], WebML) to the design of SWAs.

Acknowledgements. The first author has been funded by the FPU program (AP2007-03076) of the Spanish Ministry of Science and Innovation.

References

1. Heflin, J., Hendler, J., Luke, S.: Shoe: A knowledge representation language for internet applications. Technical report (1999)
2. Berners-Lee, T., Hendler, J., Lassila, O.: The Semantic Web. Scientific American 248(5), 34–43 (2001)
3. Koivunen, M.R., Miller, E.: W3C Semantic Web Activity. In: Hyvönen, E. (ed.) Semantic Web Kick-Off in Finland. Vision, Technologies, Research and Applications, pp. 27–43 (2002)
4. Gómez, J., Cachero, C., Pastor, O.: Conceptual Modeling of Device-Independent Web Applications. IEEE MultiMedia 8(2), 26–39 (2001)

Hybrid Algorithm for Word-Level Alignment of Parallel Texts[*]

Eduardo Cendejas, Grettel Barceló, Alexander Gelbukh, and Grigori Sidorov

Center for Computing Research, National Polytechnic Institute, Mexico City, Mexico
{ecendejasa07,gbarceloa07}@sagitario.cic.ipn.mx
http://www.gelbukh.com, http://cic.ipn.mx/~sidorov

Abstract. Given a text in two languages, word alignment task consists of identifying in the two variants of the text specific word occurrences that are mutual translations. The majority of existing text alignment systems follow either a linguistic or a statistical approach. We argue for that both approaches are insufficient when used separately, and suggest a flexible algorithm that combines statistical and linguistic techniques.

1 Introduction

Given a bilingual corpus, the text alignment task establishes a correspondence between structures, in the two languages. There are two main alignment approaches: linguistic and statistical. Linguistic approaches use linguistic information, which implies its reliance on availability of resources. Statistical approaches are based on frequencies of occurrences, though they imply a lower computational cost but usually give lower precision.

Ideally, the units (words, sentences, paragraphs) of the two texts ought to be in direct one to one correspondence. However, the alignment task is complicated by many effects that break such an ideal model. There are many reasons for which word alignment is more difficult than alignment of others units, e.g.: the degree of inflectivity of the two languages, syntactic structure of the two languages, models employed for alignment and the available linguistic resources [1,2].

2 Alignment Algorithm

The proposed alignment algorithm combines statistical and linguistic approaches to reduce the disadvantages that both approaches present when used independently. The statistical stage of the proposed system relies on three different techniques: boolean modified K-Vec algorithm, modified K-Vec algorithm with frequencies and IBM model 2. The statistical processing includes:

1. Segmentation of the input texts. The original K-Vec algorithm allows the text to be divided into small pieces or segments. Our modification allows the pieces to be paragraphs, sentences, or a specific number of words (a window).

---

[*] Work done under partial support of Mexican Government (SIP-IPN 20091587 and 20090772, CONACYT 50206-H and 83270, SNI, PIFI-IPN).

H. Horacek et al. (Eds.): NLDB 2009, LNCS 5723, pp. 293–294, 2010.

2. Generating a list of words with associated vectors. A vector contains the occurrences (boolean values or the frequency) of the word in the segments.
3. Construction of a contingency table. The vector corresponding to each word in the source language is compared to all the vectors in the translation.
4. Calculation of similarity for each pair in each table. The similarity of words is determined by means of an association test (Pointwise Mutual Information, T-score, Log-likelihood ratio and Dice coefficient).
5. Selection of the word with the greatest level of association. The other candidates in the table are discarded as translations.

If the algorithm is used in a bidirectional way, the same process is carried out interchanging the languages. The linguistic processing incorporates: (1) dictionaries, to extract lexical information, (2) lexicons with morphological information, to compare lemmas and verify grammatical categories, (3) syntactic trees, for identification of its parts – subject, predicate, etc. – and to facilitate comparisons with its counterparts in the target language, (4) semantic domains, to establish semantic relations between the meanings of a word, (5) cognates, to align words that totally or partially coincide, and (6) learning: all alignment hypotheses that can be obtained will serve as a reference for future alignment tasks.

3　Preliminary Results

We used fragments (Spanish–English) from five novels. The following table shows the obtained results. We used a modified K-Vec procedure, and we applied the cognates during the linguistic processing to reinforce the alignments.

Aligner	precision (%)	generates a dictionary?	generates alignment file?
K-Vec++	35	Not for all words	No
Uplug	45	Not as a result	Yes
GIZA++	52	Yes	Yes
Proposed algorithm	**53**	Yes	Yes

4　Conclusions

Statistical techniques provide a good starting point for the alignment processes; however, incorporation of linguistic techniques improves the quality by involving intrinsic characteristics of the involved languages. The main disadvantage of linguistic processing is the need in the linguistic resources given their limited availability. In addition, this has an impact on the algorithm speed. Nonetheless, employing databases of optimization has proved to minimize this disadvantage.

References

1. Borin, L.: You'll take the high road and i'll take the low road: Using a third language to improve bilingual word alignment. In: ACL 2000, vol. 1, pp. 97–103 (2000)
2. Mihalca, R., Pedersen, T.: An evaluation exercise for word alignment. In: HLT-NAACL 2003 Workshop on Building and using parallel texts, vol. 3, pp. 1–10 (2003)

A Hybrid Approach for Data Mart Schema Design from NL-OLAP Requirements

Fahmi Bargui, Hanene Ben-Abdallah, and Jamel Feki

Mir@cl Laboratory,
FSEG, University of Sfax Tunisia, Po Box *1088*
{fahmi.bargui,jamel.feki,hanene.benabdallah}@fsegs.rnu.tn

1 Introduction

OLAP systems remain difficult to develop for two primary reasons. The first reason is that the REQ analysis step is typically overlooked [1]. The second reason is that, even when attention is accorded to this step, REQ are described from a too technical perspective for decision makers to understand and validate [2]. However, the lack of user involvement in the specification can easily lead to the failure of the project. To ensure the involvement of decision makers in REQ specification, two solutions are offered: non technical notations for OLAP REQ expression (*cf.*, tabular format [3]) and approaches to assist decision makers in identifying their REQ (*cf.*, goal-driven approaches with either a graphical notation [1] or SQL scenarios [4], reuse of Multi-Dimensional (MD) patterns representing generic OLAP REQ [5]). Despite the assistance they offer, the proposed REQ notations and specification approaches presume that the decision makers are familiar with the MD concepts and their relationships.

Within this context, our proposed solution (Fig.1) aims at facilitating the involvement of decision makers in the REQ specification step by offering a NL based notation for expressing the OLAP REQ, and a validation approach to ensure that the specified REQ are realizable. As illustrated in Fig.1, our approach relies on a *template* for OLAP REQ identification [6]. This template is composed of elements derived from the decision making process to identify the analyzed business process, the relevant actors, the indicators formulas, and the analytical queries used for decision making.

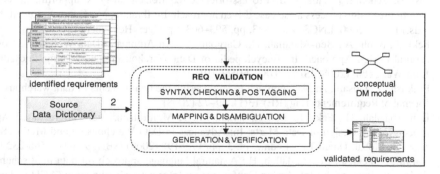

Fig. 1. Our approach for OLAP REQ acquisition, verification and validation

H. Horacek et al. (Eds.): NLDB 2009, LNCS 5723, pp. 295–296, 2010.

This template has four advantages: it has a simple structure; decision makers are familiar with its components; its analytical queries are expressed in NL and, thus, it does not presume any MD modeling knowledge; and its components can be transformed to a MD schema. In addition, to provide for an automated MD schema derivation process, we have identified empirically a linguistic pattern for the analytical queries. This pattern is used to verify the coherence of the queries, to validate the REQ with respect to a given data source, and to propose enrichments of the specified REQ with all the potential analysis information in a given data source. Thus, the final step of our REQ specification approach produces a set of coherent, validated and enriched OLAP REQ that are mapped to a given source. The mapping is vital information for the ETL procedure definition and the realization of the REQ.

2 Conclusions and Future Work

In this paper, we have presented an approach for the extraction and validation of MD concepts from OLAP REQ specified in NL. To automate this approach, we proposed a template for REQ acquisition that covers the concepts of the decision making process. In addition, this template allows a decision maker to express their analytical queries according to a linguistic pattern that we defined empirically. Once specified, the analytical queries are parsed to extract pertinent terms that could be either dimensions or parameters. To decide on the MD type of a term, our process applies a set of matching and expansion rules on the data source represented through its data dictionary. The term-data entry matching provides for: validation of the conceptual class of the query terms, verification of the functional dependencies required among the MD concepts, and traceability to the data source.

We are currently finalizing the development of a CASE toolset supporting our extraction and validation approach. In addition, we are defining the generation step of the conceptual data mart schema from validated OLAP requirements.

References

1. Giorgini, P., Rizzi, S., Garzetti, M.: GRAnd: A goal-oriented approach to requirement analysis in data warehouses. Decision Support Systems 45(1), 4–21 (2007)
2. List, B., Scheifer, J., Tjoa, A.M.: Process-oriented requirement analysis supporting the data warehouse design process a use case driven approach. In: Ibrahim, M., Küng, J., Revell, N. (eds.) DEXA 2000. LNCS, vol. 1873, pp. 593–603. Springer, Heidelberg (2000)
3. Feki, J., Nabli, A., Ben-Abdallah, H., Gargouri, F.: An Automatic Data Warehouse Conceptual Design Approach. In: Encyclopedia of Data Warehousing and Mining, 2nd edn., John Wang edn. (2008) ISBN: 978-1-60566-010-3
4. Prakash, N., Gosain, A.: An approach to engineering the requirements of data warehouses. Journal of Requirements Eng (RE) 13(1), 49–72 (2008)
5. Ben-Abdallah, H., Feki, J., Ben Abdallah, M.: A Multidimensional Pattern based Approach for the Design of Data Marts. In: Advances in Data Warehousing and Mining Series, IGI Global. David Taniar edn., vol. 3, pp. 172–192 (2009) ISBN: 978-1-60566-232-9
6. Bargui, F., Feki, J., Ben-Abdallah, H.: A natural language approach for data mart schema design. In: Proc. of 9th Int. Arabic Conference on Information Technology (ACIT), Hammamet-Tunisia, December 16-18 (2008)

Spreadsheet Information Retrieval through Natural Language

Derek Flood, Kevin McDaid, and Fergal McCaffery

Dundalk Institute of Technology
derek.flood@dkit.ie, kevin.mcdaid@dkit.ie,
fergal.mccaffery@dkit.ie

Abstract. The versatility of spreadsheet applications have enabled them to be used for a wide variety of tasks, from inventory management to financial modeling. When they are used for inventory management, extracting additional information can be a difficult task, especially for novice spreadsheet users. Typically, novice users have little experience of using the advanced data processing tools that are available in most spreadsheet applications. To address this issue, the authors propose the use of Natural Language Processing (NLP) techniques to enable such users to perform these tasks. A recent evaluation of NLP-SIR (Natural Language Processing for Spreadsheet Information Retrieval) has shown that natural language can be more effective than conventional data processing tools.

1 Introduction

Spreadsheets are a versatile software tool. Microsoft© Excel a leading spreadsheet application is estimated to be installed on over 150 million desktops worldwide[1]. They allow users to perform a wide variety of tasks such as modelling financial situations, running complex simulations and inventory management. When storing considerable amounts of data, such as in inventory management, extracting information can be a difficult task. Features such as PivotTables and Filters, included in most spreadsheet applications, can assist users in these tasks however for novice spreadsheet users these features can be difficult to use. One way to address this issue is through the use of Natural Language Processing.

Natural Language Processing (NLP) is a set of techniques which allow computers to analyze, understand, and generate languages which are used naturally. NLP has been demonstrated to simplify the way users interact with a variety of applications such as Databases[2] and Chart Generation[3].

2 NLP-SIR

NLP-SIR (Natural Language Processing for Spreadsheet Information Retrieval) is a Natural Language Interface (NLI) which allows users to perform common information retrieval tasks, such as filtering data and generating summary tables, using natural language.

The current system has some limitations. When users are referring to a column within the spreadsheet they must use the column heading as it appears in the spreadsheet. The

H. Horacek et al. (Eds.): NLDB 2009, LNCS 5723, pp. 297–298, 2010.

present version of the system does not recognise synonyms however it is hoped that future versions of the system will accommodate this. NLP-SIR is designed to interpret positive statements. If a user issues a negative statement, such as "show all golf courses except those in Boone" the system will not interpret this correctly, it will incorrectly assume the user wants to see all of the courses in Boone.

3 Evaluation

In order to evaluate the effectiveness and efficiency of the NLP-SIR system, a controlled experiment was conducted. This experiment asked 41 Novice spreadsheet users to perform ten information retrieval tasks on a given spreadsheet. This spreadsheet contained a list of all of the golf courses in the American state of Indiana. The ten tasks were divided into 3 categories of difficulty; Easy, Intermediate and Complex, based on the difficulty of performing the task through the existing interface. The participants were randomly assigned into two groups. The first group were asked to use NLP-SIR system while the second was asked to perform the same tasks using the existing Microsoft© Excel interface.

To measure the effectiveness of the system, the number of tasks successfully completed on average by each group was sought. The group who used the NLP-SIR system were able to successfully complete, on average, 6.75 of the 10 tasks where as those using the existing interface only completed 3.43 of the tasks successfully.

The efficiency of the system was measured using the average time it took participants to complete each task. It was found that those using NLP-SIR completed each category of task faster than those using Microsoft© Excel. For the Easy tasks it was found that those using NLP-SIR performed the task in 11 seconds less than those using the traditional interface. As the complexity of the task increased it was found that the time saved by the NLP-SIR participants also increased. It was found that on the complex tasks these participants saved approximately 1 minute.

In order to evaluate the flexibility of the NLP-SIR Grammar, the average number of utterances used by participants was also evaluated. It was found that on average 1.97 utterances were used by each participant to complete each task.

The results obtained to date show that the NLP-SIR is an effective and efficient interface for spreadsheet information retrieval. The experiment did however highlight some issues with the current implementation. As well as the limitations above, it was found that using a separate window for the NLP-SIR interface was irritating to participants. It is hoped that future versions of the system will be embedded in the Microsoft© Excel interface as a task pane. It would also be useful for participants to see which rows from the data contributed to the results generated by NLP-SIR.

References

[1] Kugel, R.: Scoping out the Emerging Spreadsheet Management Market. Ventana Research, Article ID: V06-68 (2006)
[2] Liddy, E., Liddy, J.: An NLP approach for improving access to statistical information for the masses. In: The FCSM 2001 Research Conference (2001) (presented at)
[3] Kato, T., Matsushita, M., Maeda, E.: Answering it with charts - Dialogue in natural language and charts. In: Presented at The 19th international conference on Computational linguistics (2002)

Characterizing Weblog Corpora

Fernando Perez-Tellez[1], David Pinto[2], John Cardiff[1], and Paolo Rosso[3]

[1] Social Media Research Group, Institute of Technology Tallaght, Dublin, Ireland
fernandoperez@itnet.ie, John.Cardiff@ittdublin.ie
[2] Benemerita Universidad Autónoma de Puebla, Mexico
dpinto@cs.buap.mx
[3] Natural Language Engineering Lab. – EliRF, Dept. Sistemas Informáticos y Computación,
Universidad Politécnica Valencia, Spain
prosso@dsic.upv.es

1 Introduction

In order to exploit the huge volume of information being published in the blogosphere, it is essential to provide techniques such as clustering, which can automatically analyze and classify their contents. However these typically can produce better results when dealing with wide domain full-text documents. In most cases however, blogs can be considered to be "short texts", i.e., they are not extensive documents and exhibit undesirable characteristics from a clustering perspective such as low frequency terms, short vocabulary size and vocabulary overlapping of some domains. Furthermore, their characteristics vary widely depending on the specific interests of the writer, their linguistic style, and the volume of texts that they produce.

In this work, we present a set of evaluation features by which we can establish the relative hardness of the clustering task, i.e., how easy or difficult it will be to accurately cluster the blog datasets. These are the shortness, domain broadness, class imbalance, stylometry, and structure. We report results obtained on corpora extracted from two popular blogging sites, Boing Boing ("B-B") and Slashdot[1]. The results are contrasted with characterizations of a number of other corpora, consisting of newspaper articles and academic papers. We can use the results to provide knowledge of the most appropriate methodology for clustering.

2 Corpora Evaluation Measures

As established in [2], it is important for the evaluation measures to be domain-independent. We focus on the following corpus characteristics: *Domain Broadness* detects if two different categories should be tagged as "wide" or "narrow". *Class Imbalance* indicates how evenly the documents are distributed across the categories. *Stylometry* helps to distinguish between writings styles of authors. *Shortness* calculates features derived from the length of a text, such as the maximum term frequency per document. *Structure measures* validate the similarity and dissimilarity of the categories of the gold standard by providing a single value which represents the structure of the document collection. By determining the degree of these measures of corpora

---

[1] Boing Boing http://boingboing.net; Slashdot http://slashdot.org. A preprocessed version of each dataset is available at http://www.dsic.upv.es/grupos/nle/downloads.html (June 2009).

H. Horacek et al. (Eds.): NLDB 2009, LNCS 5723, pp. 299–300, 2010.

we can test clustering methods in order to determine the complexity of classifying text collections of this type.

3 Experiments and Results

In Table 1, we compare the results of the evaluation of the blog corpora with four other short text corpora, consisting of scientific abstracts and newsfeeds [2].

Table 1. Features of the corpora

Corpora	Text type	Shortness	Class imbalance ranking	Stylometry	Structural ranking	Broadness measures
Cicling-2002	Scientific	Very short	4	Specific	4	Narrow
WSI-SemEval	Scientific	Very short	1	Specific	3	Narrow
R8-Training	News	Short	3	General	2	Wide
R8-Test	News	Short	2	General	1	Wide
B-B	Blog	Very short	4	General	4	Narrow
Slashdot	Blog	Very short	4	General	3	Narrow

In relation to class imbalance, both blog corpora appear well-balanced and so this will have no impact on the clustering blog process. The stylometry measure indicates a general language writing style for blog corpora as the documents were written by many different people. The structural measures confirm the best gold standard structure is for Slashdot. We validate the similarity and dissimilarity of the suggested groups or categories of the gold standard providing a single value which represents the structure of the document collection. Slashdot seems to have good structure, it is expected that Slashdot obtain much better clustering results than B-B. In relation to broadness, Slashdot is shown by all measures to be of narrower domain than B-B but both are considered narrow domain.

4 Conclusions and Further Work

In the corpora we have analyzed, our experiments indicate that the blogs were characterized as short text, with a general writing style and in a narrow domain. Having this knowledge allows us to employ a methodology for clustering which is described in [1] which takes advantage of a new self-enriching technique in order to address the challenges of clustering short texts, particularly in relation to blogs.

Acknowledgements. The work of the first author is supported by the HEA under grant PP06TA12. The work of the fourth author is supported by the CICYT TIN2006-15265-C06 research project.

References

1. Pinto, D., Rosso, P., Jiménez-Salazar, H.: UPV-SI: Word Sense Induction using Self-Term Expansion. In: 4th Workshop on Semantic Evaluations - SemEval 2007. Association for Computational Linguistics (2007)
2. Pinto, D.: On Clustering and Evaluation of Narrow Domain Short-Text Corpora, PhD dissertation, Universidad Politécnica de Valencia, Spain (2008)

Improving Full Text Search with Text Mining Tools

Scott Piao, Brian Rea, John McNaught, and Sophia Ananiadou

National Centre for Text Mining,
School of Computer Science,
The University of Manchester,
Manchester UK
{scott.piao,brian.rea,john.mcnaught,sophia.ananiadou}@manchester.ac.uk

Keywords: Information Retrieval, Full Text Search, Term extraction, Termine, Document clustering, Natural Language processing.

Today, academic researchers face a flood of information. Full text search provides an important way of finding useful information from mountains of publications, but it generally suffers from low precision, or low quality of document retrieval. A full text search algorithm typically examines every word in a given text, trying to find the query words. Unfortunately, many words in natural language are polysemous, and thus many documents retrieved using this approach are irrelevant to actual search queries. A variety of techniques have been used to mitigate this problem, such as controlled vocabularies, keywords (manually provided by authors or indexers) and phrase search, however, these techniques have their own limitations, such as low coverage of controlled vocabularies.

In our work, we attempt to improve full text search with text mining tools. we have developed a document search engine[1] based on Apache Lucene [2] at the UK National Centre for Text Mining (NaCTcM)[3] for the INTUTE Repository Search (IRS) Project [4], in which we attempt to improve the full text search by using text mining tools, particularly terminology extraction and document clustering tools.

We observe that one of the main reasons for the low precision of full text search algorithms lies in the fact that they match the query words indiscriminately against all words in a document, whether they reflect the topic of the document or not. We address this problem by constraining the search to those

---

[1] It has been deployed as a web demonstrator at url:
http://www.nactem.ac.uk/nactem_irs/doc_search
[2] Apache Lucene is a high-performance, full-featured text search engine library written in Java. See website: http://lucene.apache.org
[3] For information about NaCTeM, see website: http://www.nactem.ac.uk
[4] This project is funded by UK JISC (http://www.jisc.ac.uk). For further details, see http://www.nactem.ac.uk/intute

H. Horacek et al. (Eds.): NLDB 2009, LNCS 5723, pp. 301–302, 2010.
© Springer-Verlag Berlin Heidelberg 2010

words or terms representative of the documents. We use a terminology extraction tool, named Termine[5], which has been developed for text mining in NaCTeM. Employing the C-value termhood metric, it identifies candidate terms which are representative of a given document. This tool was integrated in the document indexing process, such that the document contents are represented by the term sets extracted by the tool. Given a query, the search is carried out on the terms. In other words, our search is for the documents in which the query occurs as a key term or part of such terms. For example, if users use query "monthly", our algorithm would only retrieve those documents in which the query word is a domain concept term or part of such terms, such as "monthly breast self-examination". In addition to improving the quality of full text retrieval, the extracted concept terms are also used to address the information overlook problem by guiding users to navigate those documents which share domain concepts, helping users to discover the underlying semantic network of documents.

Our initial evaluation of the terminology tool shows that it is effective for reducing the number of irrelevant documents retrieved while increasing the porportion of "interesting" documents. For example, when we tried the query "help", compared with standard full text search, the term-based search reduced the number of retrieved documents from 3,470 to 135. Meanwhile, when we checked the top ten documents retrieved, the number of "interesting" documents, in which this word is used in domain terms such as *self-help manual*, increased from 2 to 8.

Although our tools need further improvement and evaluation, our experiment demonstrates that, compared with the baseline full text approach, the term-based approach is capable of filtering out a significant amount of noise documents while increasing the number of "interesting" documents aming the top documents retrieved.

Furthermore, we also incorporated a document clustering package, Carrot2[6], which employs the LINGO algorithm to provide the functionality that helps overcome the information overlook problem. We use this tool to cluster top 200 retrieved documents on the fly, grouping documents under human readable labels. For example, for a query "environment", our search engine cluster the retrieved documents into groups under labels such as *Learning Environment*, *Virtual Research*, *Software Design* etc, which reflect different aspects from which the topic "environment" is addressed.

Our tool will be further improved with the development of a more efficient document clustering tool based on Termine and a customised visualisation tool.

---

[5] See: Frantzi, Katerina, Ananiadou, Sophia, Mima, Hideki: Automatic recognition of multi-word terms: the C-value/NC-value method. International Journal of Digital librarie, 3(2), pp. 117132 (2000)

[6] See Carrot2 website: http://www.carrot2.org

Cognos: A Natural Interaction Knowledge Management Toolkit

Francisco Javier Calle, Esperanza Albacete, Enrique Sánchez,
David del Valle Agudo, Jessica Rivero, and Dolores Cuadra

Computer Sc. Dep., Carlos III University of Madrid, Avda. Universidad 30, 28911 Leganés
{fcalle,ealbacet,essotes,dvalle,jrivero,dcuadra}@inf.uc3m.es

Abstract. The poster introduces a toolkit (Cognos) for assisting the analysis, annotation and management of interaction knowledge acquired from corpus. The description is focused on three of its elements: Dialogue, NL, and Communicative Acts.

1 Introduction

Cognos is a toolkit to make easier the annotation, edition and management of interaction knowledge. This toolkit consists of eight independent and integrated applications, each of which uses the same knowledge base. The global system implements a client/server architecture where the knowledge bases are allocated in a server and the knowledge consumers (diverse Knowledge Models implementations) act as clients. The toolkit includes directly new knowledge into the correspondent bases, also playing a client role. While analyzing, knowledge is added as draft, and once the knowledge analysis is finished, its state is changed and therefore it can be applied from that very moment.

The toolkit contains the following elements: a tool for the analysis and annotation of dialogue corpus (Cognos.Dial), a communicative acts edition tool (Cognos.CA), a tool for analysis and edition of relaxed grammars for natural language processing (Cognos.NL), analysis and edition tool for ontology, circumstances knowledge editor, user model editor, self-model editor, and the emotional knowledge editor. The first three are already developed (in Java v1.6), while the rest are in development phase.

All the knowledge bases are integrated and stored along with the corpus and they are implemented through relational databases in the DBMS Oracle® 11g.The corpus is stored in the same database in two ways, its textual transcriptions and the source multi-media files, which enrich the annotated knowledge and enhances its reuse possibilities. The following section is set to describe briefly some of its components: Cognos.Dial, Cognos.NL, and Cognos.CA.

2 The Pragmatic Annotation Tool

Cognos.Dial is a tool for pragmatic knowledge deep analysis of dialogue corpus. Such analysis is carried out through seven phases. These phases are: pre-segmentation (consists in bounding each sentence and discours), microanalysis (facilitates the

H. Horacek et al. (Eds.): NLDB 2009, LNCS 5723, pp. 303–304, 2010.

assignment of one or more communicative acts to each discourse), temporal realization (assists the inclusion of some prosodic information between two interventions such as silences and overlaps), segmentation (the dialogue division into independent functional units), commitment (is set to let the analyst annotating which segments alter the degree of commitment of dialogue and the way in which the change affects the commitment), operative (enables to describe involved tasks: who executes them application or external agent, inputs, outputs and the associated effects), and structural (the basic segment structure (interaction states and transitions) is automatically calculated from the rest of the knowledge. Nevertheless, this part of the tool enables the analyst to edit and modify that proposed structure).

Progress in a sample analysis is stored in the database as annotation draft. The outcome of a completed sample analysis can be stored in the knowledge base (upload). A completed sample analysis can be also saved into an XML-file, following an own format implemented through an XML-schema.

Cognos.NL tool enables to analyze and edit knowledge related with utterances and their associated communicative actions, for later Natural Language Processing. The application focuses the same knowledge bases for interpretation (through relaxed grammars) and generation. The inputs are expressions, each of which will have associated a pattern and a set of communicative acts instances. Patterns are analyzed as sets of elements that can be: sub-pattern, variable, wildcard, and token. Besides, tokens do not have discourse pieces associated.

The CA set editor (**Cognos.CA**) fulfils the need of having communicative acts made to measure for each corpus in the toolkit. Yet such aim seems to harden reusability, in fact counting on such tool enables to define compatible CA sets for different corpus, which at medium-long term should lead to an extended CA set definition, enclosing all the minor specific ones. Hence, the application lets the user to manage the communicative acts set associated to a corpus, from defining the abstract communicative acts types and categories to fixing feasible instances through category-value and category-domain constraints. Both Cognos.Dial and Cognos.NL tools only use the communicative acts defined by this application.

3 Conclusions

Cognos toolkit integrates several annotation facilities for assisting corpus analysis, reducing its cost, and enhancing results with more coherent and integrated annotated knowledge. With such aim, requirement definition phase of Cognos has been supported by the experience acquired through several national and international research projects. This work was partially supported by the Reg. Gov. of Madrid under the Research Network MAVIR (S-0505/TIC-0267), and the Spanish Education Ministry through SOPAT proj (CIT-410000-2007-12). Currently it is being improved through the THUBAN project (TIN2008-02711) also founded by Education Ministry.

The annotated knowledge is already in use by some components of an interaction system (Dialogue Model and NL engine). However, there is still much work in progress, since some parts of the toolkit are in development phase. Besides, some tools involve an analysis methodology that is also being evaluated.

Specialized Entailment Engines: Approaching Linguistic Aspects of Textual Entailment*

Elena Cabrio

FBK-Irst and University of Trento, Italy
cabrio@fbk.eu

Abstract. Textual Entailment (TE), one of the current hot topics in Computational Linguistics, has been proposed as a task to address the problem of language variability. Since TE is due to the combination of different linguistic phenomena which interact among them in a complex way, this paper proposes to experiment the use of specialized entailment engines, each addressing a specific phenomenon relevant to entailment.

Keywords: Textual Entailment, Tree Edit Distance.

1 Introduction

In 2005, the Recognizing Textual Entailment (RTE) Challenge has been launched [3], defining TE as a task for automatic systems. The task consists in developing a system that, given two text fragments (text T and hypothesis H), can determine whether the meaning of one text is entailed, i.e. can be inferred, from the other (e.g. T: *A tiger cub rejected by her mother in a zoo in Russia has been adopted by a surrogate canine mother* ⇒ H: *A dog has adopted a tiger*).

Language variability comes to light at various levels of complexity, involving almost all linguistic phenomena of natural languages (lexical, syntactic, semantic variations). Most of the systems submitted to RTE face this problem with omnicomprehensive approaches, without trying to address the problem of the interaction of the linguistic phenomena underlying the entailment relation. The idea underlying the proposed approach is to have different independent Entailment Engines (EEs), each of which able to deal with an aspect of language variability, overcoming one of the most evident limitations of most of the current systems, i.e. the impossibility of evaluating their performances on single aspects of entailment.

2 Related Work

Concerning the idea of using combined specialized EEs, the DFKI group [8] developed a system based on specialized RTE-modules for RTE4. Three main

---

* Ph.D. advisor: Bernardo Magnini, FBK-Irst, Trento, Italy. magnini@fbk.eu
  The author is carrying out part of the work described in this research proposal during an internship period at Xerox Research Centre Europe, Grenoble, France.

H. Horacek et al. (Eds.): NLDB 2009, LNCS 5723, pp. 305–308, 2010.

modules are included, to deal with temporal expressions, Named Entities and cases with two arguments for each event. Besides these precision-oriented modules, robust backup strategies have been considered. The good results obtained by this modular system let us expect that this is a promising direction to explore.

Several studies tried to analyze linguistic levels involved in the TE task. In Garoufi [5], a scheme for manual annotation of TE datasets (ARTE) is proposed, with the aim of highlighting a wide variety of entailment phenomena in the data and their distribution. An attempt to isolate the set of T-H pairs whose categorization can be accurately predicted based solely on syntactic cues has been carried out by Vanderwende et al. [7]. Clark et al. [2] highlight that the majority of entailment cases rely on significant amount of the so called "common human understanding" of lexical and world knowledge. Several existing resources and their capacity for supplying that knowledge are discussed. A definition of contradiction for TE task is provided by de Marneffe et al. [4], together with a collection of contradiction corpora.

3   Proposed Research

With the aim of coping with the problem of how linguistic aspects interact with respect to the TE task, this work proposes to experiment the use of specialized EEs, each addressing a specific phenomenon relevant to entailment (e.g. negation, modals, and active/passive forms, lexical similarity). The expected benefits aim at covering the limitations of the actual approaches: *(i)* considering complex linguistic aspects involved in TE separately aims at improving their comprehension, in order to focus on the best way to cope with them; *(ii)* adopting a modular approach would permit to evaluate the progresses of the system on single aspects of entailment; *(iii)* the creation of *ad hoc* specialized datasets aims at providing linguistic resources that can be used by whatever system to evaluate performances on linguistic phenomena basing on a common ground; *(iv)* the study on how to combine the different EEs supposes and can help in understanding the dependencies of the linguistic phenomena involved; *(v)* being able to detect entailment basing on linguistic foundations should strengthen the system. Three research directions are proposed and discussed in the following paragraphs. The general framework is based on distance between T and H and aims to be flexible enough to allow the combination of these EEs. The output of the whole system is defined by the combination of the distances produced by each specialized EE. Tree Edit Distance is the considered distance-based algorithm [6].

Designing specialized EEs. A main working hypothesis is proposed, according to which each engine is specialized for a linguistic phenomenon. In a sense, it is a reformulation of the definition of TE with respect to a phenomenon x. For a given T-H pair, the output of each module can correspond to three values: *yes* if x has been detected and is fundamental/contributes to state that there is entailment; *no* if x has been detected and is fundamental/contributes to state that there is no entailment; *unknown* in all other cases. Each pair can contain more than one phenomenon. Given the example in Section 1 and three specialized EEs (EEneg, dealing with negative polarity items; EEact/pass dealing with active/passive forms; EElex, dealing with lexical similarity), the expected output of EEneg is *unknown*, since no NPIs are present; the output of EEact/pass is *yes*, since the act/pass form is detected and there is entailment according to this phenomenon (*A has been adopted by B ⇒ B has adopted A*); the output of EElex is *yes* since the lexical similarity between T and H is over a certain threshold previously estimated (beside identical words, also the synonym *canine mother ⇒ dog* have a high lexical similarity).

Resources for specialized EEs. Two working hypothesis are proposed: *(i)* the exploitation of manually annotated resources of linguistic phenomena (e.g. ARTE [5]), to evaluate how a single module is able to detect the phenomena it is specialized for among the others; *(ii)* the creation of "artificial" datasets: the underlying criterion can be to extract T from the RTE datasets, and to create H identical to T except for the presence of the investigated phenomenon (in order to avoid other phenomena influencing the decisions of the system). Here the investigated phenomenon is isolated and the performances of the specialized engines are evaluated on a controlled setting. Intuitively, phenomena like negation are easier to isolate than others (e.g. lexical similarity), for which the dependencies on other linguistic phenomena seem to be stronger. The presented hypothesis should be tested on the EEs developed in the first phase.

Combining specialized EEs. In order to combine the results of the EEs, different work hypothesis are proposed and should be experimented: *(i)* adopting a simple voting mechanism, the judgment given by the majority of the engines will be taken as final output; *(ii)* in order to reward the modules that detect contradiction (e.g. the module dealing with negation), weighted voting mechanisms can be used (the final output of the system will be *yes* if the majority of the EEs results is *yes* and there are no negative opinions; *no* if there is at least one negative opinion; *unknown* in all other cases); *(iii)* the sum of the edit distances produced by each module can be considered as output of the overall system: a certain threshold empirically estimated over training data should be fixed on the general task, to separate the entailment cases from the non entailment ones; *(iv)* the combinations of the engines could be performed taking into consideration the dependencies of the phenomena on which the modules act. Machine Learning methods of feature analysis will be investigated to face the problem.

Current work. The feasibility of the described approach has been experimented in RTE4 [1] basing on the EDITS system (Edit Distance Textual Entailment Suite)[1], that assumes that the distance between T and H is a characteristic that separates the positive pairs, for which entailment holds, from the negative pairs, for which entailment does not hold [6]. The first module that has been implemented works on negative polarity items (NPIs) (direct licensors of negation and

---

[1] Available as open source at http://edits.fbk.eu/

antonyms). In EDITSneg, the cost schema sets specific costs for edit operations concerning negation, and the Linear Distance algorithm is used to determine the less costly sequence of edit operations to transform T into H. The underlying intuition is that assigning high costs to edit operations that involve NPIs should prevent the system from assigning positive entailment to a T-H pair in which one of the two fragments contradicts the other.

To make experiments on negation, an artificial balanced dataset has been built. Performances of EDITSneg on the specialized dataset are quite satisfying (77% acc), while on RTE4 they are lower (54% acc) (as expected, only few pairs contained NPIs). Obviously, phenomena like negation are easier to circumscribe and isolate than others. Furthermore, the dependencies of the different linguistic aspects should not be underestimated, and should be on the basis of the combination of the different EEs. At first sight, some dependencies cannot be disregarded and seem to be particularly strong.

4 Conclusions

Since different sources of knowledge are exploited by TE systems, but their peculiar contribution in solving the task seems still unclear, this work proposes an approach based on different independent EEs, each of which able to deal with an aspect of language variability. Results obtained in the participation at RTE4 let us expect that this is a promising direction to explore. Adding more specialized engines, and therefore considering a higher number of linguistic aspects, would permit to improve the current results.

References

1. Cabrio, E., Kouylekov, M., Magnini, B.: Combining Specialized Entailment Engines for RTE4. In: Proc. of TAC 2008 Workshop on TE, Gaithersburg, Maryland (2008)
2. Clark, P., Harrison, P., Thompson, J., Murray, W., Hobbs, J., Fellbaum, C.: On the Role of Lexical and World Knowledge in RTE3. In: Proc. of the ACL-PASCAL Workshop on TE and Paraphrasing, Prague, Czech Republic (2007)
3. Dagan, I., Glickman, O., Magnini, B.: The PASCAL Recognizing Textual Entailment Challenge. In: Proc. of the First PASCAL Challenges Workshop on RTE, Southampton, UK (2005)
4. De Marneffe, M.C., Rafferty, A.N., Manning, C.D.: Finding Contradictions in Text. In: Proc. of the ACL 2008, HLT Columbus, Ohio (2008)
5. Garoufi, K.: Towards a Better Understanding of Applied Textual Entailment. Master Thesis. Saarland University, Saarbrücken, Germany (2007)
6. Kouylekov, M., Magnini, B.: Combining Lexical Resources with Tree Edit Distance for Recognizing Textual Entailment. In: Quiñonero-Candela, J., Dagan, I., Magnini, B., d'Alché-Buc, F. (eds.) MLCW 2005. LNCS (LNAI), vol. 3944, pp. 217–230. Springer, Heidelberg (2006)
7. Vanderwende, L., Coughlin, D., Dolan, B.: What Syntax can Contribute in Entailment Task. In: Proc. of the First PASCAL Challenges Workshop on RTE, Southampton, UK (2005)
8. Wang, R., Neumann, G.: An Accuracy-Oriented Divide-and-Conquer Strategy. In: Proc. of the TAC 2008 Workshop on TE, Gaithersburg, Maryland (2008)

Natural Language Processing for Spreadsheet Information Retrieval

Derek Flood[*]

Dundalk institute of Technology
derek.flood@dkit.ie

Abstract. Spreadsheets can be used for a variety of tasks including inventory management. Most spreadsheet applications provide the ability to extract additional information from this data, through techniques such as filtering and PivotTables. These techniques however can be quite complex and time consuming, especially for novice spreadsheet users. We believe that these issues can be addressed through the use of Natural Language Processing, a technology that has already demonstrated its effectiveness in similar domains, such as databases and chart generation.

1 Introduction

When spreadsheets are used to analyse large amounts of data, extracting meaningful information can be a difficult task. Inbuilt features such as Filters and PivotTables can help, however these features often require multiple interactions, which can be time consuming, especially for novice spreadsheet users. Section 2 outlines this problem in more detail.

The aim of this research is to investigate and improve the usability and effectiveness of spreadsheet information retrieval. It is believed that this improvement can be achieved by simplifying the way in which users interact with spreadsheet applications. Natural Language Processing (NLP) techniques are one way this can be accomplished.

In the past, Natural Language Interfaces (NLI's), which incorporate NLP techniques, have been used to simplify the operation of computer applications. This can be done in many ways such as reducing the number of interactions required or eliminating the need for complex structured commands. Generating charts[1], Querying databases[2] and searching XML files[3] are just some of the tasks that have been simplified through the use of NLI's.

Section 3 of this paper outlines the aim of this research along with the steps that have been taken to meet this aim. In Section 4 the work that has been completed to date is outlined. This work has involved the development of a Natural Language Interface to Microsoft© Excel which allows users to perform information retrieval tasks. Section 5 details the remainder of this research which consists of evaluating the Natural language interface through a controlled experiment.

---

[*] Supervisors: Dr. Kevin Mc Daid, Dr. Fergal Mc Caffery.

H. Horacek et al. (Eds.): NLDB 2009, LNCS 5723, pp. 309–312, 2010.
© Springer-Verlag Berlin Heidelberg 2010

2 Problem Statement

Spreadsheets are used to analyse large amounts of data such as transaction or survey data. With such a large volume of data, extracting information can be a difficult task. Consider a spreadsheet containing all of the transactions that occurred in a shop in a single week. When a manager reviews this spreadsheet they would also like to be able to see more general information, such as the frequency of a single payment type or a subset of the data such as transactions over a certain value.

Inbuilt features such as filtering and PivotTables can assist in these tasks however for novice spreadsheet users these features can be difficult and time consuming. In order to look at a subset of the transaction data, the manager, using Microsoft© Excel, must first select the data they wish to filter. Next they must switch on filtering and select the column they wish to filter by. Finally they must select the criteria by which they want to filter the data.

3 Research Methodology

This research aims to investigate and improve the effectiveness and efficiency of spreadsheet information retrieval. Research in other domains, such as the Database domain[2], has demonstrated that interfaces like those found in spreadsheet applications can be replaced, with greater efficiency, with Natural Language Interfaces. In order to meet this aim a number of steps have been taken:

1. Investigate how users would form information requests to spreadsheets through a Grammar Acquisition Study, detailed in Section 4.
2. Develop a Natural language interface for spreadsheet information retrieval
3. Evaluate the Natural language Interface through a controlled experiment

RQ1 has been posed to investigate the effectiveness of a NLI to spreadsheets for information retrieval tasks in terms of time taken and the ability of users to perform more complex tasks.

- **RQ1:** To what extent can a natural language interface be used to improve information retrieval for spreadsheet users?

We hypothesize that as the complexity of the task increases, the benefits will be seen by more experienced users. The initial research will focus on novice spreadsheet users however it will be extended later to include more experienced users.

4 Work to Date

The work to date has focused on the development of the NLI. The first step in developing this interface is to construct a grammar that could be used by the system to determine the type of information that is being requested by the user.

Before constructing this grammar, an investigation was conducted into how users describe information retrieval tasks. This exercise focused on a well structured spreadsheet where each row corresponded to a single entity. The spreadsheet chosen listed the sales for a number of stores and included the subtotal, vat, total and type of payment that was used in each sale. This approach is similar to that of Zelle[4] who collected a corpus of questions relating to a geographical database.

Approximately 40 students, who are representative of novice spreadsheet users, took part in the exercise, which presented them with the spreadsheet described above and asked them to list ten questions/statements they would expect an application to be able to answer about the spreadsheet.

During the experiment 77 unique utterances were collected. These utterances along with the initial 250 collected by Zelle[4] were used to form the basis of the grammar used by the Natural Language Interface.

A number of interesting features were found within the corpus. Some of the utterances collected were not answerable with the given dataset. Additional information would be required in order to satisfy the user's request. For example one participant asked *"Show me the sales made in the south"*, however such background information, i.e. which stores were in the south, was not included in the spreadsheet.

It was also found that some of the utterances contained "Google" style queries, where the user gave only the minimum amount of words to convey their intention, for example *"Time of the transaction"*.

The resulting NLI provides users with the ability to perform a number of information retrieval tasks by simply typing in a natural language statement. These tasks include, finding the most common values, filtering the data to only rows that meet specific criteria and creating summary tables of the information, similar to Pivot-Tables. Using this interface, a manager looking at the transaction data can find the most common payment type by simply asking *"What is the most common payment type?"*

5 Future Work

As the system has now been developed, the next step is to evaluate it. This evaluation will ask participants to retrieve certain information from a spreadsheet containing a list of golf courses within the state of Indiana.

Forty participants with little experience of spreadsheets will take part in the study. These participants, as part of a module in spreadsheets, have been shown how to use the features, such as PivotTables and Filters, which are needed for the study.

Using the spreadsheet outlined above participants will be asked to complete ten tasks. These tasks will be divided into three classes of complexity; Easy, Intermediate and Complex. The tasks will require users to filter the data, such as to show only courses in the county of Boone, or to identify the number of rows that satisfy a given set of criteria, such as the number of courses of each course type.

The participants will be split into 2 groups; Group A and Group B. Group A will be asked to retrieve information from the spreadsheet using traditional means while Group B will be asked to complete the same tasks using the natural language interface. A pilot study involving a small number of participants has already shown some very promising results. Another study will use the same approach to evaluate the system for more experienced spreadsheet users.

During the experiment a VBA macro, T-CAT (Time-stamped Cell Activity Tracker) [5] will be used to record users behaviour. T-CAT records the time of each cell change and each cell selection. For the purposes of this experiment it has been modified to record the time participants start each task and the time participants use the natural language interface. It will also record the utterances that are used.

The data collected will allow for a number of metrics to be calculated. The time to complete each task will be calculated using the time a participant moves to the next task. As well as this the number of tasks successfully completed by each participant will also be recorded.

6 Conclusions

This paper proposes the use of Natural Language Processing techniques to simplify the task of spreadsheet information retrieval. Spreadsheets are used to store massive amounts of data, however extracting meaningful information from this data can be a complex and time consuming task especially for novice spreadsheet users.

Natural language processing techniques have been used in similar applications to improve the efficiency with which these applications are used. This work looks at using natural language processing to improve the way information is extracted from spreadsheets. A comparative study is outlined which aims at evaluating such a system against the current state of the art features for information retrieval.

Acknowledgements

This work is supported by the Irish Research Council for Science, Engineering and Technology (IRCSET) through the embark initiative.

References

1. Kato, T., Matsushita, M., Maeda, E.: Answering it with charts - Dialogue in natural language and charts. In: The 19th international conference on Computational linguistics (2002) (presented at)
2. Zelle, J.M., Mooney, R.J.: Learning to Parse Database Queries Using Inductive Logic Programming. In: The Thirteenth National Conference on Artificial Intelligence AAAI (1996) (presented at)
3. Woodley, A., Tannier, X., Hassler, M., Geva, S.: Natural Language Processing and XML Retrieval. In: The Australasian Language Technology Workshop, ALTW (2006) (presented at)
4. Zelle, J.M.: Using Inductive Logic Programming to Automate the Construction of Natural Language Parsers. In: Department of Computer Sciences, PhD. University of Texas, Austin (1995)
5. Bishop, B., McDaid, K.: Unobtrusive Data Acquisition for Spreadsheet Research. In: The IEEE Symposium on Visual Languages and Human-Centric Computing, VL\HCC 2008 (2008) (presented at)

Use Speech for User's Group Identification[*]

Sajad Shirali-Shahreza

Computer Engineering Department
Sharif University of Technology
Azadi Street, Tehran, Iran
shirali@ce.sharif.edu

Abstract. Speech is one of the main communication methods used by humans. So designing speech based interfaces for computer applications can ease the use of them for different users, especially disabled and elderly people. In this project, we want to design and implement a system that can identify user's group from his/her speech. We are working on two main applications: Parental Control and Completely Automated Public Turing test to tell Computers and Human Apart (CAPTCHA). In the Parental Control application, we want to distinguish between child users and adults based on human speech. This can be used to restrict child user access to adult materials such as adult websites. In the CAPTCHA application, we want to design a fully spoken CAPTCHA which can be used easily by all users, especially blind users.

Keywords: Disabled People, Parental Control, Spoken Interfaces, Support Vector Machines (SVM), Speech Recognition, Text-To-Speech (TTS).

1 Introduction

Computer is one of the most important inventions of the 20th century. The computers changed human life in their short history so that living without computers seems impossible now. The digital technologies which are developed in conjunction with the computers are now used in most of home appliances such as televisions and washing machines. In addition, new devices are created and entered into homes such as game consoles. The creation of World Wide Web (WWW) in 1993 and the expansion of the Internet during the last 15 years had also a great impact on human life. During the last decade, the Internet entered into the daily life of all peoples. Today, the people can buy goods, register for tests, and book tickets via the Internet.

Ten years ago, the main users of computers were technicians and operators and the applications were designed for them. But today the computer is used in daily life and its users are from any age and group, including children, elderly and disabled people. So the user interfaces are changing. One of goals of new user interfaces is easier usage by ordinary people. Speech is one of the primary and natural communication ways between humans. People usually use speech in their interactions. So it is a good candidate for human users to interact with computers.

---

[*] PhD Supervisor: Dr. Hassan Abolhassani, Computer Engineering Department, Sharif University of Technology, abolhassani@sharif.edu

H. Horacek et al. (Eds.): NLDB 2009, LNCS 5723, pp. 313–316, 2010.

In this project, we want to design and implement a system that can identify user's group from his/her speech. We are focused on two special applications: identifying children group against adult users group and identifying human users group against computer programs. We will describe these two application and our solutions in the following sections in more details. This project is somewhat an interdisciplinary project. The user interaction with the system is via speech, so the system must process the speech signals and require actions such as speech recognition and speech synthesis. This part of the project is a signal processing task.

The system must identify the user's group. As we will mention in the next sections, this is done by extracting a series of feature from the speech signal and then use a machine learning method such as Support Vector Machine (SVM) for identification. This part of the project is a machine learning task and involves problems such as dealing with samples having different number of features, similar to the problems in speaker verification [9]. Finally, we are designing this system for human users. So it should be user friendly and can be used easily. One of the target groups of this system are blind people. So it must be tested by human users including blind people in real situations. This part of the project is a Human Computer Interaction (HCI) task.

2 Parental Control Based on Speaker Class Verification

The first application which we are working on is parental control. In the current digital world, children can access a huge amount of multimedia information via the Internet, digital TV, game consoles and other ways. Some of the materials available on these sources such as movies containing mature scenes are for adult users and are not suitable for children. To protect children from such materials, there are a number of features named Parental Control in devices which are used by both child and adult users. We done a complete survey on the parental control methods used in current and previous digital devices in [6]. Surprisingly we found that the de facto parental control method is using a four digit password. It is an old technique, but current devices still use it.Password based parental control has a number of problems. Mainly, it requires setting up the password and entering it each time which are not desirable tasks. As a result, the feature is not used at all, as some of the surveys like [2] reports.

This problem can be expressed in another way: we want to automatically distinguish between children users and adult users by asking a question. This problem is discussed in the Human Interactive Proof (HIP) domain during last years, but no one can solve it previously. We had analyzed the problem from this perspective in [7].

The solution we propose for this problem is speech based parental control. In our method, the system asks the user to say a certain word or sentence. Then it extracts a number of features from it. For example we extract the Mel Frequency Cepstral Coefficients (MFCC) [4]. Then we use a Support Vector Machine (SVM) [1] classifier to decide whether the speech belongs to a child user or an adult user. The SVM is previously trained on with a training set. For the training set, we use a part of a data set which we created in our lab. Our data set contains 256 speakers containing 163 children and 93 adult users. Each speaker said 54 English words. The sounds are recorded in a normal room to reflect the normal conditions of home users. For this system, we choose the random word "brown" among the 54 words said by each speaker. We select 40 adult speakers and 40 child speakers (20 under 10 years old and 20 between 10-14 years old) randomly. We use 30 speakers in each group for training and the remaining 10 speakers for testing. The system can correctly identify 92.5% of test users. This system is described in more details in [6].

3   Spoken CAPTCHA: A CAPTCHA System for Blind Users

The second application is a CAPTCHA system. Today, the Internet is used to offer different services to users. Most of these services are designed for human users, but unfortunately some computer programs are designed which abuse these services. CAPTCHA (Completely Automated Public Turing test to tell Computers and Human Apart) systems are designed to automatically distinguish between human users and computer programs and block such computer programs.

The main focus of these methods is on questions that the human user can easily answer but the present computer programs are hardly likely to be able to answer.

Most of current CAPTCHA methods are using visual patterns and hence blind users cannot use them. In these methods, the image of a word with distortion and various pictorial effects is shown to the user and he is asked to type that word. Beside visual CAPTCHA methods, there are sound based CAPTCHA methods such as [3], for disabled people. In these methods, a word is said and the user must type the word. They are based on weaknesses of speech recognition systems.

In this project, we propose a new CAPTCHA method which is designed for blind people. The overall structure of spoken CAPTCHA system is shown in Figure 1. In this method, a small sound clip is played for the user and s/he is asked to say a word. Then the user response – which is a speech file – is checked to be the requested word and also not generated by a computer. This method does not require any screen and can be used easily by normal and blind users. In addition it can be used in applications which work via telephone.

In this method, like speech based parental control, we extract MFCC features and use then use a SVM to decide whether the user is a computer program or a human user. For training the system, we use two random words "Orange" and "Brown". We use 30 adult and 30 child users as training set and 10 adult and 10 child users for testing set from the data set we discussed in parental control application. For creating the synthesized voice, we use the University of Edinburgh's Festival speech synthesis systems [8]. We also gather 25 different English speaker data for Festival. We use these 25 speakers and use effects such as emphasize for pitch change, we create 76 different sounds for each word. Among them, we use 56 for training and 20 for testing. We use the word "Brown" to train system and then test it on word "Orange". The system can correctly identify 92.5% of test samples.

The main contribution of our spoken CAPTCHA system is that it uses drawbacks of computers in both speech recognition and speech synthesis, while previous sound based CAPTCHA methods use only speech recognition drawbacks. So its tests are easier for human users and more difficult for computers, in comparison to previous sound based CAPTCHA. This method is presented in more detail in [5].

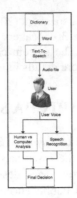

Fig. 1. The structure of our proposed Spoken CAPTCHA

4 Conclusion

In this paper, we describe our project for user group identification from speech. We created the early prototypes of the system and test them with recorded voices. Currently we are completing these prototypes for real-time test by human users. We are also working on the machine learning part to incorporate more efficient techniques such as techniques used in [9] for speaker verification.

References

1. Cortes, C., Vapnik, V.: Support Vector Networks. Mach. Learn. 20(3), 273–297 (1995)
2. Federal Communications Commission (FCC): In the Matter of Violent Television Programming and its Impact on Children, FCC 07-50, MB Docket No. 04-261 (April 2007)
3. Kochanski, G., Lopresti, D., Shih, C.: Reverse Turing Test Using Speech. In: Proc. ICSLP 2002, pp. 1357–1360. Causal Productions Pty Ltd. (2002)
4. Rabiner, L.R., Juang, B.H.: Fundamentals of Speech Recognition. Prentice-Hall, Englewood Cliffs (1993)
5. Shirali-Shahreza, S., Abolhassani, H., Sameti, H., Shirali-Shahreza, M.H.: Spoken CAPTCHA: A CAPTCHA System for Blind Users. In: Proc. CCCM 2009. IEEE, Los Alamitos (2009)
6. Shirali-Shahreza, S., Sameti, H., Shirali-Shahreza, M.: Parental Control Based on Speaker Class Verification. IEEE T. Consum. Electr. 54(3), 1244–1251 (2008)
7. Shirali-Shahreza, S., Shirali-Shahreza, M.: Identifying Child Users: Is It Possible? In: Proc. SICE 2008. IEEE Computer Society, Los Alamitos (2008)
8. Festival Speech Synthesis System,
   http://www.cstr.ed.ac.uk/projects/festival/
9. Wan, V.: Speaker Verification using Support Vector Machines, PhD Dissertation, Department of Computer Science, University of Sheffield, United Kingdom (June 2003)

Author Index